Plant Apocarotenoids

Biosynthesis, Signaling and Agricultural Applications

Edited by

Kaiser Iqbal Wani
Hainan University
China

Tariq Aftab
Aligarh Muslim University
India

WILEY

This edition first published 2026

Registered Office(s)
John Wiley & Sons, Inc., 111 River Street, Hoboken, NJ 07030, USA
John Wiley & Sons Ltd. New Era House, 8 Oldlands Way, Bognor Regis, West Sussex, PO22 9NQ, UK

For details of our global editorial offices, customer services, and more information about Wiley products visit us at www.wiley.com.

The manufacturer's authorized representative according to the EU General Product Safety Regulation is Wiley-VCH GmbH, Boschstr. 12, 69469 Weinheim, Germany, e-mail: Product_Safety@wiley.com.

Wiley also publishes its books in a variety of electronic formats and by print-on-demand. Some content that appears in standard print versions of this book may not be available in other formats.

Library of Congress Cataloging-in-Publication Data has been applied for:

Hardback ISBN: 9781394332601
Epdf ISBN: 9781394332625
Epub ISBN: 9781394332618
oBook ISBN: 9781394332632

Cover Design: Wiley
Cover Images: © ZhakYaroslav/Shutterstock, © kseniyaomega/stock.adobe.com

Contents

About the Editors

Dr. Kaiser Iqbal Wani

Kaiser Iqbal Wani has completed his Ph.D. from the Department of Botany, Aligarh Muslim University, Aligarh, India. He is currently working as a post-doctoral fellow in the National Key Laboratory for Tropical Crop Breeding, School of Breeding and Multiplication (Sanya Institute of Breeding and Multiplication), Hainan University. He is a recipient of a prestigious national fellowship CSIR/UGC JRF and SRF during his Ph.D. Besides, he has also qualified in some other national-level exams like the Graduate Aptitude Test in Engineering (GATE) in life sciences and JKSET. His research is focused on molecular, proteomic, physiological, and biotechnological studies on medicinal and horticultural plants under adverse environmental conditions. Kaiser Iqbal has a number of research publications to his credit in peer-reviewed international journals. He has authored a book published by Springer Nature and edited a book on strigolactones published by Wiley-Scrivener. Besides, he also has a number of chapters in edited volumes with international publishers.

Dr. Tariq Aftab

Tariq Aftab received his Ph.D. in the Department of Botany at Aligarh Muslim University, India, and is currently an Assistant Professor there. He is the recipient of a prestigious Leibniz-DAAD fellowship from Germany, Raman Fellowship from the Government of India, and Young Scientist Awards from the State Government of Uttar Pradesh (India) and Government of India. After completing his doctorate, he has worked as Research Fellow at National Bureau of Plant Genetic Resources, New Delhi and as Post-doctorate Fellow at Jamia Hamdard, New Delhi, India. Dr. Aftab also worked as Visiting Scientist at Leibniz Institute of Plant Genetics and Crop Plant Research (IPK), Gatersleben, Germany, and in the Department of Plant Biology, Michigan State University, USA. He is a member of various scientific associations from India and abroad.

He has edited 20 books with international publishers, including Elsevier Inc., Springer Nature and CRC Press (Taylor & Francis Group), co-authored several book chapters, and published over 100 research papers in peer-reviewed international journals. His research interests include physiological, proteomic, and molecular studies on medicinal and crop plants.

List of Contributors

Abdullah
Department of Botany
Aligarh Muslim University
Aligarh
Uttar Pradesh
India
National Key Laboratory for Tropical Crop Breeding
School of Breeding and Multiplication (Sanya Institute of Breeding and Multiplication)
Hainan University
Sanya
Hainan
China

Tariq Aftab
Department of Botany
Aligarh Muslim University
Aligarh
Uttar Pradesh
India

Zishan Ahmad
Bamboo Research Institute
School of Life Sciences
Nanjing Forestry University
Nanjing
Jiangsu
China

Adriana Delgado Alvarado
Colegio de Postgraduados
Campus Puebla
Cholula
PU
Mexico

Nauf Al Araimi
Department of Applied Biotechnology
University of Technology and Applied Sciences
Sur
Oman

Nasheeman Ashraf
Plant Molecular Biology and Biotechnology Division
CSIR-Indian Institute of Integrative Medicine
Srinagar
Jammu and Kashmir
India
Academy of Scientific and Innovative Research (AcSIR)
Ghaziabad
Uttar Pradesh
India

Muhammad Azam
Department of Plant Science
School of Agriculture and Biology
Shanghai Jiao Tong University
Shanghai
China

Hebert Jair Barrales-Cureño
Instituto Tecnológico de Estudios Superiores de Zamora
Tecnológico Nacional de México
La Piedad
MC
Mexico

Braulio Edgar Herrera Cabrera
Colegio de Postgraduados
Campus Puebla
Cholula
PU
Mexico

Vinay Kumar Dhiman
Department of Food Science and Technology
Khalsa College
Amritsar
Punjab
India

Vivek Kumar Dhiman
Centre for Herbal Pharmacology and Environmental Sustainability
Chettinad Hospital and Research Institute
Chettinad Academy of Research and Education
Kelambakkam
Tamil Nadu
India

Alejandra Hernández Garcia
Universidad Michoacana de San Nicolás de Hidalgo
Morelia
MC
Mexico

Rafael Salgado Garciglia
Universidad Michoacana de San Nicolás de Hidalgo
Morelia
MC
Mexico

Berta Gonçalves
Centre for the Research and Technology of Agro-Environmental and Biological Sciences (CITAB)
Institute for Innovation
Capacity Building and Sustainability of Agri-Food Production (Inov4Agro)
Department of Biology and Environment
School of Life and Environmental Sciences (ECVA)
University of Trás-os-Montes e Alto Douro
Quinta de Prados
Vila Real
Portugal

Antonio Granell
Instituto de Biología Molecular y Celular de Plantas
CSIC-Universidad Politécnica de València
Valencia
Spain

Aabida Gul
Department of Plant Molecular Biology & Biotechnology
CSIR-Indian Institute of Integrative Medicine
Srinagar
Jammu and Kashmir
India

Chao Guo
National Key Laboratory of Cotton Bio-breeding and Integrated Utilization
State Key Laboratory of Crop Stress Adaptation and Improvement
Henan Joint International Laboratory for Crop Multi-Omics Research
School of Life Sciences
Henan University
Kaifeng
Henan
China

Uzma Jan
Department of Plant Molecular Biology & Biotechnology
CSIR-Indian Institute of Integrative Medicine
Srinagar
Jammu and Kashmir
India

Jianhua Jia
State Key Laboratory of Component-based Chinese Medicine
Tianjin University of Traditional Chinese Medicine
Tianjin
China
Haihe Laboratory of Modern Chinese Medicine
Tianjin
China

Kun-Peng Jia
National Key Laboratory of Cotton Bio-breeding and Integrated Utilization
State Key Laboratory of Crop Stress Adaptation and Improvement
Henan Joint International Laboratory for Crop Multi-Omics Research
School of Life Sciences
Henan University
Kaifeng
Henan
China
Sanya Institute of Henan University
Sanya
Hainan
China

Shamshad Ahmad Khan
Department of Applied Biotechnology
University of Technology and Applied Sciences
Sur
Oman

Rodrigo Ledesma-Amaro
Department of Bioengineering
Imperial College Centre for Synthetic Biology
Bezos Centre for Sustainable Protein
UKRI Engineering Biology Mission Hub on Microbial Food Centre for Synthetic Biology
Imperial College London
London
UK

Weiqiang Li
Jilin Da'an Agro-Ecosystem National Observation Research Station
Changchun Jingyuetan Remote Sensing Experiment Station
State Key Laboratory of Black Soils Conservation and Utilization
Northeast Institute of Geography and Agroecology
Chinese Academy of Sciences
Changchun
Jilin
China

Jiali Liu
State Key Laboratory of Component-based Chinese Medicine
Tianjin University of Traditional Chinese Medicine
Tianjin
China
Haihe Laboratory of Modern Chinese Medicine
Tianjin
China

Maria Lobato-Gómez
Instituto de Biología Molecular y Celular de Plantas
CSIC-Universidad Politécnica de València
Valencia
Spain

Muhammad Aamir Manzoor
Key Laboratory for Quality Regulation of Tropical Horticultural Crops of Hainan Province
School of Tropical Agriculture and Forestry, Hainan University
Haikou
Hainan
China
National Key Laboratory for Tropical Crop Breeding
School of Breeding and Multiplication (Sanya Institute of Breeding and Multiplication)
Hainan University
Sanya
Hainan
China

Ishrat Mehmood
Department of Biophysics
Panjab University
Chandigarh
India

Yuchen Miao
National Key Laboratory of Cotton Bio-breeding and Integrated Utilization
State Key Laboratory of Crop Stress Adaptation and Improvement
Henan Joint International Laboratory for Crop Multi-Omics Research
School of Life Sciences
Henan University
Kaifeng
Henan
China
Sanya Institute of Henan University
Sanya
Hainan
China

Francisco Javier Verduzco Miramón
Instituto Tecnológico de Estudios Superiores de Zamora
Tecnológico Nacional de México
La Piedad
MC
Mexico

Anshu Mishra
Maharishi School of Pharmaceutical Sciences
Lucknow
Uttar Pradesh
India

Ghulam Murtaza
School of Agriculture
Yunnan University
Kunming
Yunnan
China

Jewel Jameeta Noor
Department of Plant Molecular Biology & Biotechnology
CSIR-Indian Institute of Integrative Medicine
Srinagar
Jammu and Kashmir
India

Aamir Hussain Pala
Department of Plant Molecular Biology & Biotechnology
CSIR-Indian Institute of Integrative Medicine
Srinagar
Jammu and Kashmir
India

Nayaab Shahir Pandit
School of Computational and Integrative Sciences (SC&IS)
Jawaharlal Nehru University
New Delhi
India

Anu Pandita
Vatsalya Clinic
New Delhi
India

Deepu Pandita
Government Department of School Education
Jammu
Jammu and Kashmir
India

Muhammad Waheed Riaz
Department of Agronomy
Shandong Agricultural University
Taian
Shandong
China

Irfan Ali Sabir
Department of Plant Science
School of Agriculture and Biology
Shanghai Jiao Tong University
Shanghai
China

Najwa Shabir
Plant Molecular Biology and Biotechnology Division
CSIR-Indian Institute of Integrative Medicine
Srinagar
Jammu and Kashmir
India
Academy of Scientific and Innovative Research (AcSIR)
Ghaziabad
Uttar Pradesh
India

Mohsin Shafi
Genomic Biology Lab
Department of Biosciences
Jamia Millia Islamia
New Delhi
India

Iftikhar Hussain Shah
Department of Plant Science
School of Agriculture and Biology
Shanghai Jiao Tong University
Shanghai
China

Adnan Shakeel
Department of Botany
Central University of Kashmir
Ganderbal

Mehnaz Shakeel
Department of Plant Molecular Biology & Biotechnology
CSIR-Indian Institute of Integrative Medicine
Srinagar
Jammu and Kashmir
India

Alam Sher
Department of Agronomy
Shandong Agricultural University
Taian
Shandong
China

Devendra Singh
Maharishi School of Pharmaceutical Sciences
Lucknow
Uttar Pradesh
India
Faculty of Biotechnology, Institute of Biosciences and Technology
Shri Ramswaroop Memorial University
Barabanki
Uttar Pradesh
India

Cheng Song
College of Biological and Pharmaceutical Engineering
West Anhui University
Lu'an
Anhui
China

Sudarshna
Department of Bioscience and Bioengineering
Indian Institute of Technology Jodhpur
Jodhpur
Rajasthan
India

Muhammad Usman
Department of Plant Science
School of Agriculture and Biology
Shanghai Jiao Tong University
Shanghai
China

Luis Germán López Valdez
Universidad Autónoma Chapingo
Texcoco
EM
Mexico

Priyanka Verma
(Ex-Pool Scientist) Division of Biochemical Sciences
CSIR-National Chemical Laboratory (NCL)
Pune
Maharashtra
India

Sunil Kumar Verma
Faculty of Biotechnology, Institute of Biosciences and Technology
Shri Ramswaroop Memorial University
Barabanki
Uttar Pradesh
India

Alice Vilela
Chemistry Research Centre—Vila Real (CQ-VR)
Department of Agronomy
School of Agrarian and Veterinary Sciences (ECAV)
University of Trás-os-Montes e Alto Douro
Quinta de Prados
Vila Real
Portugal

Kaiser Iqbal Wani
Key Laboratory for Quality Regulation of Tropical Horticultural Crops of Hainan Province
School of Tropical Agriculture and Forestry
Hainan University
Haikou
Hainan
China
National Key Laboratory for Tropical Crop Breeding
School of Breeding and Multiplication (Sanya Institute of Breeding and Multiplication)
Hainan University
Sanya
Hainan
China

Xiaohui Yan
State Key Laboratory of Component-based Chinese Medicine
Tianjin University of Traditional Chinese Medicine
Tianjin
China
Haihe Laboratory of Modern Chinese Medicine
Tianjin
China

Peng Zhong
State Key Laboratory of Component-based Chinese Medicine
Tianjin University of Traditional Chinese Medicine
Tianjin
China
Haihe Laboratory of Modern Chinese Medicine
Tianjin
China

Tingan Zhou
Department of Natural Product Biosynthesis
Max Planck Institute for Chemical Ecology
Jena
Germany

Preface

Apocarotenoids, derived from the oxidative cleavage of carotenoids, constitute a diverse group of metabolites with critical functions in plant biology. They are not only essential for pigmentation, aroma, and flavor but also serve as signaling molecules, growth regulators, and mediators of plant–environment interactions. In recent years, research on apocarotenoids has gained significant momentum, owing to their pivotal roles in plant growth and development, biotic and abiotic stress responses, symbiotic associations, and their emerging potential in agricultural biotechnology.

This volume, *Plant Apocarotenoids: Biosynthesis, Signaling and Agricultural Applications*, has been conceived to provide a comprehensive and up-to-date account of advances in this expanding field. The book is organized into five sections, each addressing a specific thematic area. The introductory section outlines the fundamentals of apocarotenoid biosynthesis, signaling pathways, and regulatory mechanisms. The second section highlights key apocarotenoids, including strigolactones, zaxinone, β-cyclocitral, and β-ionone, with emphasis on their biological significance and potential applications. A special section is devoted to saffron apocarotenoids, reflecting their unique evolutionary trajectory, biosynthetic regulation, and biotechnological relevance. The fourth section explores the physiological functions of apocarotenoids, particularly in plant defense responses, stress tolerance, and symbiotic interactions. The final section addresses applied aspects, focusing on metabolic engineering approaches aimed at enhancing apocarotenoid production for agricultural improvement and sustainability.

This volume brings together contributions from leading experts in the field, providing an integrated perspective that spans from fundamental biochemical and molecular insights to translational applications in crop science. It is envisaged that the book will serve as a valuable reference for researchers, graduate students, and professionals in plant biology, crop physiology, molecular biology, and agricultural biotechnology.

We sincerely acknowledge the efforts of all contributing authors for their scholarly chapters and dedication, which have greatly enriched this volume. We also express our gratitude to our respective institutions for their encouragement and support throughout the preparation of this book. Finally, we extend our appreciation to the publishing team for their assistance and professionalism in bringing this work to completion.

Kaiser Iqbal Wani
Tariq Aftab
(Editors)

Section I

Introduction and Fundamentals

1

General Overview of Apocarotenoids

Anu Pandita[1] *and Deepu Pandita*[2]*

[1] *Vatsalya Clinic, New Delhi, India*
[2] *Government Department of School Education, Jammu, Jammu and Kashmir, India*

*Corresponding author: deepupandita@gmail.com

1.1 Introduction

Apocarotenoid carbonyl products are degradation products resulting from the enzymatically catalyzed oxidative cleavage and/or nonenzymatic degradation of carotenoids and terpenoid compounds containing fewer than 40 carbon (C) atoms. The apocarotenoid terpenoids can be found in bacteria, fungi, plants, and some animals. Many apocarotenoids have been identified due to the vast array of natural carotenoids.

These result from the degradation of carotenoids, occurring by the loss of C atoms in the side chain of the carotenoid structure. The gained apocarotenoids have shorter fragments than precursor carotenoids due to cleavage at one or both ends (Beltran and Stange 2016; Barrero et al. 2017; Dzib-Cauich et al. 2020). The oxidative cleavage of carotenoids can be site-specific (enzymatic) or non-site-specific (nonenzymatic) (Dzib-Cauich et al. 2020).

The carotenoid-specific cleavage oxygenases (CCOs) are nonheme iron-type enzymes that cleave double bonds in conjugated C chain of carotenoids, resulting in the formation of specific apocarotenoids (Ahrazem et al. 2016). CCOs consist of carotenoid cleavage dioxygenases (CCDs) and 9-*cis*-epoxydioxygenases (Beltran and Stange 2016). CCDs containing the retinal segment epithelial membrane protein (RPE65) domain, which is responsible for binding Fe^{2+}, typically function by the incorporation of oxygen (O_2) atoms between adjacent C atoms along the conjugated carotenoid's backbone (Hou et al. 2016). Furthermore, nonenzymatic oxidation exposes carotenoids to reactive oxygen species (ROS).

The apocarotenoids can be classified as volatile, semi-volatile, or nonvolatile on the basis of their chemical structures. Additionally, these can be categorized into trisporoids, cyclofarnesoids, and methylhexanoids depending on the C atom number in their structure. The word apocarotenoids searched in PubMed showed 82 papers in 2022 (Figure 1.1).

Plant Apocarotenoids: Biosynthesis, Signaling and Agricultural Applications, First Edition. Edited by Kaiser Iqbal Wani and Tariq Aftab.
Published 2026 by John Wiley & Sons Ltd.

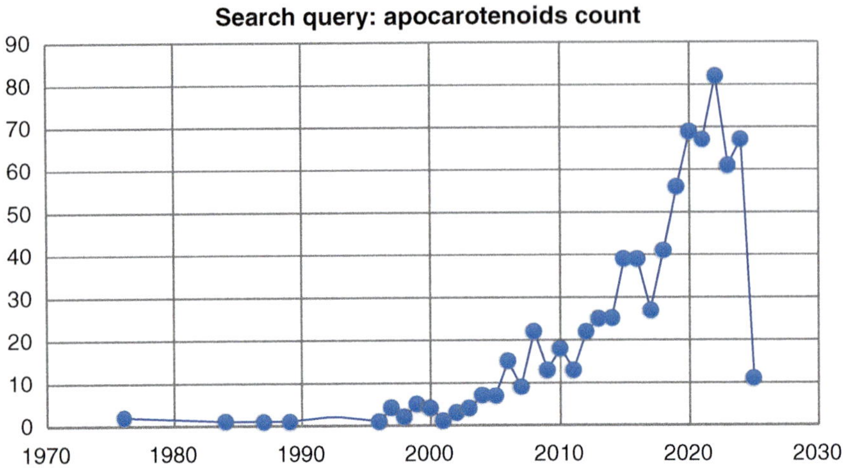

Figure 1.1 Results of search word "apocarotenoids" by year in PubMed, with the highest papers (82) in 2022.

1.2 Apocarotenoids Sources

Plants, fungi, and bacteria all contain apocarotenoids (Ahrazem et al. 2016). The key enzymes β-carotene 15,15′-oxygenase (CMOI) or BCO1 and β-carotene 9′,10′-oxygenase (CMOII) or β-carotene 9′,10′-dioxygenase, or BCO2 regulate the conversion of carotenoids into apocarotenoids (Dzib-Cauich et al. 2020).

Apocarotenoids within plant cells are found in a variety of plastids, including amyloplasts, chloroplasts, leucoplasts, intermediate forms of chloroplasts (etioplasts), and chromoplasts. Apocarotenoids are found in various plant parts such as seeds, fruits, barks, flowers, and roots (Beltran and Stange 2016; Mi et al. 2018), and interspecies fungi such as *Phycomyces blakesleeanus* (Medina et al. 2011; Barrero et al. 2017), *Arbuscular mycorrhiza* (Fester et al. 2002), *Blakeslea trispora* Thaxter (1914), and *Mucor mucedo* Linnaeus (1753) (Barrero et al. 2017) also contain apocarotenoids. Arbuscular mycorrhizal maize roots contain two types of apocarotenoids, that is, mycorradicin and blumenol C cellobioside (Fester et al. 2002). Apocarotenoids such as trisporic acid, trisporin, trisporol, cyclofarnesine T, (2S,8S,E)-cyclofarnesoid, and (2S,8R,E)-cyclofarnesoid have been isolated from various cultures of *P. blakesleeanus*, Burgeff 1925, *M. mucedo*, and *B. trispora* (Barrero et al. 2017). Apocarotenoids have diverse and different functions in a plant, which are as follows (Figure 1.2) (Beltran and Stange 2016; Mi et al. 2018; Havaux 2020; Bo et al. 2023):

Figure 1.2 Functions of apocarotenoids in plants.

1. act as repellents, chemoattractants, growth simulators, and inhibitors.
2. provide colors to flowers, fruits, and seeds that support plant reproduction and protection from pathogen infections.

3. attract animals for flower pollination and seed dispersal.
4. increase the photooxidative stress tolerance.
5. upregulation of lipid peroxidation.
6. play role in plant's response to drought and other abiotic stress conditions.
7. establishing plant's shoot and root architecture.
8. increase the osmolyte accumulation.
9. promote cell division in root meristems and stimulate lateral root branching.
10. take part in intraspecies communication.

1.3 Classification of Apocarotenoids

Some 850 naturally occurring carotenoids have been reported so far (Maoka 2020). The high number of carotenoids leads to a wide variety of apocarotenoids. Volatile apocarotenoids consist of β-ionone (also known as (E)-β-ionone or *trans*-β-ionone), pseudoionone (citrylideneacetone), β-cyclocitral, *cis*-6-geranylacetone (*trans*-geranylacetone), 2-hydroxy-β-ionone (4-(5-hydroxy-2,6,6-trimethyl-1-cyclohexen-1-yl)-3-buten-2-one), neryl acetate ((Z)-3,7-dimethylocta-2,6-dien-1-yl acetate), β-damascenone (*trans*-β-damascenone), geranyl acetate ([(2E)-3,7-dimethylocta-2,6-dienyl] acetate), geranyl acetone ((5E)-6,10-dimethylundeca-5,9-dien-2-one), and 6-methyl-5-hepten-2-one (sulcatone) (Rodríguez-Concepción and Welsch 2020; Shi et al. 2020). Farnesylacetone ((5E,9E)-6,10,14-trimethylpentadeca-5,9,13-trien-2-one) is one example of semi-volatile linear $C_{18}H_{30}O$ apocarotenoid found in *Solanum lycopersicum* (D'Ambrosio et al. 2023). Non-volatile apocarotenoids identified in *S. lycopersicum* contain apocarotenoids, known as the aglycons (genin). There are six glycosylated aglycons in *S. lycopersicum* reported, including hydroxy-β-cyclocitral, hydroxy-β-damascenone 1 (3-hydroxy-β-damascone), hydroxy-β-damascenone 2, hydroxy-β-ionone, 3-hydroxy-5,6-epoxy-β-ionone (HEBI), and 6-hydroxy-3-oxo-α-ionone (Vomifoliol) (D'Ambrosio et al. 2023). Additionally, six glycosylated hydroxy-apocarotenoids are detected and quantified in *S. lycopersicum*, which are hydroxy-β-cyclocitral-glucoside (picrocrocin), hydroxy-β-damascenone-glucoside, hydroxy-β-ionone-glucoside 1, 3-hydroxy-5,6-epoxy-β-ionone-glucoside 1, 3-hydroxy-5,6-epoxy-β-ionone-glucoside 2, and hydroxy-β-ionone-glucoside 2 (D'Ambrosio et al. 2023). In the meantime, nonvolatile apocarotenoids include β-apo-8′-carotenal (Apocarotenal), β-apo-10′-carotenal (10′-apo-β-carotenal), β-apo-12′-carotenal (12′-apo-β,ψ-carotenal), β-apo-14′-carotenal, β-apo-15-carotenal, and β-apo-13-carotenone (Mi et al. 2018). Moreover, β-carotene-5,6-epoxide (5,6-epoxy-β,β-carotene), β-carotene-5,8-epoxide, mutachrome (Citroxanthin), and aurochrome (3′-methyl-phthalanilic acid) are other examples of nonvolatile apocarotenoids (Singh et al. 2015). The scheme of the classification of the apocarotenoids is demonstrated in Figure 1.3.

Apocarotenoids are further divided into three groups based on the C number in their structures: trisporoids, cyclofarnesoids, and methylhexanoids (Medina et al. 2011; Barrero et al. 2017). Trisporic acid (1,3-dimethyl-2-[(1E,3E)-3-methylocta-1,3-dienyl]-4-oxocyclohex-2-ene-1-carboxylic acid), trisporin, trisporol, and methyl trisporate B (methyl 1,3-dimethyl-2-[(1E,3E)-3-methyl-7-oxoocta-1,3-dienyl]-4-oxocyclohex-2-ene-1-carboxylate) are trisporoids isolated from different cultures of *P. blakesleeanus* Burgeff, 1925, *Choanephora trispora*, and *Mucor coprophilus* (Barrero et al. 2017). The second apocarotenoid group has 15 C

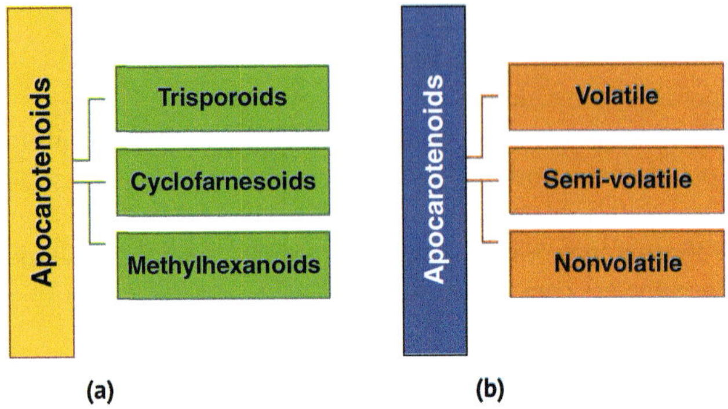

Figure 1.3 Classes of apocarotenoids based on (a) chemical structure and (b) C number.

atoms and is now called cyclofarnesoids (apotrisporoids). Cyclofarnesine T ((4S)-5-hydroxy-4-(hydroxymethyl)-3-(5-hydroxy-3-methylpenta-1,3-dienyl)-2,4-dimethylcyclohex-2-en-1-one), (2S,8S,E)-cyclofarnesoid, and (2S,8R,E)-cyclofarnesoid are cyclofarnesoids isolated from *P. blakesleeanus*, while (3S,5S,8R,E)-cyclofarnesoid U is isolated from *C. trispora*. The methylhexanoid apocarotenoids are the third type with seven C atoms and 4-dihydrocyclofarnesine T is one example of methylhexanoids isolated from *P. blakesleeanus* (Barrero et al. 2017). Research suggests that the three groups of apocarotenoids originate from separate fragments of β-carotene (Medina et al. 2011).

1.4 Structural Diversity of Apocarotenoids

The structures and molecular formulas of various apocarotenoids were derived from chemspider.com and pubchem.com (Figure 1.4). Some examples of apocarotenoids, their structures, and their sources are tabulated in Table 1.1.

1.5 Significance of Apocarotenoids

A variety of carotenoids used as precursors have an impact on the variety of apocarotenoids produced. The α-ionone, β-ionone, β-cyclocitral, and β-damascenone have been estimated to contribute as much as 8% to the aroma profile of Valencia orange juice, and up to 78% of the total floral aroma (Mahattanatawee et al. 2005). β-Ionone is the most significant volatile compound that contributes to the flavor of *S. lycopersicum*, commonly known as tomato fruit. The β-cyclocitral that is identified along with β-ionone is the second most significant volatile compound that contributes to the flavor of fruits of both *S. lycopersicum* and *Citrullus lanatus* containing β-carotene. Besides, α-ionone is detected only in an orange-fleshed *S. lycopersicum* mutant Delta, which accumulates ζ-carotene (Baldwin et al. 2000; Lewinsohn et al. 2005). The β-ionone has also been identified as one of the major components of leaves of the *Lawsonia inermis, C. lanatus*, and Moroccan herbs (*Cistus monspeliensis* or narrow-leaved cistus, *Myrtus communis* or true myrtle, and *Cistus ladanifer* L. or gum rockrose) and roots of the orange and purple carrot (*Daucus carota* L. or wild carrot), and imparts fragrance in the flowers of *Boronia megastigma* Bartl. (brown boronia),

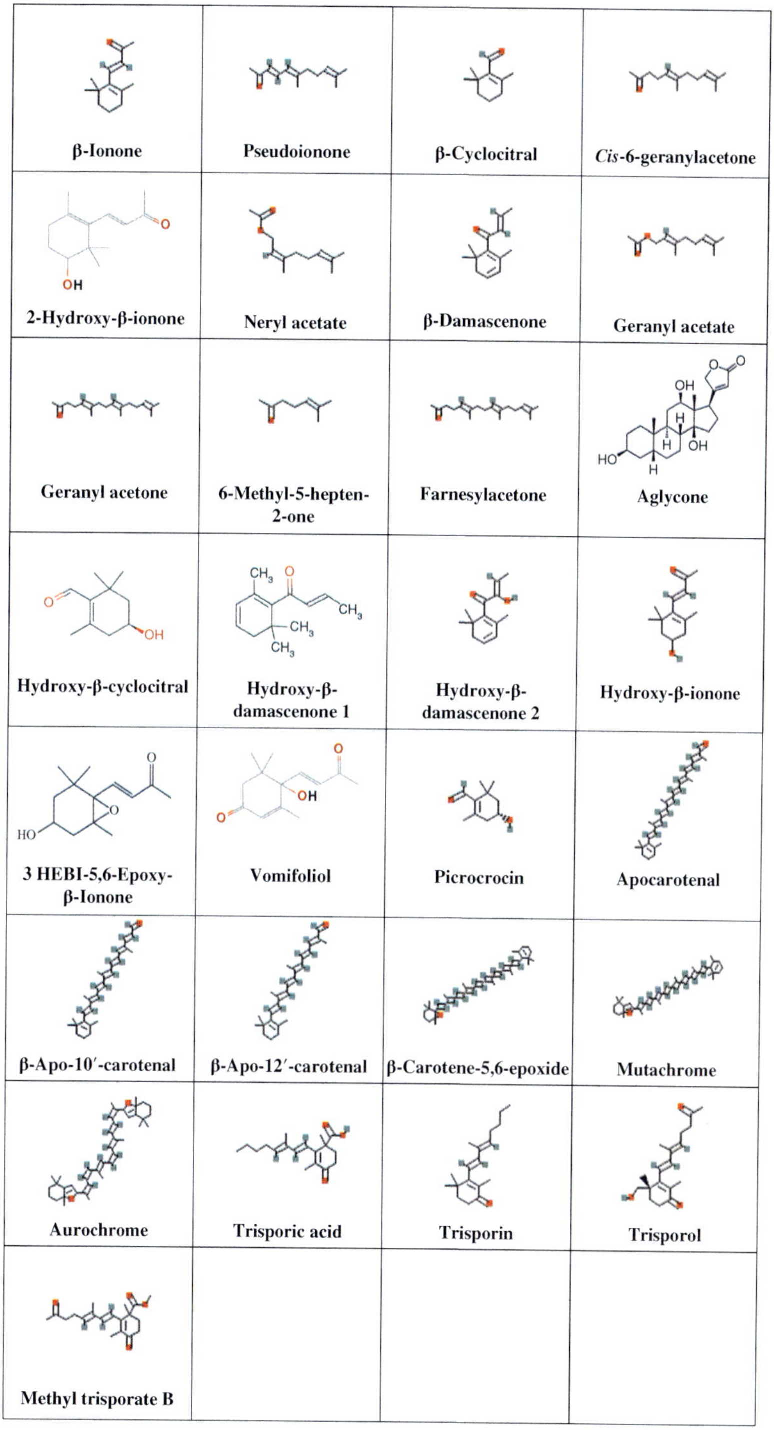

Figure 1.4 Structures of some apocarotenoids.

Table 1.1 Structural diversity and sources of various apocarotenoids.

Apocarotenoids	Molecular formula	Structure	Sources	References
β-Ionone	$C_{13}H_{20}O$	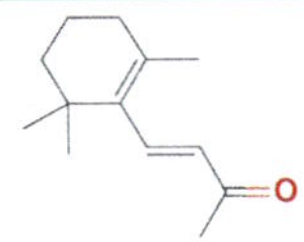	*Osmanthus fragrans* Lour. *Petunia × hybrida* *Rosa bourboniana* *Lawsonia inermis* L. *Myrtus communis* L. *Cistus ladanifer* L. *Montpellier cistus* L. *Cascabela thevetia* (L.) Lippold *Daucus carota* *Solanum lycopersicum* *Malus domestica* *Prunus domestica* *Cucumis melo* L. *Rubus idaeus* *Prunus armeniaca*	Paparella et al. (2021)
Pseudoionone	$C_{13}H_{20}O$	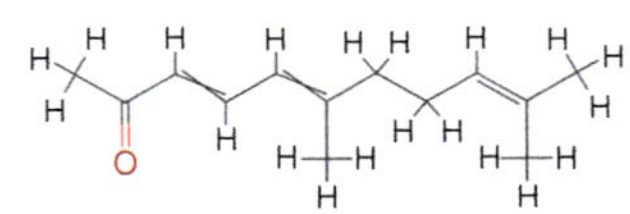	*Solanum lycopersicum* *Citrullus lanatus*	Simkin (2021)
β-Cyclocitral	$C_{10}H_{16}O$	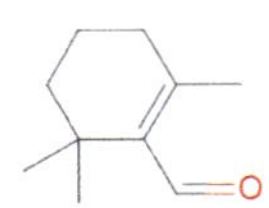	*Cyanidium caldarium* *Camellia sinensis* (L.) Kuntze *Piper nigrum* L. *Solanum lycopersicum* L. *Oryza sativa* L. *Auxenochlorella pyrenoidosa* *Arabidopsis thaliana* L.	Faizan et al. (2022)

β-Damascenone	$C_{13}H_{18}O$	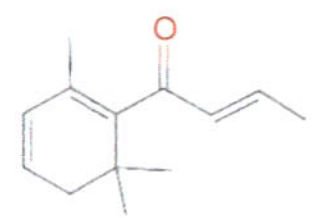	*Rosa* spp. *Vitis vinifera*	Tomasino and Bolman (2021)
Geranyl acetone	$C_{13}H_{22}O$	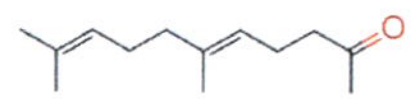	*Solanum lycopersicum* *Vitis vinifera* *Cucumis melo* L. *Petunia* spp.	Zheng et al. (2021), Shi et al. (2020), Saad et al. (2019)
Crocin	$C_{44}H_{64}O_{24}$	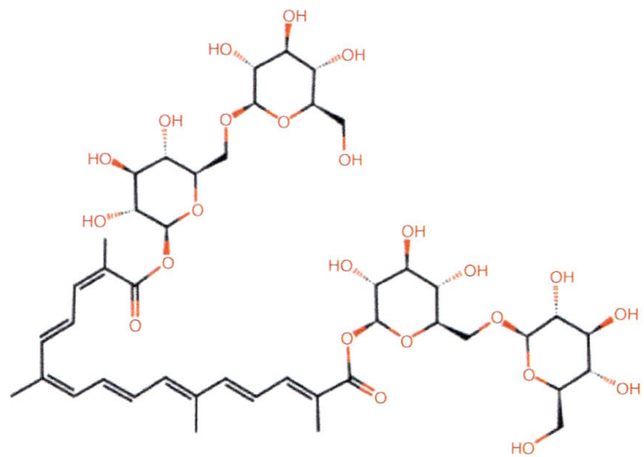	*Crocus sativus* *Gardenia jasminoides* *Buddleja davidii* *Nyctanthes arbor-tristis*	Zheng et al. (2021), Morote et al. (2023)
β-Citraurin	$C_{30}H_{40}O_2$	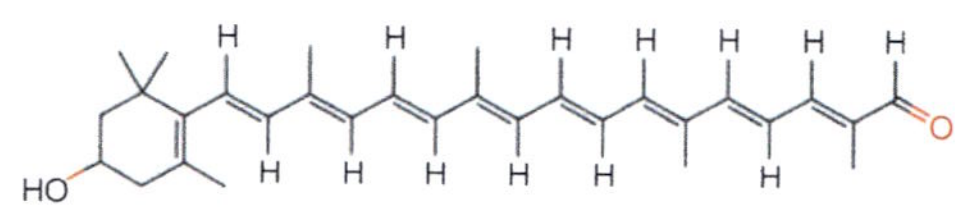	*Citrus* spp.	Rodríguez-Concepción and Welsch (2020)

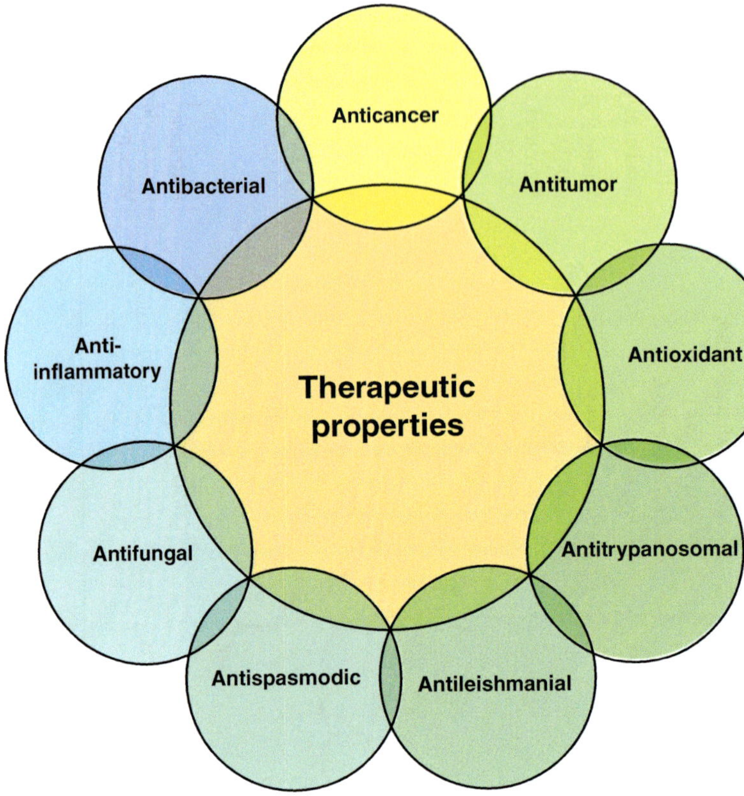

Figure 1.5 Pharmacological activities of β-ionone.

Petunia × *hybrida* hort. ex E. Vilm., and *Rosa* × *damascena* Mill. (Damask rose) (Cooper et al. 2003; Simkin et al. 2004; Oyedeji et al. 2005; Ibdah et al. 2006; Huang et al. 2009; Viuda-Martos et al. 2011; Yahyaa et al. 2013).

Apocarotenoids exhibit a variety of biological activities, including antispasmodic, anticancer, antitumor, antioxidant, antibacterial, antifungal, antitrypanosomal, anti-inflammatory, and antileishmanial (Neuhaus et al. 2009; Saad et al. 2019; Paparella et al. 2021; Zheng et al. 2021). As shown in Figure 1.5, β-ionone has pharmacological activities like antibacterial, antifungal, and antileishmanial (Suryawanshi et al. 2007; Sharma et al. 2012). β-Ionone actively inhibits proliferation of *Escherichia coli* and *Candida albicans*. β-Ionone also affects the growth of *Aspergillus flavus* and sporulation of *A. flavus* and *Aspergillus parasiticus* (Wilson et al. 1981; Griffin et al. 1999). The β-ionone is anti-inflammatory and cancer-preventing (Balbi et al. 2006; Ansari and Emami 2016; Dong et al. 2019). β-Ionone suppresses breast cancer cell proliferation (Dong et al. 2019), prostate cancer cell growth (Xie et al. 2019), induces apoptosis in murine B16 melanoma cells, human leukemia, and suppresses the proliferation of human colon adenocarcinoma cell lines (Mo and Elson 1999), human colon cancer (Janakiram et al. 2008), and human gastric adenocarcinoma (Dong et al. 2013).

The bixin and crocin apocarotenoids (Beltran and Stange 2016) are the two most important colorants and are used as flavors in the food and cosmetic industries (Kapoor and

Ramamoorthy 2021; Zheng et al. 2021). Crocin and picrocrocin are water-soluble apocarotenoids that are derived from the cleavage of zeaxanthin at 7,8/7′,8′ positions by a specific CCD enzyme (CsCCD2L) (Ahrazem et al. 2022). These hydrophilic pigments occur in *Crocus sativus*, and in reduced amounts in *Buddleja* (*Buddlea*) and *Gardenia* (Bukhari et al. 2018; Ahrazem et al. 2022). Bixin is a polyene derived from the enzymatic cleavage of the central part of a C_{40} lycopene, which communicates bright red color to the specific tissues of *Bixa orellana* L. (Kapoor and Ramamoorthy 2021).

1.6 Conclusion

The carotenoids serve as precursors for the synthesis of apocarotenoids, which may be volatile, semi-volatile, or nonvolatile compounds based on their chemical structures, and trisporoids, cyclofarnesoids, and methylhexanoids based on the number of C atoms contained in their structure. The apocarotenoids are antispasmodic, anticancer, antitumor, antioxidant, antibacterial, antifungal, antitrypanosomal, anti-inflammatory, and antileishmanial and find use in the food and cosmetic industries.

References

Ahrazem, O., Diretto, G., Luis Rambla, J. et al. (2022). Engineering high levels of saffron apocarotenoids in tomato. *Horticulture Research* 9, uhac074. https://doi.org/10.1093/hr/uhac074.

Ahrazem, O., Gómez-Gómez, L., Rodrigo, M.J. et al. (2016). Carotenoid cleavage oxygenases from microbes and photosynthetic organisms: features and functions. *International Journal of Molecular Sciences* 17(11), 1781. https://doi.org/10.3390/ijms17111781.

Ansari, M. and Emami, S. (2016). β-Ionone and its analogs as promising anticancer agents. *European Journal of Medicinal Chemistry* 123, 141–154.

Balbi, A., Anzaldi, M., Mazzei, M. et al. (2006). Synthesis and biological evaluation of novel heterocyclic ionone-like derivatives as anti-inflammatory agents. *Bioorganic & Medicinal Chemistry* 14, 5152–5160.

Baldwin, E.A., Scott, J.W., Shewmaker, C.K. et al. (2000). Flavor trivia and tomato aroma: biochemistry and possible mechanisms for control of important aroma components. *Journal of Horticultural Sciences* 35, 1013–1022.

Barrero, A.F., Del Pino, M.M.H., Arteaga, J.F. et al. (2017). Occurrence and chemical synthesis of apocarotenoids from mucorales: a review. *Natural Product Communications* 12(5), 733–741. https://doi.org/10.1177/1934578x1701200518.

Beltran, J.C.M. and Stange, C. (2016). Apocarotenoids: a new carotenoid-derived pathway. In: *Subcellular Chemistry: Vol. 79: Carotenoids in Nature* (ed. C. Stange), 239–272. Springer. https://doi.org/10.1007/978-3-319-39126-7_9.

Bo, R., Zhan, Y., Wei, S. et al. (2023). Tea tree oil nanoliposomes: optimization, characterization, and antibacterial activity against *Escherichia coli* in vitro and in vivo. *Poultry Science* 102(1), 102238. https://doi.org/10.1016/j.psj.2022.102238.

Bukhari, S.I. Manzoor, M. and Dhar, M.K. (2018). A comprehensive review of the pharmacological potential of *Crocus sativus* and its bioactive apocarotenoids. *Biomedicine & Pharmacotherapy* 98, 733–745. https://doi.org/10.1016/j.biopha.2017.12.090.

Cooper, C.M., Davies, N.W. and Menary, R.C. (2003). C-27 apocarotenoids in the flowers of *Boronia megastigma* (Nees). *Journal of Agricultural and Food Chemistry* 51, 2384–2389.

D'Ambrosio, C., Stigliani, A.L., Rambla, J.L. et al. (2023). A xanthophyll-derived apocarotenoid regulates carotenogenesis in tomato chromoplasts. *Plant Science* 328, 111575. https://doi.org/10.1016/j.plantsci.2022.111575.

Dong, H.W., Wang, K., Chang, X.X. et al. (2019). Beta-ionone-inhibited proliferation of breast cancer cells by inhibited COX-2 activity. *Archives of Toxicology* 93, 2993–3003.

Dong, H.W., Zhang, S., Sun, W.G. et al. (2013). β-Ionone arrests cell cycle of gastric carcinoma cancer cells by a MAPK pathway. *Archives of Toxicology* 87, 1797–1808.

Dzib-Cauich, J., Us-Camas, R. and Rivera Madrid, R. (2020). Natural sources of apocarotenoids and their applications. In: *Biology, Chemistry and Applications of Apocarotenoids* (eds. S. Ramamoorthy, R.R. Madrid and C. George Priya Doss), 11–39. CRC Press. https://doi.org/10.1201/9780429344206-2.

Faizan, M., Tonny, S.H., Afzal, S. et al. (2022). β-Cyclocitral: emerging bioactive compound in plants. *Molecules* 27(20), 6845. https://doi.org/10.3390/molecules27206845.

Fester, T., Hause, B., Schmidt, D. et al. (2002). Occurrence and localization of apocarotenoids in arbuscular mycorrhizal plant roots. *Plant & Cell Physiology* 43(3), 256–265. https://doi.org/10.1093/pcp/pcf029.

Griffin, S.G., Wyllie, S.G., Markham, J.L. et al. (1999). The role of structure and molecular properties of terpenoids in determining their antimicrobial activity. *Flavour and Fragrance Journal* 14, 322–332.

Havaux, M. (2020). β-Cyclocitral and derivatives: emerging molecular signals serving multiple biological functions. *Plant Physiology and Biochemistry* 155, 35–41. https://doi.org/10.1016/j.plaphy.2020.07.032.

Hou, X., Rivers, J., León, P. et al. (2016). Synthesis and function of apocarotenoid signals in plants. *Trends in Plant Science* 21(9), 792–803. https://doi.org/10.1016/j.tplants.2016.06.001.

Huang, F.C., Horváth, G., Molnár, P. et al. (2009). Substrate promiscuity of RdCCD1, a carotenoid cleavage oxygenase from *Rosa damascena*. *Phytochemistry* 70, 457–464.

Ibdah, M., Azulay, Y., Portnoy, V. et al. (2006). Functional characterization of CmCCD1, a carotenoid cleavage dioxygenase from melon. *Phytochemistry* 67, 1579–1589.

Janakiram, N.B., Cooma, I., Mohammed, A. et al. (2008). β-Ionone inhibits colonic aberrant crypt foci formation in rats, suppresses cell growth, and induces retinoid X receptor-α human colon cancer cells. *Molecular Cancer Therapeutics* 7, 181–190.

Kapoor, L. and Ramamoorthy, S. (2021). Strategies to meet the global demand for natural food colorant bixin: a multidisciplinary approach. *Journal of Biotechnology* 338, 40–51. https://doi.org/10.1016/j.jbiotec.2021.07.007.

Lewinsohn, E., Sitrit, Y., Bar, E. et al. (2005). Carotenoid pigmentation affects the volatile composition of tomato and watermelon fruits, as revealed by comparative genetic analyses. *Journal of Agricultural and Food Chemistry* 53, 3142–3148.

Mahattanatawee, K., Rouseff, R., Valim, M.F. et al. (2005). Identification and aroma impact of norisoprenoids in orange juice. *Journal of Agricultural and Food Chemistry* 53, 393–397.

Maoka, T. (2020). Carotenoids as natural functional pigments. *Journal of Natural Medicines* 74(1), 1–16. https://doi.org/10.1007/s11418-019-01364-x.

Medina, H.R., Cerdá-Olmedo, E. and Al-Babili, S. (2011). Cleavage oxygenases for the biosynthesis of trisporoids and other apocarotenoids in *Phycomyces*. *Molecular Microbiology* 82(1), 199–208. https://doi.org/10.1111/j.1365 2958.2011.07805.x.

Mi, J., Jia, K.P., Wang, J.Y. et al. (2018). A rapid LC-MS method for qualitative and quantitative profiling of plant apocarotenoids. *Analytica Chimica Acta* 1035, 87–95. https://doi.org/10.1016/j.aca.2018.07.002.

Mo, H. and Elson, C.E. (1999). Apoptosis and cell-cycle arrest in human and murine tumor cells are initiated by isoprenoids. *The Journal of Nutrition* 129, 804–813.

Morote, L., Lobato-Gómez, M., Ahrazem, O. et al. (2023). Crocins-rich tomato extracts showed enhanced protective effects *in vitro*. *Journal of Functional Foods* 101, 105432. https://doi.org/10.1016/j.jff.2023.105432.

Neuhaus, E.M., Zhang, W., Gelis, L. et al. (2009). Activation of an olfactory receptor inhibits proliferation of prostate cancer cells. *Journal of Biological Chemistry* 284, 16218–16225.

Oyedeji, A.O., Ekundayo, O. and Koenig, W.A. (2005). Essential oil composition of *Lawsonia inermis* L. leaves from Nigeria. *Journal of Essential Oil Research* 17, 403–404.

Paparella, A., Shaltiel-Harpaza, L. and Ibdah, M. (2021). β-Ionone: its occurrence and biological function and metabolic engineering. *Plants* 10(4), 754. https://doi.org/10.3390/plants10040754.

Rodríguez-Concepción, M. and Welsch, R. (2020). *Methods in Molecular Biology: Vol. 2083: Plant and Food Carotenoids*. Humana. https://doi.org/10.1007/978-1-4939-9952-1_8.

Saad, S.B., Ibrahim, M.A., Jatau, I.D. et al. (2019). Trypanostatic activity of geranylacetone: mitigation of *Trypanosoma congolense*-associated pathological pertubations and insight into the mechanism of anaemia amelioration using *in vitro* and *in silico* models. *Experimental Parasitology* 201, 49–56. https://doi.org/10.1016/j.exppara.2019.04.011.

Sharma, V., Singh, G., Kaur, H. et al. (2012). Synthesis of β-ionone derived chalcones as potent antimicrobial agents. *Bioorganic & Medicinal Chemistry Letters* 22, 6343–6346.

Shi, J., Cao, C., Xu, J. et al. (2020). Research advances on biosynthesis, regulation, and biological activities of apocarotenoid aroma in horticultural plants. *Journal of Chemistry* 2020, 2526956. https://doi.org/10.1155/2020/2526956.

Simkin, A.J. (2021). Carotenoids and apocarotenoids in planta: their role in plant development, contribution to the flavour and aroma of fruits and flowers, and their nutraceutical benefits. *Plants* 10, 2321. https://doi.org/10.3390/plants10112321.

Simkin, A.J., Underwood, B.A., Auldridge, M. et al. (2004). Circadian regulation of the PhCCD1 carotenoid cleavage dioxygenase controls emission of β-ionone, a fragrance volatile of petunia flowers. *Plant Physiology* 136, 3504–3514.

Singh, A., Ahmad, S. and Ahmad, A. (2015). Green extraction methods and environmental applications of carotenoids—a review. *RCS Advances* 5, 62358–62393.

Suryawanshi, S.N., Bhat, B.A., Pandey, S. et al. (2007). Chemotherapy of leishmaniasis. Part VII: synthesis and bioevaluation of substituted terpenyl pyrimidines. *European Journal of Medicinal Chemistry* 42, 1211–1217.

Tomasino, E. and Bolman, S. (2021). The potential effect of β-ionone and β-damascenone on sensory perception of pinot noir wine aroma. *Molecules* 26(5), 1288. https://doi.org/10.3390/molecules26051288.

Viuda-Martos, M., Sendra, E., Pérez-Alvarez, J.A. et al. (2011). Identification of flavonoid content and chemical composition of the essential oils of Moroccan herbs: myrtle (*Myrtus communis* L.), rockrose (*Cistus ladanifer* L.) and Montpellier cistus (*Cistus monspeliensis* L.). *Journal of Essential Oil Research* 23, 1–9.

Wilson, D.M., Gueldner, R.C., Mckinney, J.K. et al. (1981). Effect of β-ionone on *Aspergillus flavus* and *Aspergillus parasiticus* growth, sporulation, morphology and aflatoxin production. *Journal of the American Oil Chemists' Society* 58, A959–A961.

Xie, H., Liu, T., Chen, J. et al. (2019). Activation of PSGR with β-ionone suppresses prostate cancer progression by blocking androgen receptor nuclear translocation. *Cancer Letters* 453, 193–205.

Yahyaa, M., Bar, E., Dubey, N.K. et al. (2013). Formation of norisoprenoid flavor compounds in carrot (*Daucus carota* L.) roots: characterization of a cyclic-specific carotenoid cleavage dioxygenase 1 gene. *Journal of Agricultural and Food Chemistry* 61, 12244–12252.

Zheng, X., Yang, Y. and Al-Babili, S. (2021). Exploring the diversity and regulation of apocarotenoid metabolic pathways in plants. *Frontiers in Plant Science* 12, 787049. https://doi.org/10.3389/fpls.2021.787049.

2

Apocarotenoid Biosynthesis, Signaling, and Regulatory Mechanisms in Plants

Sunil Kumar Verma[1], Devendra Singh[1], and Anshu Mishra[2]*

[1] *Faculty of Biotechnology, Institute of Biosciences and Technology, Shri Ramswaroop Memorial University, Barabanki, Uttar Pradesh, India*
[2] *Maharishi School of Pharmaceutical Sciences, Lucknow, Uttar Pradesh, India*

*Corresponding author: devendrasingh.ibst@srmu.ac.in

2.1 Introduction

Carotenoids are isoprenoid compounds that are lipophilic and comprise a polyene backbone with 3–11 conjugated double bonds. Plants, photosynthetic bacteria, and algae, in addition to several non-photosynthetic microorganisms, can all synthesize them from scratch (Hirschberg 2001; Fraser and Bramley 2004; DellaPenna and Pogson 2006; Moise et al. 2014; Rodriguez-Concepcion et al. 2018; Zheng et al. 2020a; Verma et al. 2024a). Although humans cannot produce carotenoids from scratch, they are nonetheless necessary for provitamin A and as significant antioxidants, especially in the event that food derived from animals is unavailable (Fraser and Bramley 2004; DellaPenna and Pogson 2006; Giuliano 2014; Nisar et al. 2015; Zheng et al. 2020b; Singh et al. 2023, 2024b).

Carotenoids give flowers, fruits, seeds, and tubers in several crops and various plant species their vibrant red-to-yellow colors (Yuan et al. 2015; Giuliano 2017; Rathva et al. 2024). These colors draw in animal pollinators as well as distributors and are a significant feature for consumers, indicating the plant products' quality, thereby promoting their marketability. More significantly, because they shield the photosynthetic machinery from photooxidative damage and aid in light harvesting, carotenoids are necessary for plant photosynthesis (Hashimoto et al. 2016; Moreno et al. 2021; Sharma et al. 2023b). Due to their conjugated double bonds, they are vulnerable to a variety of nonspecific oxidative breakdown processes, including co-oxidation mediated by lipoxygenase (LOX), photooxidation, and targeted cleavage of specific double bonds by enzymes (Zheng et al. 2020a; Moreno et al. 2021). The outputs of cleavage resulting from these reactions are called "apocarotenoids."

Plant Apocarotenoids: Biosynthesis, Signaling and Agricultural Applications, First Edition. Edited by Kaiser Iqbal Wani and Tariq Aftab.
Published 2026 by John Wiley & Sons Ltd.

Numerous apocarotenoids with a range of physicochemical characteristics and biological roles are produced by the several double bonds found in the carotenoid backbone and distinct carotenoids (Rodriguez-Concepcion et al. 2018). Pigments and volatiles that give fruits, flowers, and vegetables their color and scent are also members of the compound family of apocarotenoids (Felemban et al. 2019). These substances include precursors of significant phytohormones that have a role in plant development, growth, and resistance to diseases and herbivores, such as strigolactones (SLs) and abscisic acid (ABA) (Jia et al. 2018; Wang et al. 2020; Watkins and Pogson 2020).

β-Carotene cleavage oxygenases 1 and 2 (BCO1 and BCO2) are enzymes found in humans that can break down certain carotenoids into a limited group of apocarotenoids, such as vitamin A (Harrison 2005; von Lintig 2010). Humans are unable to produce many apocarotenoids that are good for health, like safranal and crocins (Frusciante et al. 2014; Ahrazem et al. 2016). Since apocarotenoids are essential to both humans and plants, a great deal of research has been done to understand their biological roles and their biosynthetic processes (Gao et al. 2019; Zheng et al. 2019; Zoccali et al. 2021). Understanding the apocarotenoid regulation production and the genetic regulation of apocarotenoid diversity will be aided by this information (Fiorilli et al. 2019). These days, the development of multi-omics methods and global profiling techniques, including transcriptomics, metabolomics, and genomics, has proven to be quite beneficial in examining the genetic underpinnings behind the variance in apocarotenoid levels. The manufacture of signaling molecules and hormones produced from apocarotenoid pigments, in addition to their function in plant development and growth, is the subject of several recent reviews (Felemban et al. 2019; Wang et al. 2020; Dhiman et al. 2024b). This chapter will primarily address the latest developments in our knowledge of the metabolism, regulation, and genetic process behind the apocarotenoid volatiles and pigment diversity. We also show how comprehending the principles underlying the diversity of apocarotenoids may speed up molecular breeding and next-generation apocarotenoids' metabolic engineering based on CRISPR for improved agricultural yields (Moreno et al. 2021; Sunya et al. 2022; Chauhan et al. 2024; Dhiman et al. 2024a; Sharma et al. 2024a).

2.2 The Significance of Apocarotenoids and Their Functions

Plant metabolites that are biologically significant include signaling molecules, pigments, volatiles, and phytohormones, as seen in Figure 2.1.

Carotenoids undergo oxidative cleavage to yield these metabolites, both by enzymatic and nonenzymatic means, and by subsequent metabolism of the apocarotenoids that are so produced (Table 2.1). For biologically significant plant apocarotenoids, the examples are the phytohormones SLs and ABA, which play crucial roles in several physiological, developmental, and plant interactions with biotic environmental factors. Widely found in fungi, algae, cyanobacteria, plants, and mammals, ABA is a common component of many living organisms (Wang et al. 2020). It is important for the development of seed dormancy, the growth of roots and shoots, the control of stomatal closure, as well as several other developmental and physiological processes (Sharma et al. 2024b; Singh et al. 2024b, d; Verma et al. 2024a). It also plays a significant part in the plant's response to both biotic and abiotic stressors (Chen et al. 2020; Moreno et al. 2021). SLs were first acknowledged as the

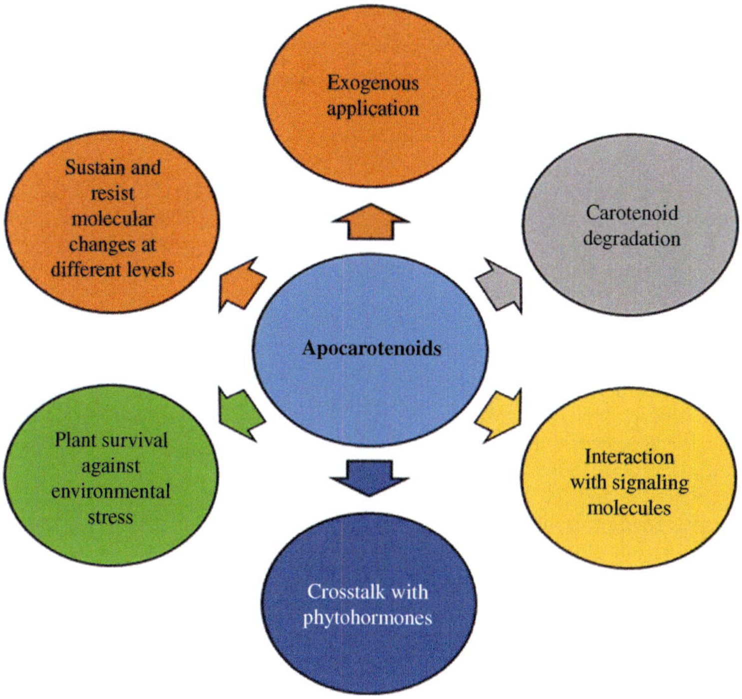

Figure 2.1 Significance of apocarotenoids and their functions.

rhizospheric chemical signals derived from the host that trigger germination of seeds in root parasite plants like *Striga hermonthica*, which pose a danger to the world's food supply (Xie et al. 2010; Jamil et al. 2021). Subsequent research on arbuscular mycorrhization has shown that SLs are the signals derived from plants that cause hyphal branching in arbuscular mycorrhization fungi, hence facilitating host colonization (Cook et al. 1966; Lanfranco et al. 2018; Wang et al. 2020). Plant SLs promote internode elongation, thicken stems, and prevent adventitious root formation and shoot branching. Moreover, SLs play a role in the biotic and abiotic stress responses, in addition to the senescence of leaves and floral organs (Decker et al. 2017; Wang et al. 2020). In addition to these two phytohormones formed from apocarotenoid compounds, other carotenoid-derived plant growth regulators and signaling molecules have been discovered through current investigations into the synthesis and biological roles of apocarotenoids (Moreno et al. 2021). For instance, zaxinone is a growth regulator that influences rice's hormone homeostasis while promoting rice growth. Zaxinone and its mimics can be applied exogenously to improve rice growth and combat root parasitic weeds. This resulted in an increase in the rice seedlings' crown root growth and a reduction in *Striga* infestation in a cultivar of rice that is susceptible to *Striga*. Zaxinone growth-promoting effects in rice are probably due to increased uptake and metabolism of sugar, and cytokinin content modulation is probably the cause of zaxinone's impact on root cell division activity in the root meristem (Al-Babili and Bouwmeester 2015; Wang et al. 2019a). A recent investigation on hydroponically cultivated seedlings of *Arabidopsis* revealed that, in disparity to rice, zaxinone treatment enhanced the SL and ABA levels in roots without enhancing growth. This finding implies that different plant species may have

Table 2.1 Plant apocarotenoids and their derivatives: biosynthesis and biological roles.

Apocarotenoid	Biological functions	CCD enzyme(s) involved	References
Zaxinone	A naturally occurring growth regulator that is necessary for the appropriate rice root development and growth. In *Arabidopsis*, it controls the production of SL and ABA and participates in mycorrhization.	ZAS (CCD10)	Ablazov et al. (2020), Wang et al. (2019, 2021)
Crocetin	Primarily seen in the crimson stigmas of gardenia fruits and crocus flowers. It is a naturally occurring apocarotenoid dicarboxylic acid. In both animal models and cell culture systems, it demonstrates strong antitumorigenic properties.	*Crocus* CCD2, *Gardenia* CCD4a, *Buddleja* CCD4.1/ CCD4.3	Frusciante et al. (2014), Ahrazem et al. (2016, 2017)
SLs	Phytohormones are involved in many developmental processes: rhizospheric signaling molecules that cause AMF's hyphal branching and root parasite plant seed germination. To develop the arbuscular mycorrhizal (AM) symbiosis, the latter is required.	CCD7, CCD8	Zhang et al. (2014)
3-Hydroxy-carlactone	The key stage in the SL production.	CCD7, CCD8	Yoneyama et al. (2020)
Carlactone	A forerunner of SLs that are still unknown.	CCD7, CCD8	Bruno et al. (2014)
Anchorage	A naturally occurring plant metabolite and apocarotenoid dialdehyde that is essential to the *Arabidopsis* anchor roots development.	Unknown	Jia et al. (2019)
ABA	A phytohormone is involved in several developmental processes, such as seed dormancy, as well as the plant's response to abiotic and biotic stress.	NCEDs	Tan et al. (2003), Moreno et al. (2021)
Isoanchorage	An anchorage isomer that prevents *Arabidopsis* primary root development.	Unknown	Jia et al. (2021a)
β-Ionone	A volatile cyclic apocarotenoid that smells fruity or violet. It has a strong repelling effect on flea beetles and spider mites.	CCD1/CCD4/CCD7	Cáceres et al. (2016), Liang et al. (2021)

Apocarotenoid	Biological functions	CCD enzyme(s) involved	References
β-CC	A volatile apocarotenoid that cycles and tastes grassy or tobacco-like. While being a signaling molecule, it controls the oxidative stress response, encourages the growth of *Arabidopsis* roots, and creates resistance in plants to *Spodoptera littoralis* as a result of herbivore-triggered signals.	*Citrus* CCD4b	Zheng et al. (2021), Dickinson et al. (2019), Gao et al. (2019), Mitra et al. (2021)
α-Ionone	A volatile cyclic apocarotenoid that smells fruity or violet. It has a strong repelling effect on flea beetles and spider mites and draws pollinators. Against flower thrips in the West.	CCD1/CCD4	Murata et al. (2020), Liang et al. (2021)
6-Methyl-5-hepten-2-one	An essential component of fruit and flower scent, linear volatile apocarotenoid.	CCD1/CCD4	Liang et al. (2021), Shi et al. (2020), Pandey et al. (2022)
6-Damascenone	A delicious, honey-like cyclic volatile apocarotenoid. It has an extremely low odor threshold for detection. It enhances the scent of flowers and fruits.	Unknown	
Loliolide or (—)-Loliolide	A naturally occurring regulatory metabolite that encourages allelochemical formation in the interactions between barnyard grass and rice, governs the plant's defense response against herbivores.	Unknown	Murata et al. (2019), Li et al. (2020)
Bixin	An apocarotenoid dicarboxylic monomethyl ester that gives *Bixa orellana* seeds their crimson hue. It has anti-inflammatory and antinociceptive properties and is utilized in the food and cosmetics industries.	*Bixa* CCD4	Bouvier et al. (2003), Pacheco et al. (2019)
Safranal	One of the main volatile ingredients of crocus stigma is an apocarotenoid with a strong scent.	*Crocus* CCD2	Rezaee and Hosseinzadeh (2013), Diretto et al. (2019)
Picrocrocin	A 3-OH-β-cyclocitral 6-D-glucoside. It is the ancestor of safranal and causes the crocus stigmas' bitter flavor. It is utilized in food, cosmetics, and medical industries and has pharmacological effects, including a decrease in the growth of melanoma.	*Crocus* CCD2	

(Continued)

Table 2.1 (Continued)

Apocarotenoid	Biological functions	CCD enzyme(s) involved	References
β-Citraurinene	Citrus fruit's red peel is caused by this C30 red apocarotenoid pigment.	*Citrus* CCD4b	Ma et al. (2013), Rodrigo et al. (2013), Zheng et al. (2019)
β-Citraurin	Citrus fruit's red peel is caused by the C30 red apocarotenoid pigment.	*Citrus* CCD4b	
Crocins	Water-soluble apocarotenoids found in nature that are composed of crocetin glycosides. They are mostly present in the crimson stigmas of gardenia fruit and crocus flowers. Their pharmacological effects are diverse, encompassing anti-inflammatory, antiaging, analgesic, and neuroprotective properties.	*Crocus* CCD2, *Gardenia* CCD4a, *Buddleja* CCD4.1/CCD4.3	Frusciante et al. (2014), Ahrazem et al. (2016, 2017), Bukhari et al. (2018)

different zaxinone activities and functions, which are most likely related to their capacity to form symbioses with arbuscular mycorrhizal fungi (AMFs) (Wang et al. 2019; Ablazov et al. 2020; Moreno et al. 2021).

Dialdehyde compounds are produced when carotenoids undergo oxidative breakage at multiple double bonds: antenene, a C10-dialdehyde produced from carotenoids, specifically promotes *Arabidopsis* anchor roots development and growth. By modifying auxin homeostasis, this form of root arises from the collet area or the root hypocotyl junction. Employing carotenoid biosynthesis-impaired mutants and pharmacologic carotenoid biosynthesis inhibitors, as well as developing an LC–MS-based method for dialdehyde extraction and detection from carotenoid, anchorage's authenticity as a naturally occurring carotenoid-derived metabolite was verified (Jia et al. 2019; Mi et al. 2019). When anchoring was discovered, it became clear how important dialdehydes made from carotenoid pigments are to plant development.

The volatile apocarotenoid known as β-Cyclocitral (β-CC) is produced when carotenoids, for instance β-carotene, oxidize. It is found in wildlife and is frequently found containing fungi, plants, and cyanobacteria. Research revealed that β-CC functions as a signaling molecule, controlling the expression of genes that respond to oxidative stress and helping plants adapt to high light levels (Ramel et al. 2012). As a stress signal that increases drought tolerance, β-cycle citric acid (β-CCA) most likely uses a different signaling pathway than β-CC (D'Alessandro et al. 2019). Compared to its ancestor β-CC, β-CCA is more suited for agricultural relevance and easier to utilize to boost crop resistance to drought by spraying or irrigation due to its water-soluble nature (Havaux 2020). β-CC functions as a growth regulator for the root as well as its part in the oxidative stress response. In tomato, *Arabidopsis*, and rice, exogenous administration of β-CC accelerated primary root growth. Additionally, research has demonstrated that β-CC lessens the negative effects of salinity on rice root growth (Dickinson et al. 2019). According to a recent study, β-CC directly inhibits 1-deoxy-D-xylulose-5-phosphate synthase (DXS), an enzyme that limits the MEP pathway rate, thereby downregulating the flux of MEP pathway, which is in charge of plastid isoprenoid biosynthesis. This results in plant resistance alongside *Spodoptera littoralis* as a signal triggered by herbivores, and it also directly inhibits the MEP pathway flux (Mitra et al. 2021). Herbivore resistance or responsiveness is also influenced by other volatile apocarotenoids. For instance, in *Brassica napus*, the herbivore-induced defensive volatile β-ionone is generated in canola and has a potent repellent effect on flea beetles and spider mites (Murata et al. 2020; Verma et al. 2023). Loliolide, a C11-terpene lactone, acts as an endogenous signaling molecule that, presumably independently of JA signaling, increases resistance to a variety of herbivores, including the western flower thrips and common cutworms. Loliolide decreases the survival rate of common cutworm and two-spotted spider mite larvae without endangering the herbivores themselves (Murata et al. 2019). In addition, it was revealed that loliolide, also known as (—)-loliolide, functions as a signaling molecule that promotes the synthesis of allelochemicals such as tricin and momilactone B to facilitate the coordination of allelopathic interactions between barnyard grass and rice (Kong et al. 2018; Li et al. 2020).

Apocarotenoid volatiles are known for their significant scent components found in many horticultural plant fruits and flowers, in addition to their biological roles within plants. When it comes to people's perception of fragrance, cyclic apocarotenoids—like β-CC, β-ionone, and β-damascenone—have greater effects and lower odor thresholds than linear volatiles (Shi et al. 2020). In light of this, cyclic apocarotenoids are acknowledged as being more significant volatiles with regard to the fragrant aroma of fruits and flowers (Goff and Klee 2006; Shi et al. 2020). The fragrance industry has widely employed C13 cyclic β-damascenone as a flavoring ingredient because of its appealing honey-like, flowery, and fruity smell. The scent of the cyclic volatiles β-ionone and α-ionone is fruity, woodsy, and violet-like, whereas the scent of β-CC is grassy, woodsy, and tobacco-like. They are both frequently utilized as flavoring compounds in perfumes. Furthermore, these volatile apocarotenoids serve as efficient attracting factors for pollinators and dispersers of seeds (Liang et al. 2021).

Apocarotenoids are chemicals that give fruits and flowers their color and have antioxidant properties over and above their aroma. In fact, a number of derivatives of long-chain apocarotenoids have a significant positive influence on human health and are employed as premium food coloring agents, cosmetics, and antioxidants. Citrus fruits contain two important C30 apocarotenoid pigments: β-citraurinene and β-citraurin (3-hydroxy-β-apo-8′-carotenal). In certain citrus species, the appealing red pigmentation of the fruit peel is caused by their hyper-accumulation (Zheng et al. 2019). Bixin (C25) is an additional naturally occurring apocarotenoid pigment that gives *Bixa orellana* seeds their red hue. It is a common color additive used in the cosmetic and food industry, taken from natural sources (Giuliano et al. 2003). These effects were mediated by a decrease in neutrophil migration in rats utilized in simulated wound testing. All-*trans*-lycopene is broken down into bixin aldehyde at the start of the route. Saffron is a costly spice on the planet and one of the first naturally occurring food additions (Srivastava et al. 2024). The saffron flower stigma contains apocarotenoid C10, safranal; pigment C20, crocins; and 3-OH-β-cyclocitral glycoside, picrocrocin. The saffron red color is produced by crocetins and crocins, while the bitter flavor and strong scent are caused by picrocrocin and safranal (Frusciante et al. 2014). Many countries employ these health-promoting apocarotenoids in cosmetic and food products. It has also been demonstrated that saffron apocarotenoids have a number of pharmacological properties, including anti-inflammatory, antiaging, analgesic, and neuroprotective properties (Bukhari et al. 2018).

2.3 The Structural Diversity of Plant Apocarotenoids and Their Metabolic Basis

2.3.1 Formation of a Variety of Apocarotenoids by Site-specific Oxidative Tailored Carotenoids by Several CCDs

Nonheme iron-dependent carotenoid cleavage dioxygenase (CCD) enzymes found in plants, animals, and fungi at specific locations normally catalyze the carotenoid's enzymatic cleavage. In the last few decades, various plant CCD kinds have been recognized, each with unique characteristics. CCD8, CCD4, CCD7, and CCD1, in addition to 9-*cis*-epoxycarotenoid dioxygenase (NCED), are the four different forms of CCD enzymes found in *Arabidopsis*. Five NCEDs—NCED5, NCED9, NCED2, NCED3, and NCED6—are found in *Arabidopsis*, whereas a single enzyme represents each of the other categories (Auldridge et al. 2006; Walter and Strack 2011). The cytosol receives this apocarotenoid product from plastids, the most plant cleavage enzymes, and carotenoid production. Short-chain dehydrogenase (SDR) then converts it to abscisic aldehyde (González-Guzmán et al. 2002). An aldehyde oxidase (AAO) that is dependent on molybdenum oxidizes the resulting abscisic aldehyde to create ABA (Seo et al. 2000). 3-OH-β-apo-11-carotenal, β-apo-11-carotenal, 9-*cis*-3-OH-β-apo-11-carotenal, and 9-*cis*-β-apo-11-carotenal, β-apo-11-carotenoids (C15), have been found very recently to perform biological processes akin to those of ABA, such as causing the activation of ABA-responsive genes and seed dormancy. Additional feeding trials utilizing labeled apocarotenoids, in conjunction with an examination of transcriptional and physiological reactions, demonstrated that plants are capable of producing ABA from C15 β-apo-11-carotenoids. This indicates an alternative mechanism for ABA biosynthesis that does not rely on zeaxanthin oxidase (Jia et al. 2021b). Further investigation into the biological importance of this pathway's isomerization, hydroxylation, and oxidation of C15 β-apo-11-carotenoids will be crucial given the lack of clarity surrounding the genes implicated in these processes in this novel ABA biosynthesis pathway. The biological functions and cleavage activities of the remaining four *Arabidopsis* CCDs, CCD4, CCD7, CCD1, and CCD8, vary (Zheng et al. 2020a).

The cleavage of apocarotenoid and carotenoid substrates at distinct double bonds is achieved by plant CCD1 enzymes, resulting in diverse apocarotenoid ketones and aldehydes production with varied chain lengths (Schwartz et al. 2001; Vogel et al. 2008; Ilg et al. 2009, 2014). The substrate selectivity and loosened cleavage site of these enzymes set them apart. Important horticultural crops' fruits and flowers have aromas and flavors that are attributed to certain volatile chemicals derived from the CCD1 cleavage. For example, pseudoionone, β-ionophore, β-ionone; and geranylacetone, α-ionone are among the C14 dialdehyde and C13 ketone products that arise from CCD1-catalyzed C9–C10. Many plant species, like tomatoes, have fruits or flowers that contain these volatiles, which are aroma and flavor (Simkin et al. 2004a), grape (Mathieu et al. 2005), melon (Ibdah et al. 2006), and petunia (Simkin et al. 2004b). Additionally, it was demonstrated that CCD1 cleaves lycopene at the C5′–C6′ and/or C5–C6 position to produce MHO, a significant tomato fragrance component (Vogel et al. 2008). However, because CCD1 enzymes are cytoplasm-localized, they appear to cleave apocarotenoids—carotenoids that have already been broken down—rather than the carotenoid substrates found in plastids. This theory is reinforced by the finding that overexpressing rice CCD1 in Golden Rice did not significantly reduce the amount of carotenoid in the endosperm (Ilg et al. 2009, 2010).

CCD2, a unique CCD type found only in the genera *Crocus* and *Freesia* of the family Iridaceae, is located in plastids but shares a close kinship with the cytoplasmic CCD1 subfamily (Fang et al. 2020). The cleavage of zeaxanthin's C7′—C8′ and C7—C8 double bonds by *Crocus* CCD2 has been observed to yield crocetin dialdehyde (C20) and 3-OH-β-cyclocitral (Frusciante et al. 2014; Ahrazem et al. 2016). These compounds are then transformed into crocins and safranal, respectively, and influence the color and scent of saffron stigma (Demurtas et al. 2018; Diretto et al. 2019).

The fourth class of plastid-localized CCDs is CCD4. Their activity regulates carotenoid concentration and apocarotenoid pigment consolidation (Verma et al. 2024a). The substrate specificity, along with the regioselectivity of CCD4 enzymes, differed throughout plant species. Bicyclic carotenoids in *Arabidopsis* and potatoes have been demonstrated to be cleaved at the C9′—C10′ or C9—C10 double bond by CCD4; this leads to the creation of C13 volatiles and C27 apocarotenoids, which are assumed to be further broken down into smaller metabolites. The carotenoid concentration in a variety of plant species' tissues is reduced by enzymatic activity regarding CCD4 (Gonzalez-Jorge et al. 2013; Zhang et al. 2015; Watanabe et al. 2018; Han et al. 2019a). *Citrus* CCD4b, contrariwise, targets a different site to form C30 apocarotenoid pigments, like β-citraurin, along with C10 volatiles. This location is the C7′—C8′ or C7—C8 double bond found in zeaxanthin, β-cryptoxanthin, and β-carotene (Ma et al. 2013; Rodrigo et al. 2013; Zheng et al. 2015, 2019, 2021). Citrus fruit-specific C30 apocarotenoid pigments, β-citraurinene and β-citraurol, are assumed to be produced in citrus peels by *Citrus* CCD4b as well (Zheng et al. 2021). In fact, the primary long-chain apocarotenoids in citrus callus that overexpresses *CCD4b* are C30 apocarotenoids, confirming the central functions of CCD4b in the making of C30 pigments in citrus peel (Zheng et al. 2019, 2021). *Buddleja davidii*'s BdCCD4.1 and BdCCD4.3, which exhibit similar enzymatic activity as *Crocus* CCD2, break zeaxanthin's C7′—C8′ and C7—C8 double bonds both *in vitro* and *in vivo* to produce 3-hydroxy-β-cyclocitral and crocetin dialdehyde (Ahrazem et al. 2017). Another form of CCD4 from *Gardenia jasminoides* was found in the latest report. It cleaves many substrates, such as lycopene, zeaxanthin, and β-carotene, at C7′—C8′ and C7—C8 double bonds (Xu et al. 2020).

The new CCD type known as zaxinone synthase (ZAS), or CCD10, has just recently been discovered in maize and rice. It is absent in *Arabidopsis* (Zhong et al. 2020). The 3-OH-apocarotenoids with varying chain lengths, for instance, C30 all-*trans*-3-OH-apo-8′-carotenal, C27 all-*trans*-3-OH-apo-10′-carotenal, and C25 all-*trans*-3-OH-apo-12′-carotenal, were demonstrated to be cleaved by the rice ZAS. The resulting C18-ketone product, zaxinone, is illustrated as a signaling molecule that regulates the development and growth of plants. Non-mycorrhizal plants, for instance, *Brassica* species, don't have this kind of CCDs, suggesting the enzyme presence is correlated with the capability to form a mycorrhizal symbiosis (Wang et al. 2019).

2.3.2 Oxidative Cleavage Processes Without Specificity in the Apocarotenoids Formation

Apart from CCDs, other enzymes also mediate carotenoid degradation. For instance, LOXs utilize carotenoids as co-substrates, resulting in their nonspecific oxidative cleavage. *In vitro* studies have shown that polyunsaturated fatty acids such as arachidonic, linolenic, and linoleic acids can facilitate this process. Furthermore, research has demonstrated that

LOX disruption or downregulation reduces carotenoid degradation during pasta production and limits carotenoid loss in Golden Rice during storage. These findings underscore the critical role of LOXs in the oxidative cleavage of carotenoids (Carrera et al. 2007; Gayen et al. 2015). Recently discovered tomato LOX, known as *TomLoxC*, is in charge of producing apocarotenoid volatiles, for example, in tomato fruits' β-CC and β-ionone, utilizing quantitative trait locus mapping, pan-genome analysis along with functional analysis (Gao et al. 2019).

Apart from enzymatic degradation processes that lack specificity, it can cause nonspecific destructive oxidation of carotenoids (Ramel et al. 2012; Hou et al. 2016). This can start metabolic pathways that result in different volatile compounds and/or signaling molecule production. For instance, all-trans β-carotene is attacked by singlet oxygen that forms in chloroplasts, especially photosystem (PS) II, in high-light conditions (Srivastava et al. 2024; Verma et al. 2024b). As a result, different apocarotenoid aldehydes, ketones, epoxides, endoperoxides, and lactones are formed, including the signaling molecule β-CC and growth regulator. Apart from the potential role of CCD1 enzymes, linear volatile apocarotenoids, for instance, citral, geranylacetone, pseudoionone, MHO, and farnesylacetone, are formed throughout unspecific oxidative breakdown of precursors of carotenoid pigments (Simkin et al. 2004a; Moreno et al. 2021).

2.4 Molecular Regulation of Apocarotenoid Biosynthesis

Plant apocarotenoids are produced in different amounts based on the tissue and developmental stage, as well as environmental factors that cause nonenzymatic carotenoid cleavage. The genesis of various apocarotenoids is carefully regulated at several stages.

2.4.1 Transcriptional Regulation

Numerous investigations have revealed that certain *CCD* genes exhibit tissue/organ-specific expression patterns that are also dependent on environmental factors and growth/developmental phases. Several transcriptional regulators control the gene expression connected to the synthesis of apocarotenoid substances in the fruits and flowers of horticultural crops. For example, it has been proven that the sweet osmanthus (*Osmanthus fragrans*) transcription factors *OfWRKY3* and *OfERF61* unambiguously attach to W-box and CAACA box sites in *OfCCD4* promoter, increasing the level of *OfCCD4* expression (Han et al. 2019b). Given that it is produced by the *OfCCD4* cleavage reaction, it is deemed that two transcription factors positively influence the β-ionone biosynthesis, an essential aromatic component of sweet osmanthus petals. It has been demonstrated that the concentrations of various volatile apocarotenoid compounds in grapes, such as β-ionone, β-damascenone, and MHO, positively correlate with the *VvCCD4b* expression level (Huang et al. 2009; Han et al. 2016, 2019b). A transcription factor belonging to the MADS family called *VvMADS4* directly binds to the *VvCCD4b* promoter, thereby repressing its expression (Meng et al. 2020). Sometimes, the transcription-related variables of genes responsible for apocarotenoid biosynthesis also boost the gene expression responsible for producing the carotenoid substrates. It has been demonstrated, for example, that citrus *CsMADS6* directly binds to and activates the *Citrus* CCD1 promoters, in addition to genes that synthesize carotenoid pigments, including *LCYB1* (*lycopene β-cyclase 1*), *PSY* (*phytoene synthase*), and *PDS* (Lu et al. 2018). Recent research revealed that citrus *CsERF061* diametrically binds to the *NCED3, CCD4*, and *CCD1* promoters, accompanied by seven gene promoters enmeshed in the carotenoid biosynthesis: *PSY1, PDS, carotene isomerase* (*CRTISO*), *LCYB2, LCYB1, zeaxanthin epoxidase* (*ZEP*), and *β-carotene hydroxylase* (*BCH*). This suggests that *CsERF061* regulates apocarotenoid and carotenoid metabolism on multiple targets.

The transcription factor *Crocus*-derived *CsULT1* (*Ultrapetala1*) positively regulates the *PSY, PDS, BCH, CCD2*, and *CCD4b* gene expression (Ashraf et al. 2015), thereby modulating carotenoid biosynthesis and their enzymatic cleavage into apocarotenoids. The formation of volatile apocarotenoids by unspecific enzymatic means is facilitated by the *LOXC* gene, whose expression level was significantly elevated in tomato upon *SlMYB75* overexpression, an MYB-type transcription factor. MYBPLANT and MYBPZM cis-elements can be directly bound by SlMYB75 in the *LOXC* promoter, according to a yeast one-hybrid experiment. Additional dual-luciferase experiments demonstrated SlMYB75's function in *LOXC* promoter activation (Jian et al. 2019).

Noteworthy is the fact that while these transcription factors control the gene expression level involved in carotenoid and/or apocarotenoid compound biosynthesis, more is needed about how these factors affect the amount and makeup of apocarotenoid compounds. To verify these transcription factors' role in apocarotenoid formation, more apocarotenoid pigment and/or volatile quantification by LC–MS/GC–MS profiling ought to be carried out in plants that overexpress or lack these transcription factors (Mani et al. 2017; Verma et al. 2024c).

As might be expected, transcriptional factors in several tissues as well as organs also influence the production of hormones produced from apocarotenoids. For instance, the CCD7 promoter can bind to *basic pentacysteine 1* (*BPC1*), and in the *Malus baccata* roots, temporary overexpression of *BPC1* reduced the promoter activity of *CCD7*. Upon binding to the *NCED3* and *NCED2* promoters, the transcription factor CDF4 from *Arabidopsis*, a protein belonging to the DOF family, promotes the expression of these genes and acts as a progressive regulator of ABA production and leaf senescence. It has been demonstrated that the *Arabidopsis* NAC transcription factor ATAF1 binds to the *NCED3* promoter *in vivo* and activates its expression, raising the ABA levels (Yue et al. 2015).

2.4.2 Post-translational Regulation

Another way to manage the enzymatic activity of CCD in plants and the amount and makeup of apocarotenoids is post-translational regulation. PGM48, an *Arabidopsis* plastoglobule (PG)-localized metallopeptidase, progressively controls leaf senescence (Dhiman et al. 2022; Singh et al. 2022a; Chauhan et al. 2023; Singh and Verma 2023; Tyagi et al. 2024). The sole protease found in PG proteomes is PGM48. PG-localized CCD4 reduced during leaf senescence. *PGM48* overexpression in *Arabidopsis* significantly lowered CCD4 protein levels. Furthermore, it was demonstrated that PGM48 and CCD4 can interact directly, indicating that CCD4 may be a PGM48 substrate. Though its exact function in the apocarotenoid synthesis is nevertheless unknown, PGM48's direct association and co-localization with CCD4 have shed fresh light on the post-translational control of CCD enzymes. There is still a need to do more apocarotenoid profiling and thorough characterization of PGM48 peptidase activity (Bhuiyan et al. 2016; Bhuiyan and van Wijk 2017; Singh et al. 2022b).

A cysteine-rich DnaJ-like protein called plastid-localized ORANGE (OR) acts as a post-translational regulator of PSY by stabilizing the enzyme and enhancing its activity. This regulatory function contributes significantly to carotenoid pigment accumulation (Zhou et al. 2015). Recent research indicates sweet potato OR protein, or IbOR, may have an impact on carotenoid cleavage, which is crucial for maintaining carotenoid homeostasis. This interaction has been demonstrated with both the established target, *PSY*, and a novel target, *CCD4*. The interaction effect on the amount or activity of CCD4 protein is yet unclear, though (Park et al. 2016; Chayut et al. 2017).

2.4.3 Epigenetic Modification

Another strategy used by plants to control apocarotenoid production is epigenetic regulation. In the yellow-flowered *Oncidium* cultivar "Gower Ramsey" (GR) and the white-flowered cultivar "Jade," methylation analysis of the *OgCCD1* promoter revealed significantly higher DNA methylation levels in GR. This elevated methylation is associated with the suppression of *OgCCD1* expression, resulting in higher carotenoid accumulation in GR's yellow floral tissues (Chiou et al. 2010). When 5-azacytidine, a DNA methyltransferase inhibitor, was applied to citrus callus, the amount of carotenoid content significantly decreased while the *CpCCD1* gene was significantly upregulated (Xu et al. 2017). The 5-azacytidine impact did not demethylate the *CpCCD1* promoter areas evaluated in this investigation, indicating that methylation alterations may indirectly control *CpCCD1* expression. Yellow-colored GR cultivar *Oncidium* promoter region OgCCD1 exhibited a higher degree of DNA methylation as compared to "Jade," which led to *OgCCD1* down-regulation in GR and a decrease in the amount of carotenoids converted in floral tissue into a colorless apocarotenoid. Furthermore, by directly modifying the methylation levels of the SlNCED promoter region, METHYLTRANSFERASE1 (SlMET1) facilitates the *SlNCED* gene expression and influences the production of ABA (Yao et al. 2020).

2.5 Genetic Perspectives on Apocarotenoid Natural Variations

Numerous agricultural plants, particularly those grown for human consumption, exhibit significant differences in the amounts and kinds of apocarotenoids found in various species and/or natural accessions within a single species. There is a significant ecological adaptation in the amounts of MHO and β-ionone that accumulate in different papaya accessions (Jing et al. 2015). Among the carrot accessions, orange (Nairobi) and red (Rothild), yellow (Yellowstone), purple (Purple Haze), and white (Creme de Lite), β-ionone, α-ionone, and farnesylacetone also show notable natural variation. According to the odor activity values (OVA) method, two of the main fragrance molecules for fruit quality in apricot fruit are β-ionone and dihydro-β-ionone (Yahyaa et al. 2013). The white apricot cultivar pulp, like "Luntaixiaobaixing" and "Baixing," had much larger contents of dihydro-β-ionone and β-ionone as compared to yellow cultivars, "Hongyuxing" and "Danxing," which even had higher carotenoid content. This discovery implies that the quantity of carotenoid precursors, additionally controlled by genetics, is the variation in apocarotenoid concentration of different apricot species (Xi et al. 2016).

The integration of diverse omics datasets, including transcriptomics, metabolomics, and genomics, has been made possible by technological developments in data mining, metabolomics, and high-throughput sequencing. Exploiting the natural variety of apocarotenoid pigments and volatiles has been made easier by the multi-omics technique in conjunction with functional and genetic research. This has opened the door to metabolism, function, and evolutionary history of plant apocarotenoids (Singh et al. 2024b).

For example, from the genome sequence records of 725 tomato accessions, to create a tomato pan-genome, an unusual variant in farmed tomatoes was discovered using comparative analysis utilizing this created tomato pan-genome. The *TomLoxC* gene, which encodes 13-LOX, has a ~4-kb substitution in its promoter region that defines this allele. Subsequent examination of RNA-Seq data revealed that levels of the *TomLoxC* gene expression are much greater in accessions that possess the unusual allele that was recorded present in the "Heinz 1706" genome than in other accessions. These studies support the significance of an alternate pathway for apocarotenoid production via an unspecific oxidation mechanism and they further establish that *LOX* acts as a genetic driver for natural variation in apocarotenoid creation. Apocarotenoid volatiles are vital constituents of tomato fruit quality and flavor (Tieman et al. 2017; Gao et al. 2019). However, this pan-genome-captured *TomLoxC* promoter allele shows substantial negative selection during domestication. This might be because qualities related to scent quality should have been taken into consideration in favor of modern breeding's major focus on tomato productivity, stress resistance, and long shelf life (Sharma et al. 2023a).

Apart from volatiles, a number of recent investigations have profiled apocarotenoid pigments in fruits and associated natural variations in numerous significant horticultural crops, like the species of *Citrus* and *Capsicum*. Citrus fruits vary in color depending on the cultivar (Sangma et al. 2022; Singh et al. 2024c). The formation of C30 apocarotenoid pigments, including β-citraurin, is essential for the peel pigmentation of mature citrus fruit, in addition to carotenoid quantity and composition (Ma et al. 2013; Rodrigo et al. 2013; Zheng et al. 2015, 2019, 2021). In fact, in an F1 pseudo-testcross population resulting from a cross between red-peeled *Citrus reticulate* and yellow-peeled *Poncirus trifoliata*, in the contents of C30 apocarotenoid pigments, namely β-citraurinene and β-citraurin, a significant association was seen between their ratio to the aggregate citrus peel carotenoids and color index of the progenies with red peels (Zheng et al. 2019). A 5′ cis-regulatory mutation of *Citrus CCD4b* (*CitCCD4b*) is a key genetic predictor of natural variation in accretion of the C30 pigments β-citraurinene and β-citraurin, according to BSR, eQTL, and mQTL-seq analysis. Enzyme CCD4b contributes to the β-citraurinene synthesis, indicating a new process in β-citraurin aldehyde group modification. Citrus peels' red coloration is caused by the high accumulation of C30 apocarotenoid pigments, which are linked to the occurrence of a particular SNP in MITE of the *CitCCD4b* promoter and the *CitCCD4b* high expression level in progenies of the pseudo-testcross population, in addition to 115 distinct citrus accessions. *Citrus changshanensis* red-peeled mutant, also known as "Huyou," had 100 times higher levels of *CitCCD4b* expression and a significantly higher content of C30 apocarotenoid β-citraurin than the yellow-peeled ordinary "Huyou." These results also showed a correlation with the supposed enhancer SNP in the *CitCCD4b* promoter that was previously mentioned. This finding implies that *CitCCD4b* expression variation plays a dual role in fruit scent and pigmentation synthesis, in addition to the natural variance of volatiles and apocarotenoid pigments among various citrus species (Luan et al. 2020).

The CCD4 enzyme cleaves carotenoids into C27 apocarotenoids, which are further catabolized into smaller, colorless metabolites. However, mutations in the coding sequence of *CCD4* have been shown to reduce its activity, thereby enhancing carotenoid accumulation in various tissues or organs of crops such as *Brassica* and peach (Mi and Al-Babili 2019). According to recent studies, CRISPR/Cas9-mediated disruption of the *CCD4* gene led to a significant increase in carotenoid content and visible pigmentation changes: in *Ipomoea nil*, the petals developed a pale-yellow phenotype, while in a white-rooted carrot variety, the taproots exhibited a yellow phenotype due to carotenoid accumulation (Watanabe et al. 2018; Prasad et al. 2024). According to reports, the CCD4 enzymes are also in charge of the apocarotenoid volatiles synthesis in tissues and organs, including peach fruit (Falchi et al. 2013; Li et al. 2021; Sharma et al. 2023c).

Beyond pigments and volatiles, apocarotenoid-derived hormones such as ABA exhibit natural variation among *Arabidopsis* accessions. For example, under low water potential, the accession "Shahdara" (Sha) accumulates significantly less ABA than "Columbia" (Col) or "*Landsberg erecta*" (*L. er*), suggesting genotypic differences in stress-responsive apocarotenoid metabolism. The *NCED3* variant is liable for variance in ABA accumulation among the Ler and Sha accessions, as shown by complementation tests and genetic study of Sha × Ler recombinant inbred lines (RILs) (Verma et al. 2024c). The degree of tillering, along with the unique vulnerability to *Striga* infection of the rice cultivars "Bala" and "Azucena," can be attributed to their significant differences in SL biosynthesis. Rearranging a 51–59-kbp region between 28.9 and 29 Mbp on chromosome 1 is a substantial QTL. A chromosomal rearrangement leading to the deletion of two *AtMAX1* orthologs has been linked to decreased SL accumulation in 367 rice cultivars, highlighting a structural genomic basis for natural variation in SL biosynthesis (Cardoso et al. 2014; Singh et al. 2024a).

2.6 Conclusion and Prospects

Besides their importance as building blocks for plant hormones, their function as growth regulators and signaling molecules, apocarotenoids have significant roles as antioxidants, medicines, fragrances, and pigments. They are also the results of carotenoid degradation. This chapter has covered the biological role of apocarotenoid chemicals, which are crucial for plant growth, stress response, and biotic interactions, including significance for human health. A summary of the control of newly characterized biosynthetic pathways is also provided. Further investigation demonstrates that when administered exogenously, apocarotenoids influence plant growth and development along with biotic and abiotic stress tolerances in a wide variety of plant species, hence serving as regulatory metabolites in biology. However, uncertainty remains on how naturally occurring variations of these apocarotenoids relate to specific plant phenotypes. The reasons behind the numerous functions and varying activities of apocarotenoids with somewhat different structures in plants, over and above the contradicting effects of some apocarotenoids like zaxinone on distinct plant species, remain unclear. We suggest that investigating the biochemical and genetic causes of apocarotenoid metabolic variation/diversity would hasten the investigation of the bioactive apocarotenoid pathway, which is presently unknown or unclear. This could help to clarify the roles that these substances play in crop end-phenotypes and the currently unclear regulations and changes. The molecular underpinnings and natural genetic variation mechanisms of only a small number of apocarotenoid volatiles and pigments have been thoroughly studied to date; phytohormones derived from apocarotenoid pigments along with other important apocarotenoids. We believe that apocarotenoids MS profiling in conjunction with multi-omics techniques, such as genome-wide association (GWA), RNA-seq analysis, and QTL mapping, will open up new avenues for QTL discovery and the identification of the genes that underlie them in a variety of significant crops. The progression in this field will allow for the identification of associations between DNA markers and apocarotenoid traits for marker-assisted development of agronomic traits in crops, such as fruit and flower color and aroma. Additionally, it will offer fresh targets for CRISPR editing, enabling low-cost, effective, stable, and environmentally friendly methods of enhancing high-value apocarotenoids in response to future needs for greater crop quality.

References

Ablazov, A., Mi, J., Jamil, M. et al. (2020). The apocarotenoid zaxinone is a positive regulator of strigolactone and abscisic acid biosynthesis in Arabidopsis roots. *Frontiers in Plant Science* 11, 578. https://doi.org/10.3389/fpls.2020.00578.

Ahrazem, O., Diretto, G., Argandoña, J. et al. (2017). Evolutionarily distinct carotenoid cleavage dioxygenases are responsible for crocetin production in *Buddleja davidii*. *The Journal of Experimental Biology* 68, 4663–4677. https://doi.org/10.1093/jxb/erx277.

Ahrazem, O., Rubio-Moraga, A., Berman, J. et al. (2016). The carotenoid cleavage dioxygenase CCD2 catalysing the synthesis of crocetin in spring crocuses and saffron is a plastidial enzyme. *New Phytologist* 209, 650–663. https://doi.org/10.1111/nph.13609.

Al-Babili, S. and Bouwmeester, H.J. (2015). Strigolactones, a novel carotenoid-derived plant hormone. *Annual Review of Plant Biology* 66, 161–186. https://doi.org/10.1146/annurev-arplant-043014-114759.

Ashraf, N., Jain, D. and Vishwakarma, R.A. (2015). Identification, cloning and characterization of an ultrapetala transcription factor *CsULT1* from crocus: a novel regulator of apocarotenoid biosynthesis. *BMC Plant Biology* 15, 25. https://doi.org/10.1186/s12870-015-0423-7.

Auldridge, M.E., Block, A., Vogel, J.T. et al. (2006). Characterization of three members of the Arabidopsis carotenoid cleavage dioxygenase family demonstrates the divergent roles of this multifunctional enzyme family. *The Plant Journal* 45, 982–993. https://doi.org/10.1111/j.1365-313X.2006.02666.x.

Bhuiyan, N.H., Friso, G., Rowland, E. et al. (2016). The plastoglobule-localized metallopeptidase PGM48 is a positive regulator of senescence in *Arabidopsis thaliana*. *The Plant Cell* 28, 3020–3037. https://doi.org/10.1105/tpc.16.00745.

Bhuiyan, N.H. and van Wijk, K.J. (2017). Functions and substrates of plastoglobule-localized metallopeptidase PGM48. *Plant Signaling & Behavior* 12, 3020–3037. https://doi.org/10.1080/15592324.2017.1331197.

Bouvier, F., Dogbo, O. and Camara, B. (2003). Biosynthesis of the food and cosmetic plant pigment bixin (annatto). *Science* 300, 2089–2091. https://doi.org/10.1126/science.1085162.

Bruno, M., Hofmann, M., Vermathen, M. et al. (2014). On the substrate-and stereospecificity of the plant carotenoid cleavage dioxygenase 7. *FEBS Letters* 588, 1802–1807. https://doi.org/10.1016/j.febslet.2014.03.041.

Bukhari, S.I., Manzoor, M. and Dhar, M. (2018). A comprehensive review of the pharmacological potential of *Crocus sativus* and its bioactive apocarotenoids. *Biomedicine & Pharmacotherapy* 98, 733–745. https://doi.org/10.1016/j.biopha.2017.12.090.

Cáceres, L., Lakshminarayan, S., Yeung, K.-C. et al. (2016). Repellent and attractive effects of α-, β-, and dihydro-β-ionone to generalist and specialist herbivores. *Journal of Chemical Ecology* 42, 107–117. https://doi.org/10.1007/s10886-016-0669-z.

Cardoso, C., Zhang, Y., Jamil, M. et al. (2014). Natural variation of rice strigolactone biosynthesis is associated with the deletion of two *MAX1* orthologs. *Proceedings of the National Academy of Sciences of the United States of America* 111, 2379–2384. https://doi.org/10.1073/pnas.1317360111.

Carrera, A., Echenique, V., Zhang, W. et al. (2007). A deletion at the *Lpx-B1* locus is associated with low lipoxygenase activity and improved pasta color in durum wheat (*Triticum turgidum* ssp. *durum*). *Journal of Cereal Science* 45, 67–77. https://doi.org/10.1016/j.jcs.2006.07.001.

Chauhan, D., Prasad, G., Singh, D. et al. (2024). Role of next-generation sequencing technology in crop improvement. In: *Smart Breeding* (eds. K. Chandra, S. Chand, R.P. Saini and R. Sharma), 371–390. Apple Academic Press.

Chauhan, D., Singh, D., Pandey, H. et al. (2023). Impact of transcription factors in plant abiotic stress: a recent advancement for crop improvement. In: *Plant Transcription Factors: Contribution in Development, Metabolism, and Environmental Stress* (eds. V. Srivastava, S. Mishra, S. Mehrotra and S.K. Upadhyay), 271–286. Academic Press.

Chayut, N., Yuan, H., Ohali, S. et al. (2017). Distinct mechanisms of the ORANGE protein in controlling carotenoid flux. *Plant Physiology* 173, 376–389. https://doi.org/10.1104/pp.16.01256.

Chen, K., Li, G.J., Bressan, R.A. et al. (2020). Abscisic acid dynamics, signaling, and functions in plants. *Journal of Integrative Plant Biology* 62, 25–54. https://doi.org/10.1111/jipb.12899.

Chiou, C.-Y., Pan, H.-A., Chuang, Y.-N. et al. (2010). Differential expression of carotenoid-related genes determines diversified carotenoid coloration in floral tissues of *Oncidium* cultivars. *Planta* 232, 937–948. https://doi.org/10.1007/s00425-010-1222-x.

Cook, C., Whichard, L.P., Turner, B. et al. (1966). Germination of witchweed (*Striga lutea* Lour.): isolation and properties of a potent stimulant. *Science* 154, 1189–1190. https://doi.org/10.1126/science.154.3753.1189.

D'Alessandro, S., Mizokami, Y., Legeret, B. et al. (2019). The apocarotenoid β-cyclocitric acid elicits drought tolerance in plants. *iScience* 19, 461–473. https://doi.org/10.1016/j.isci.2019.08.003.

Decker, E.L., Alder, A., Hunn, S. et al. (2017). Strigolactone biosynthesis is evolutionarily conserved, regulated by phosphate starvation and contributes to resistance against phytopathogenic fungi in a moss, *Physcomitrella patens*. *New Phytologist* 216, 455–468. https://doi.org/10.1111/nph.14506.

DellaPenna, D. and Pogson, B.J. (2006). Vitamin synthesis in plants: tocopherols and carotenoids. *Annual Review of Plant Biology* 57, 711–738. https://doi.org/10.1146/annurev.arplant.56.032604.144301.

Demurtas, O.C., Frusciante, S., Ferrante, P. et al. (2018). Candidate enzymes for saffron crocin biosynthesis are localized in multiple cellular compartments. *Plant Physiology* 177, 990–1006. https://doi.org/10.1104/pp.17.01815.

Dhiman, V.K., Singh, D., Dhiman, V.K. et al. (2024a). Multi-omics: an advanced bioinformatics approach for crop improvement in agriculture. In: *Bioinformatics for Plant Research and Crop Breeding* (ed. J.-T. Chen), 75–98. John Wiley & Sons.

Dhiman, V.K., Singh, D., Dhiman, V.K. et al. (2024b). Insects as biotic stress in agriculture. In: *Plant-Microbe Interaction and Stress Management* (eds. P. Singh Chauhan, S.K. Tewari and S. Misra), 193–210. Springer Nature Singapore.

Dhiman, V.K., Singh, D., Pandey, H. et al. (2022). Mycoremediation of agricultural waste for the cultivation of edible mushroom. In: *Environmental Contamination Remediation and Management: Sustainable Management of Environmental Contaminants* (ed. T. Aftab), 471–482. Springer. https://doi.org/10.1007/978-3-031-08446-1_18.

Dickinson, A.J., Lehner, K., Mi, J. et al. (2019). β-Cyclocitral is a conserved root growth regulator. *Proceedings of the National Academy of Sciences of the United States of America* 116, 10563–10567. https://doi.org/10.1073/pnas.1821445116.

Diretto, G., Ahrazem, O., Rubio-Moraga, Á. et al. (2019). UGT709G1: a novel uridine diphosphate glycosyltransferase involved in the biosynthesis of picrocrocin, the precursor of safranal in saffron (*Crocus sativus*). *New Phytologist* 224, 725–740. https://doi.org/10.1111/nph.16079.

Falchi, R., Vendramin, E., Zanon, L. et al. (2013). Three distinct mutational mechanisms acting on a single gene underpin the origin of yellow flesh in peach. *The Plant Journal* 76, 175–187. https://doi.org/10.1111/tpj.12283.

Fang, Q., Li, Y., Liu, B. et al. (2020). Cloning and functional characterization of a carotenoid cleavage dioxygenase 2 gene in safranal and crocin biosynthesis from *Freesia hybrida*. *Plant Physiology and Biochemistry* 154, 439–450. https://doi.org/10.1016/j.plaphy.2020.06.035.

Felemban, A., Braguy, J., Zurbriggen, M.D. et al. (2019). Apocarotenoids involved in plant development and stress response. *Frontiers in Plant Science* 10, 1168. https://doi.org/10.3389/fpls.2019.01168.

Fiorilli, V., Wang, J.Y., Bonfante, P. et al. (2019). Apocarotenoids: old and new mediators of the arbuscular mycorrhizal symbiosis. *Frontiers in Plant Science* 10, 1186. https://doi.org/10.3389/fpls.2019.01186.

Fraser, P.D. and Bramley, P.M. (2004). The biosynthesis and nutritional uses of carotenoids. *Progress in Lipid Research* 43, 228–265. https://doi.org/10.1016/j.plipres.2003.10.002.

Frusciante, S., Diretto, G., Bruno, M. et al. (2014). Novel carotenoid cleavage dioxygenase catalyzes the first dedicated step in saffron crocin biosynthesis. *Proceedings of the National Academy of Sciences of the United States of America* 111, 12246–12251. https://doi.org/10.1073/pnas.1404629111.

Gao, L., Gonda, I., Sun, H. et al. (2019). The tomato pan-genome uncovers new genes and a rare allele regulating fruit flavor. *Nature Genetics* 51, 1044–1051. https://doi.org/10.1038/s41588-019-0410-2.

Gayen, D., Ali, N., Sarkar, S.N. et al. (2015). Down-regulation of lipoxygenase gene reduces degradation of carotenoids of golden rice during storage. *Planta* 242, 353–363. https://doi.org/10.1007/s00425-015-2314-4.

Giuliano, G. (2014). Plant carotenoids: genomics meets multi-gene engineering. *Current Opinion in Plant Biology* 19, 111–117. https://doi.org/10.1016/j.pbi.2014.05.006.

Giuliano, G. (2017). Provitamin A biofortification of crop plants: a gold rush with many miners. *Current Opinion in Biotechnology* 44, 169–180. https://doi.org/10.1016/j.copbio.2017.02.001.

Giuliano, G., Rosati, C. and Bramley, P.M. (2003). To dye or not to dye: biochemistry of annatto unveiled. *Trends Biotechnology* 21, 513–516. https://doi.org/10.1016/j.tibtech.2003.10.001.

Goff, S.A. and Klee, H.J. (2006). Plant volatile compounds: sensory cues for health and nutritional value? *Science* 311, 815–819. https://doi.org/10.1126/science.1112614.

González-Guzmán, M., Apostolova, N., Bellés, J.M. et al. (2002). The short-chain alcohol dehydrogenase ABA2 catalyzes the conversion of xanthoxin to abscisic aldehyde. *The Plant Cell* 14, 1833–1846. https://doi.org/10.1105/tpc.002477.

Gonzalez-Jorge, S., Ha, S.H., Magallanes-Lundback, M. et al. (2013). *Carotenoid cleavage dioxygenase 4* is a negative regulator of β-carotene content in *Arabidopsis* seeds. *The Plant Cell* 25, 4812–4826. https://doi.org/10.1105/tpc.113.119677.

Han, F., Cui, H., Zhang, B. et al. (2019a). Map-based cloning and characterization of BoCCD4, a gene responsible for white/yellow petal color in *B. oleracea*. *BMC Genomics* 20, 242. https://doi.org/10.1186/s12864-019-5596-2.

Han, Y., Wang, H., Wang, X. et al. (2019b). Mechanism of floral scent production in *Osmanthus fragrans* and the production and regulation of its key floral constituents, β-ionone and linalool. *Horticulture Research* 6, 106. https://doi.org/10.1038/s41438-019-0189-4.

Han, Y., Wu, M., Cao, L. et al. (2016). Characterization of OfWRKY3, a transcription factor that positively regulates the carotenoid cleavage dioxygenase gene OfCCD4 in *Osmanthus fragrans*. *Plant Molecular Biology* 91, 485–496. https://doi.org/10.1007/s11103-016-0483-6.

Harrison, E.H. (2005). Mechanisms of digestion and absorption of dietary vitamin A. *Annual Review of Nutrition* 25, 87–103. https://doi.org/10.1146/annurev.nutr.25.050304.092614.

Hashimoto, H., Uragami, C. and Cogdell, R.J. (2016). Carotenoids and photosynthesis. *Carotenoids in Nature* 79, 111–139. https://doi.org/10.1007/978-3-319-39126-7_4.

Havaux, M. (2020). β-Cyclocitral and derivatives: emerging molecular signals serving multiple biological functions. *Plant Physiology and Biochemistry* 155, 35–41. https://doi.org/10.1016/j.plaphy.2020.07.032.

Hirschberg, J. (2001). Carotenoid biosynthesis in flowering plants. *Current Opinion in Plant Biology* 4, 210–218. https://doi.org/10.1016/S1369-5266(00)00163-1.

Hou, X., Rivers, J., León, P. et al. (2016). Synthesis and function of apocarotenoid signals in plants. *Trends in Plant Science* 21, 792–803. https://doi.org/10.1016/j.tplants.2016.06.001.

Huang, F.-C., Molnár, P. and Schwab, W. (2009). Cloning and functional characterization of carotenoid cleavage dioxygenase 4 genes. *Journal of Experimental Botany* 60, 3011–3022. https://doi.org/10.1093/jxb/erp137.

Ibdah, M., Azulay, Y., Portnoy, V. et al. (2006). Functional characterization of *CmCCD1*, a carotenoid cleavage dioxygenase from melon. *Phytochemistry* 67, 1579–1589. https://doi.org/10.1016/j.phytochem.2006.02.009.

Ilg, A., Beyer, P. and Al-Babili, S. (2009). Characterization of the rice carotenoid cleavage dioxygenase 1 reveals a novel route for geranial biosynthesis. *The FEBS Journal* 276, 736–747. https://doi.org/10.1111/j.1742-4658.2008.06820.x.

Ilg, A., Bruno, M., Beyer, P. et al. (2014). Tomato carotenoid cleavage dioxygenases 1A and 1B: relaxed double bond specificity leads to a plenitude of dialdehydes, mono-apocarotenoids and isoprenoid volatiles. *FEBS Open Bio* 4, 584–593. https://doi.org/10.1016/j.fob.2014.06.005.

Ilg, A., Yu, Q., Schaub, P. et al. (2010). Overexpression of the rice carotenoid cleavage dioxygenase 1 gene in Golden Rice endosperm suggests apocarotenoids as substrates in planta. *Planta* 232, 691–699. https://doi.org/10.1007/s00425-010-1205-y.

Jamil, M., Kountche, B.A. and Al-Babili, S. (2021). Current progress in *Striga* management. *Plant Physiology* 185, 1339–1352. https://doi.org/10.1093/plphys/kiab040.

Jia, K.-P., Baz, L. and Al-Babili, S. (2018). From carotenoids to strigolactones. *Journal of Experimental Botany* 69, 2189–2204. https://doi.org/10.1093/jxb/erx476.

Jia, K.-P., Dickinson, A.J., Mi, J. et al. (2019). Anchorene is a carotenoid-derived regulatory metabolite required for anchor root formation in *Arabidopsis*. *Science Advances* 5, eaaw6787. https://doi.org/10.1126/sciadv.aaw6787.

Jia, K.-P., Mi, J., Ablazov, A. et al. (2021a). Iso-anchorene is an endogenous metabolite that inhibits primary root growth in Arabidopsis. *The Plant Journal* 107, 54–66. https://doi.org/10.1111/tpj.15271.

Jia, K.-P., Mi, J., Ali, S. et al. (2021b). An alternative, zeaxanthin epoxidase-independent abscisic acid biosynthetic pathway in plants. *Molecular Plant* 15, 151–166. https://doi.org/10.1016/j.molp.2021.09.008.

Jian, W., Cao, H., Yuan, S. et al. (2019). SlMYB75, an MYB-type transcription factor, promotes anthocyanin accumulation and enhances volatile aroma production in tomato fruits. *Horticulture Research* 6, 22. https://doi.org/10.1038/s41438-018-0098-y.

Jing, G., Li, T., Qu, H. et al. (2015). Carotenoids and volatile profiles of yellow-and red-fleshed papaya fruit in relation to the expression of carotenoid cleavage dioxygenase genes. *Postharvest Biology and Technology* 109, 114–119. https://doi.org/10.1016/j.postharvbio.2015.06.006.

Kong, C.-H., Zhang, S.-Z., Li, Y.-H. et al. (2018). Plant neighbor detection and allelochemical response are driven by root-secreted signaling chemicals. *Nature Communications* 9, 3867. https://doi.org/10.1038/s41467-018-06429-1.

Lanfranco, L., Fiorilli, V., Venice, F. et al. (2018). Strigolactones cross the kingdoms: plants, fungi, and bacteria in the arbuscular mycorrhizal symbiosis. *Journal of Experimental Botany* 69, 2175–2188. https://doi.org/10.1093/jxb/erx432.

Li, T., Deng, Y.J., Liu, J.X. et al. (2021). DcCCD4 catalyzes the degradation of α-carotene and β-carotene to affect carotenoid accumulation and taproot color in carrot. *The Plant Journal* 108, 1116–1130. https://doi.org/10.1111/tpj.15524.

Li, L.-L., Zhao, H.-H. and Kong, C.-H. (2020). (—)-Loliolide, the most ubiquitous lactone, is involved in barnyardgrass-induced rice allelopathy. *Journal of Experimental Botany* 71, 1540–1550. https://doi.org/10.1093/jxb/erz497.

Liang, M.-H., He, Y.-J., Liu, D.-M. et al. (2021). Regulation of carotenoid degradation and production of apocarotenoids in natural and engineered organisms. *Critical Reviews in Biotechnology* 41, 531–534. https://doi.org/10.1080/07388551.2021.1873242.

Lu, S., Zhang, Y., Zhu, K. et al. (2018). The citrus transcription factor CsMADS6 modulates carotenoid metabolism by directly regulating carotenogenic genes. *Plant Physiology* 176, 2657–2676. https://doi.org/10.1104/pp.17.01830.

Luan, Y., Fu, X., Lu, P. et al. (2020a). Molecular mechanisms determining the differential accumulation of carotenoids in plant species and varieties. *Critical Reviews in Plant Sciences* 39, 125–139. https://doi.org/10.1080/07352689.2020.1768350.

Ma, G., Zhang, L., Matsuta, A. et al. (2013). Enzymatic formation of β-citraurin from β-cryptoxanthin and zeaxanthin by carotenoid cleavage dioxygenase 4 in the flavedo of citrus fruit. *Plant Physiology* 163, 682–695. https://doi.org/10.1104/pp.113.223297.

Mani, P., Sharma, H., Gautam, D. et al. (2017). Hydroxyapatite (HA) attenuate TiO_2 toxicity in bio-system triggering *E. coli* and mouse bone marrow mono-nuclear cells (BMMNC'S). *International Journal of Life Science and Pharma Research* 7, 46–57.

Mathieu, S., Terrier, N., Procureur, J.M. et al. (2005). A carotenoid cleavage dioxygenase from *Vitis vinifera* L.: functional characterization and expression during grape berry development in relation to C13-norisoprenoid accumulation. *Journal of Experimental Botany* 56, 2721–2731. https://doi.org/10.1093/jxb/eri265.

Meng, N., Wei, Y., Gao, Y. et al. (2020). Characterization of transcriptional expression and regulation of carotenoid cleavage dioxygenase 4b in grapes. *Frontiers in Plant Science* 11, 483. https://doi.org/10.3389/fpls.2020.00483.

Mi, J. and Al-Babili, S. (2019). To color or to decolor: that is the question. *Molecular Plant* 12, 1173–1175. https://doi.org/10.1016/j.molp.2019.07.007.

Mi, J., Jia, K.-P., Balakrishna, A. et al. (2019). A highly sensitive SPE derivatization–UHPLC–MS approach for quantitative profiling of carotenoid-derived dialdehydes from vegetables. *Journal of Agricultural and Food Chemistry* 67, 5899–5907. https://doi.org/10.1021/acs.jafc.9b01749.

Mitra, S., Estrada-Tejedor, R., Volke, D.C. et al. (2021). Negative regulation of plastidial isoprenoid pathway by herbivore-induced β-cyclocitral in *Arabidopsis thaliana*. *Proceedings of the National Academy of Sciences of the United States of America* 118, e2008747118. https://doi.org/10.1073/pnas.2008747118.

Moise, A.R., Al-Babili, S. and Wurtzel, E.T. (2014). Mechanistic aspects of carotenoid biosynthesis. *Chemical Reviews* 114, 164–193. https://doi.org/10.1021/cr400106y.

Moreno, J.C., Mi, J., Alagoz, Y. et al. (2021). Plant apocarotenoids: from retrograde signaling to interspecific communication. *The Plant Journal* 105, 351–375. https://doi.org/10.1111/tpj.15102.

Murata, M., Kobayashi, T. and Seo, S. (2020). α-Ionone, an apocarotenoid, induces plant resistance to western flower thrips, *Frankliniella occidentalis*, independently of jasmonic acid. *Molecules* 25, 17. https://doi.org/10.3390/molecules25010017.

Murata, M., Nakai, Y., Kawazu, K. et al. (2019). Loliolide, a carotenoid metabolite, is a potential endogenous inducer of herbivore resistance. *Plant Physiology* 179, 1822–1833. https://doi.org/10.1104/pp.18.00837.

Nisar, N., Li, L., Lu, S. et al. (2015). Carotenoid metabolism in plants. *Molecular Plant* 8, 68–82. https://doi.org/10.1016/j.molp.2014.12.007.

Pacheco, S.D.G., Gasparin, A.T., Jesus, C.H.A. et al. (2019). Antinociceptive and anti-inflammatory effects of bixin, a carotenoid extracted from the seeds of *Bixa orellana*. *Planta Medica* 85, 1216–1224. https://doi.org/10.1055/a-1008-1238.

Pandey, H., Singh, D., Dhiman, V.K. et al. (2022). Extremophiles and their application in bioremediation. In: *Physiology, Genomics, and Biotechnological Applications of Extremophiles* (eds. A.B. Gunjal, R. Thombre and J.A. Parray), 188–206. IGI Global.

Park, S., Kim, H.S., Jung, Y.J. et al. (2016). Orange protein has a role in phytoene synthase stabilization in sweetpotato. *Scientific Reports* 6, 33563. https://doi.org/10.1038/srep33563.

Prasad, G., Chauhan, D., Pandey, H. et al. (2024). Recent advances in CRISPR-Cas for climate-resilient horticulture in fruits crops. In: *Climate-Resilient Agriculture: A Molecular Perspective* (eds. A. Singh and S. Pandey), 415–431. Apple Academic Press.

Ramel, F., Birtic, S., Ginies, C. et al. (2012). Carotenoid oxidation products are stress signals that mediate gene responses to singlet oxygen in plants. *Proceedings of the National*

Academy of Sciences of the United States of America 109, 5535–5540. https://doi.org/10.1073/pnas.1115982109.

Rathva, H., Pandey, A.K., Suthar, K. et al. (2024). Genetic relatedness analysis in sapota using SSR markers. *Ecological Genetics and Genomics* 31, 100234.

Rezaee, R. and Hosseinzadeh, H. (2013). Safranal: from an aromatic natural product to a rewarding pharmacological agent. *Iranian Journal of Basic Medical Sciences* 16, 12–26.

Rodrigo, M.J., Alquezar, B., Alos, E. et al. (2013). A novel carotenoid cleavage activity involved in the biosynthesis of *Citrus* fruit-specific apocarotenoid pigments. *Journal of Experimental Botany* 64, 4461–4478. https://doi.org/10.1093/jxb/ert260.

Rodriguez-Concepcion, M., Avalos, J., Bonet, M.L. et al. (2018). A global perspective on carotenoids: metabolism, biotechnology, and benefits for nutrition and health. *Progress in Lipid Research* 70, 62–93. https://doi.org/10.1016/j.plipres.2018.04.004.

Sangma, R.R., Manpoong, C., Sharma, A. et al. (2022). Performance of black rice (*Oryza sativa*) varieties grown in Namsai district of Arunachal Pradesh, India. *Research on Crops* 23(1), 11–14.

Schwartz, S.H., Qin, X. and Zeevaart, J.A. (2001). Characterization of a novel carotenoid cleavage dioxygenase from plants. *Journal of Biological Chemistry* 276, 25208–25211. https://doi.org/10.1074/jbc.M102146200.

Seo, M., Peeters, A.J., Koiwai, H. et al. (2000). The *Arabidopsis* aldehyde oxidase 3 (*AAO3*) gene product catalyzes the final step in abscisic acid biosynthesis in leaves. *Proceedings of the National Academy of Sciences of the United States of America* 97, 12908–12913. https://doi.org/10.1073/pnas.220426197.

Sharma, A., Hazarika, M., Heisnam, P. et al. (2023a). Influence of storage conditions, packaging, post-harvest technology, nanotechnology and molecular approaches on shelf life of microgreens. *Journal of Agriculture and Food Research* 14, 100835.

Sharma, A., Hazarika, M., Heisnam, P. et al. (2024a). Controlled environment ecosystem: a cutting-edge technology in speed breeding. *ACS Omega* 9(27), 29114–29138.

Sharma, A., Hazarika, M., Heisnam, P. et al. (2024b). Spontaneous formation, gene regulation of *Trichoderma* and slow decomposition in cocopeat. *Next Sustainability* 4, 100051.

Sharma, A., Pandey, H., Manpoong, C. et al. (2023b). Biometric analysis, biosynthetic pathway and multipurpose uses of buckwheat local varieties in Namsai district of eastern Himalayas in India. *Trends in Food Science & Technology* 136, 251–267.

Sharma, A., Pandey, H., Nampoothiri Devadas, V.A.S. et al. (2023c). Regulations and factors affecting of light emitting diodes artificial light in crop production under soilless culture. *ACS Agricultural Science & Technology* 3(8), 616–630.

Shi, J., Cao, C., Xu, J. et al. (2020). Research advances on biosynthesis, regulation, and biological activities of apocarotenoid aroma in horticultural plants. *Journal of Chemotherapy* 2020, 2526956. https://doi.org/10.1155/2020/2526956.

Simkin, A.J., Schwartz, S.H., Auldridge, M. et al. (2004a). The tomato *carotenoid cleavage dioxygenase 1* genes contribute to the formation of the flavor volatiles β-ionone, pseudoionone, and geranylacetone. *The Plant Journal* 40, 882–892. https://doi.org/10.1111/j.1365-313X.2004.02263.x.

Simkin, A.J., Underwood, B.A., Auldridge, M. et al. (2004b). Circadian regulation of the PhCCD1 carotenoid cleavage dioxygenase controls emission of β-ionone, a fragrance volatile of petunia flowers. *Plant Physiology* 136, 3504–3514. https://doi.org/10.1104/pp.104.049718.

Singh, D., Pandey, H., Dhiman, V.K. et al. (2022a). Assessment of environmental impacts of metal/metalloid pollution on plants. In: *Metals Metalloids Soil Plant Water Systems: Phytophysiology and Remediation Techniques* (eds. T. Aftab and K. Hakeem), 217–232. Academic Press.

Singh, D., Pandey, H., Thakur, K. et al. (2022b). Aquaporins and their functions in water transportation in different plant species. *Environmental Sustainability* 5(4), 443–456.

Singh, D., Sharma, A., Verma, S.K. et al. (2024a). Impact of nanoparticles on plant physiology, nutrition, and toxicity: a short review. *Next Nanotechnology* 6, 100081.

Singh, D., Singh, A. and Kumar, S. (2023). Probiotics: friend or foe to the human immune system. *Bulletin of the National Research Centre* 47(1), 126.

Singh, D. and Verma, S.K. (2023). Impacts of particulate matter pollution on plants. In: *New Frontiers in Plant-Environment Interactions: Innovative Technologies and Developments* (ed. T. Aftab), 483–500. Springer Nature Switzerland.

Singh, D., Verma, S.K. and Kumar, S. (2024b). Genetic manipulation for secondary metabolite production. In: *Developments in Applied Microbiology and Biotechnology: Secondary Metabolites and Biotherapeutics* (eds. A. Kumar and S. Kumar), 161–188. Academic Press.

Singh, D., Verma, S.K. and Shyam, P. (2024c). Identification and purification of plant secondary metabolite as medicinal raw materials. In: *Developments in Applied Microbiology and Biotechnology: Secondary Metabolites and Biotherapeutics* (eds. A. Kumar and S. Kumar), 9–38. Academic Press.

Singh, D., Verma, S.K., Wani, K.I. et al. (2024d). An overview of metal and metalloid toxicity and signaling in plants. In: *Signaling and Communication in Plants: Metals and Metalloids in Plant Signaling* (ed. T. Aftab), 1–17. Springer.

Srivastava, S., Singh, D., Verma, S.K. et al. (2024). Recent advancements in cancer vaccines: a systematic review. *Vacunas (English Edition)* 25(1), 97–108. https://doi.org/10.1016/j.vacune.2024.02.011.

Sunya, H., Sharma, A., Manpoong, C. et al. (2022). Evaluation of improved varieties of rice (*Oryza sativa* L.) under environmental condition of Namsai, Arunachal Pradesh, India. *Ecology, Environment and Conservation* 28, 907–911.

Tan, B.-C., Joseph, L.M., Deng, W.-T. et al. (2003). Molecular characterization of the *Arabidopsis* 9-*cis* epoxycarotenoid dioxygenase gene family. *The Plant Journal* 35, 44–56. https://doi.org/10.1046/j.1365-313X.2003.01786.x.

Tieman, D., Zhu, G., Resende, M.F. et al. (2017). A chemical genetic roadmap to improved tomato flavor. *Science* 355, 391–394. https://doi.org/10.1126/science.aal1556.

Tyagi, S., Dhiman, V.K., Dhiman, V.K. et al. (2024). Plant defense strategies and biomarkers against heavy metal-induced stress: a comprehensive review. *ACS Agricultural Science & Technology* 4(2), 129–143.

Verma, S., Anjali, A., Diksha, D. et al. (2023). Recent advancements in skin tissue engineering in the application of nanotechnology. *Research Journal of Biotechnology* 18, 127–136. https://doi.org/10.25303/1802rjbt1270136.

Verma, S.K., Kumar, P., Mishra, A. et al. (2024a). Green nanotechnology: illuminating the effects of bio-based nanoparticles on plant physiology. *Biotechnology for Sustainable Materials* 1(1), 1. https://doi.org/10.1186/s44316-024-00001-2.

Verma, S., Pandey, M., Khare, R. et al. (2024b). A review on non-small cell lung cancer. *Vacunas* 25, 239–253. https://doi.org/10.1016/j.vacun.2023.12.003.

Verma, S., Pandey, M., Sharma, A. et al. (2024c). Exploring Ayurveda: principles and their application in modern medicine. *Bulletin of the National Research Centre* 48, 77. https://doi.org/10.1186/s42269-024-01231-0.

Vogel, J.T., Tan, B.-C., McCarty, D.R. et al. (2008). The carotenoid cleavage dioxygenase 1 enzyme has broad substrate specificity, cleaving multiple carotenoids at two different bond positions. *Journal of Biological Chemistry* 283, 11364–11373. https://doi.org/10.1074/jbc.M710106200.

von Lintig, J. (2010). Colors with functions: elucidating the biochemical and molecular basis of carotenoid metabolism. *Annual Review of Nutrition* 30, 35–56. https://doi.org/10.1146/annurev-nutr-080508-141027.

Walter, M.H. and Strack, D. (2011). Carotenoids and their cleavage products: biosynthesis and functions. *Natural Product Reports* 28, 663–692. https://doi.org/10.1039/c0np00036a.

Wang, J.Y., Alseekh, S., Xiao, T. et al. (2021). Multi-omics approaches explain the growth-promoting effect of the apocarotenoid growth regulator zaxinone in rice. *Communications Biology* 4, 1222. https://doi.org/10.1038/s42003-021-02740-8.

Wang, J.Y., Haider, I., Jamil, M. et al. (2019). The apocarotenoid metabolite zaxinone regulates growth and strigolactone biosynthesis in rice. *Nature Communications* 10, 810. https://doi.org/10.1038/s41467-019-08461-1.

Wang, J.Y., Lin, P.-Y. and Al-Babili, S. (2020). On the biosynthesis and evolution of apocarotenoid plant growth regulators. *Seminars in Cell & Developmental Biology* 109, 3–11. https://doi.org/10.1016/j.semcdb.2020.07.007.

Watanabe, K., Oda-Yamamizo, C., Sage-Ono, K. et al. (2018). Alteration of flower colour in *Ipomoea nil* through CRISPR/Cas9-mediated mutagenesis of *carotenoid cleavage dioxygenase 4*. *Transgenic Research* 27, 25–38. https://doi.org/10.1007/s11248-017-0051-0.

Watkins, J.L. and Pogson, B.J. (2020). Prospects for carotenoid biofortification targeting retention and catabolism. *Trends in Plant Science* 25, 501–512. https://doi.org/10.1016/j.tplants.2019.12.021.

Xi, W., Zheng, H., Zhang, Q. et al. (2016). Profiling taste and aroma compound metabolism during apricot fruit development and ripening. *International Journal of Molecular Sciences* 17, 998. https://doi.org/10.3390/ijms17070998.

Xie, X., Yoneyama, K. and Yoneyama, K. (2010). The strigolactone story. *Annual Review of Phytopathology* 48, 93–117. https://doi.org/10.1146/annurev-phyto-073009-114453.

Xu, Z., Pu, X., Gao, R. et al. (2020). Tandem gene duplications drive divergent evolution of caffeine and crocin biosynthetic pathways in plants. *BMC Biology* 18, 63. https://doi.org/10.1186/s12915-020-00795-3.

Xu, J., Wang, X., Cao, H. et al. (2017). Dynamic changes in methylome and transcriptome patterns in response to methyltransferase inhibitor 5-azacytidine treatment in citrus. *DNA Research* 24, 509–522. https://doi.org/10.1093/dnares/dsx021.

Yahyaa, M., Bar, E., Dubey, N.K. et al. (2013). Formation of norisoprenoid flavor compounds in carrot (*Daucus carota* L.) roots: characterization of a cyclic-specific carotenoid cleavage dioxygenase 1 gene. *Journal of Agricultural and Food Chemistry* 61, 12244–12252. https://doi.org/10.1021/jf404085k.

Yao, M., Chen, W., Kong, J. et al. (2020). *METHYLTRANSFERASE1* and ripening modulate vivipary during tomato fruit development. *Plant Physiology* 183, 1883–1897. https://doi.org/10.1104/pp.20.00499.

Yoneyama, K., Akiyama, K., Brewer, P.B. et al. (2020). Hydroxyl carlactone derivatives are predominant strigolactones in *Arabidopsis*. *Plant Direct* 4, e00219. https://doi.org/10.1002/pld3.219.

Yuan, H., Zhang, J., Nageswaran, D. et al. (2015). Carotenoid metabolism and regulation in horticultural crops. *Horticulture Research* 2, 15036. https://doi.org/10.1038/hortres.2015.36.

Yue, Z., Liu, H. and Ma, F. (2015). The *Malus carotenoid cleavage dioxygenase* 7 is involved in stress response and regulated by basic *pentacysteine* 1. *Scientia Horticulturae* 192, 264–270. https://doi.org/10.1016/j.scienta.2015.06.027.

Zhang, B., Liu, C., Wang, Y. et al. (2015). Disruption of a *CAROTENOID CLEAVAGE DIOXYGENASE 4* gene converts flower colour from white to yellow in *Brassica* species. *New Phytologist* 206, 1513–1526. https://doi.org/10.1111/nph.13335.

Zhang, Y., van Dijk, A.D., Scaffidi, A. et al. (2014). Rice cytochrome P450 MAX1 homologs catalyze distinct steps in strigolactone biosynthesis. *Nature Chemical Biology* 10, 1028. https://doi.org/10.1038/nchembio.1660.

Zheng, X., Giuliano, G. and Al-Babili, S. (2020a). Carotenoid biofortification in crop plants: *citius, altius, fortius*. *Biochimica et Biophysica Acta (BBA)—Molecular and Cell Biology of Lipids* 1865, 158664. https://doi.org/10.1016/j.bbalip.2020.158664.

Zheng, X., Kuijer, H.N. and Al-Babili, S. (2020b). Carotenoid biofortification of crops in the CRISPR era. *Trends in Biotechnology* 39, 857–860. https://doi.org/10.1016/j.tibtech.2020.12.003.

Zheng, X., Mi, J., Deng, X. et al. (2021). LC–MS-based profiling provides new insights into apocarotenoid biosynthesis and modifications in citrus fruits. *Journal of Agricultural and Food Chemistry* 69, 1842–1851. https://doi.org/10.1021/acs.jafc.0c06893.

Zheng, X., Xie, Z., Zhu, K. et al. (2015). Isolation and characterization of *carotenoid cleavage dioxygenase 4* genes from different citrus species. *Molecular Genetics and Genomics* 290, 1589–1603. https://doi.org/10.1007/s00438-015-1016-8.

Zheng, X., Zhu, K., Sun, Q. et al. (2019). Natural variation in *CCD4* promoter underpins species-specific evolution of red coloration in citrus peel. *Molecular Plant* 12, 1294–1307. https://doi.org/10.1016/j.molp.2019.04.014.

Zhong, Y., Pan, X., Wang, R. et al. (2020). *ZmCCD10a* encodes a distinct type of carotenoid cleavage dioxygenase and enhances plant tolerance to low phosphate. *Plant Physiology* 184, 374–392. https://doi.org/10.1104/pp.20.00378.

Zhou, X., Welsch, R., Yang, Y. et al. (2015). *Arabidopsis* OR proteins are the major posttranscriptional regulators of phytoene synthase in controlling carotenoid biosynthesis. *Proceedings of the National Academy of Sciences of the United States of America* 112, 3558–3563. https://doi.org/10.1073/pnas.1420831112.

Zoccali, M., Giuffrida, D., Salafia, F. et al. (2021). Apocarotenoids profiling in different *Capsicum* species. *Food Chemistry* 334, 127595. https://doi.org/10.1016/j.foodchem.2020.127595.

3

Carotenoid-derived Diapocarotenoids: From Pigments to Plant Growth Regulators

*Chao Guo[1], Weiqiang Li[2], Yuchen Miao[1,3], and Kun-Peng Jia[1,3]**

[1] *National Key Laboratory of Cotton Bio-breeding and Integrated Utilization, State Key Laboratory of Crop Stress Adaptation and Improvement, Henan Joint International Laboratory for Crop Multi-Omics Research, School of Life Sciences, Henan University, Kaifeng, Henan, China*
[2] *Jilin Da'an Agro-Ecosystem National Observation Research Station, Changchun Jingyuetan Remote Sensing Experiment Station, State Key Laboratory of Black Soils Conservation and Utilization, Northeast Institute of Geography and Agroecology, Chinese Academy of Sciences, Changchun, Jilin, China*
[3] *Sanya Institute of Henan University, Sanya, Hainan, China*

*Corresponding author: kunpengjia_0515@126.com

3.1 Introduction

Carotenoids are vital lipophilic isoprenoids that play a significant role in human health, serving as precursors to vitamin A and acting as efficient antioxidants (Harrison 2005; Voutilainen et al. 2006). These pigments are synthesized by a diverse range of organisms, including all photosynthetic entities such as plants, algae, and cyanobacteria, as well as various non-photosynthetic fungi and prokaryotes (Moise et al. 2014; Rosas-Saavedra and Stange 2016; Rodriguez-Concepcion et al. 2018). In plants, carotenoids serve critical functions in photosynthesis. They act as accessory light-harvesting pigments, absorbing light primarily in the blue–green region and transferring this energy to chlorophylls. This function not only enhances photosynthesis but also provides photoprotection to photosystems against singlet oxygen damage resulting from excessive light exposure (Polívka and Frank 2010; Hashimoto et al. 2016). Beyond their involvement in light absorption and protection, carotenoids are precursors for biologically significant metabolites, particularly plant hormones such as abscisic acid (ABA) and strigolactones (SLs), along with various signaling molecules like β-cyclocitral, zaxinone, and retinal (Schwartz et al. 1997; Al-Babili and Bouwmeester 2015; Jia et al. 2018; Dickinson et al. 2021; Wang et al. 2021).

Structurally, carotenoids are characterized by a common C40 backbone composed of conjugated double bond system with 3–11 double bonds capable of adopting different

Plant Apocarotenoids: Biosynthesis, Signaling and Agricultural Applications, First Edition. Edited by Kaiser Iqbal Wani and Tariq Aftab.
Published 2026 by John Wiley & Sons Ltd.

stereo-configurations, which makes them particularly susceptible to oxidative cleavage (Moise et al. 2014). The oxidative cleavage of carotenoids is primarily catalyzed by carotenoid cleavage dioxygenases (CCDs) or triggered nonenzymatically by reactive oxygen species (ROS) (Giuliano et al. 2003; Walter and Strack 2011; Ramel et al. 2012; Nisar et al. 2015). In addition, lipoxygenases were also reported to mediate this process (Porta and Rocha-Sosa 2002). The oxidative cleavage of carotenoids leads to the generation of apocarotenoids, a class of monocarbonyl products derived from carotenoid cleavage, which could act as the direct or precursors of above-mentioned carotenoid-derived bioactive metabolites (Moreno et al. 2021).

Diapocarotenoids (DIAPOs), a class of linear carotenoid oxidation products characterized by two aldehyde functional groups, can be formed through the further oxidative cleavage of apocarotenoids (Mi et al. 2019). Generally recognized as metabolite intermediates in carotenoid metabolism, DIAPOs have recently garnered interest due to their biological significance. Notably, two specific DIAPOs, crocin dialdehyde and bixin aldehyde, are precursors to bixin and crocin, pigments found in annatto fruits and saffron flowers, respectively (Bouvier et al. 2003a; Frusciante et al. 2014). Recent research has provided growing evidence that DIAPOs may serve as biologically important metabolites in both plants and animals. For example, it has been reported that anchorene and isoanchorene, two C10 DIAPOs, function as endogenous bioactive metabolites that regulate root development in *Arabidopsis* and rice (Jia et al. 2019, 2021; Ke et al. 2022, 2024). Additionally, a C14 DIAPO and its structural isomer have been shown to inhibit the activity of nuclear factor kappa B in bone and cancer cells (Linnewiel et al. 2009; Linnewiel-Hermoni et al. 2014).

In this review chapter, we will explore the origin, nomenclature, and identification of carotenoid-derived DIAPOs in plants, elucidating their biological functions in pigmentation and plant growth regulation, with a particular focus on the research progress on anchorene.

3.2 Nomenclature and Identification of Carotenoid-derived DIAPOs in Plants

DIAPOs can theoretically be produced through the oxidative cleavage of all carotenoids. Because all all-trans carotenoids share a conjugated double bond system, DIAPOs could be named based on the specific positions of oxidative cleavage within these carotenoids (Figure 3.1a). For example, when considering all-*trans*-β-carotene, cleaving at the 11,12 and 11′,12′ double bonds yields 12,12′-diapocarotene-12,12′-dial, commonly known as anchorene (Jia et al. 2019). This specificity in cleavage results in distinct DIAPO products that may contain the same or different carbon numbers. A summary of most potential DIAPOs (with more than seven carbons) derived from all-*trans*-β-carotene is provided in Figure 3.1a, b. Additionally, DIAPOs can undergo further isomerization or modification to generate products such as cis isomers, dialcohols, and diacids, with derivatives named according to the nature of these modifications (Figure 3.1c).

Initially identified through high-performance liquid chromatography (HPLC) and liquid chromatography–mass spectrometry (LC–MS) in products of CCD-catalyzed enzymatic assays, DIAPOs are typically produced as side products in considerable quantities, facilitating their detection (Scherzinger et al. 2006; Alder et al. 2008; Ilg et al. 2009, 2014).

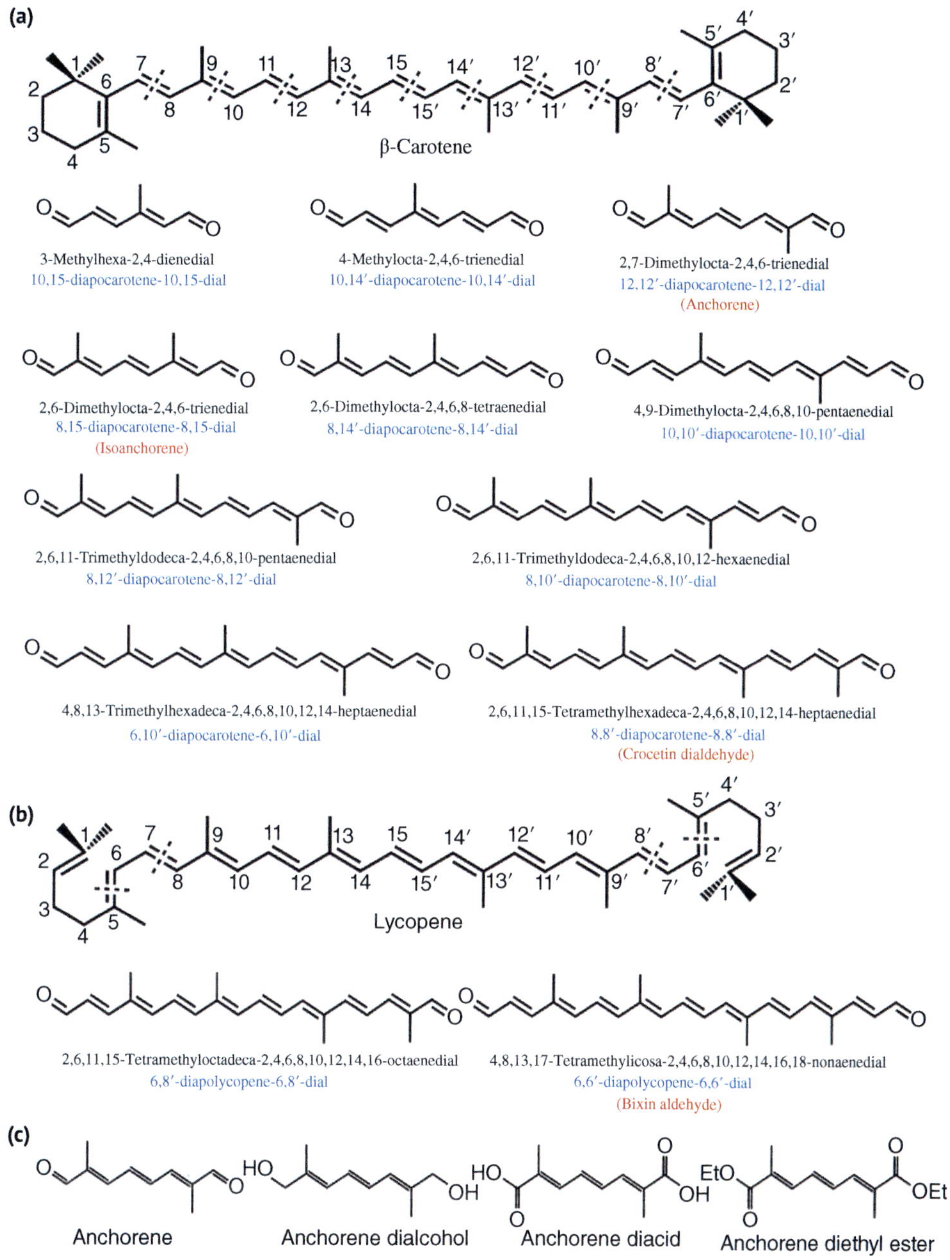

Figure 3.1 Potential carotenoid-derived DIAPOs and their derivatives. Potential DIAPOs derived from (a) carotene and (b) lycopene. (c) Potential DIAPO derivatives using anchorene as a reference. In panels (a) and (b), the DIAPOs named based on their chemical structures, carotenoid origin, and biological functions are marked by black, blue, and red, respectively.

For instance, CCD1 can cleave both monocyclic and bicyclic carotenoids at various double bond locations, resulting in a diverse range of DIAPOs (Ilg et al. 2009, 2014). However, endogenous DIAPOs are generally unstable and exist in low abundance within plants, resulting in poor ionization efficiency in MS, which complicates their direct detection via LC–MS. To address these challenges, Mi and co-workers developed a chemical derivatization-based ultra-HPLC method combined with a hybrid quadrupole-Orbitrap (UHPLC-Q-Orbitrap) mass spectrometer specifically for analyzing carotenoid-derived DIAPOs from plant sources (Mi et al. 2019). The reagent used for derivatization, *N*2,*N*2,*N*4,*N*4-tetraethyl-6-hydrazineyl-1,3,5-triazine-2,4-diamine, effectively converts the aldehyde groups in DIAPOs into stable structures, enhancing their stability and detection sensitivity.

Using the derivatization-based UHPLC-Q-Orbitrap mass spectrometer, both anchorene and isoanchorene were identified in the root extracts of *Arabidopsis* and rice, confirming their endogenous presence in these plants (Jia et al. 2019, 2021). Notably, in *Arabidopsis* seedlings treated with NF (a carotenoid biosynthesis inhibitor), the levels of anchorene and isoanchorene significantly decreased, further indicating their carotenoid origin. Subsequently, Mi and co-workers detected various lengths of carbon chain DIAPOs at concentrations as low as 0.05 pg/mg in dry material from several vegetables, including pepper, carrot, and tomato (Mi et al. 2020). These findings collectively suggest that DIAPOs are widely present across different plant species.

3.3 Biosynthesis of Carotenoid-derived DIAPOs in Plants

3.3.1 CCDs in Carotenoid Metabolism

The extended conjugated double bond system in carotenoids makes them particularly susceptible to cleavage, either enzymatically by CCDs or nonenzymatically through ROS, to produce apocarotenoids and DIAPOs (Giuliano et al. 2003; Walter and Strack 2011; Ramel et al. 2012; Nisar et al. 2015). CCDs are a widespread family of nonheme iron enzymes found in bacteria, fungi, plants, and animals (Sui et al. 2013; Ahrazem et al. 2016). Plant CCD enzymes are categorized into six subfamilies based on substrate specificity, stereospecificity, and cleavage sites: CCD1, CCD4, CCD7, CCD8, 9-*cis*-epoxycarotenoid cleavage dioxygenases (NCEDs), and zaxinone synthase (ZAS) (Giuliano et al. 2003; Wang et al. 2019). Among these, NCEDs play a critical role in both classical and nonclassical synthetic pathways for ABA (Schwartz et al. 1997). They cleave the C11–C12 double bond in 9-*cis*-violaxanthin and 9′-*cis*-neoxanthin to produce xanthoxin, the essential ABA precursor, and serve as the rate-limiting enzymes in *de novo* ABA synthesis. The CCD7 and CCD8 subfamilies have been identified as key enzymes involved in the biosynthesis of SLs (Jia et al. 2018). CCD7 is a stereospecific enzyme that cleaves 9-*cis*-β-carotene, leading to the generation of β-ionone and 9-*cis*-configured β-apo-10′-carotenal, which is further cleaved by CCD8 to produce the key SL precursor, carlactone. Members of the ZAS subfamily have recently been reported to cleave apo-10′-zeaxanthinal to produce an apocarotenoid zaxinone, a newly identified plant growth regulator in rice (Wang et al. 2019). CCD1 is notable for its ability to cleave a variety of linear, monocyclic, and bicyclic carotenoid and apocarotenoid substrates at multiple positions (Ilg et al. 2014). This versatility allows CCD1 to function as a scavenger, breaking down apocarotenoids into a wide range of products. Similarly, CCD4 enzymes primarily target the C9—C10 or C7—C8 double bonds in bicyclic carotenoids, influencing carotenoid content and pigmentation across different plant organs and tissues (Zhang et al. 2015; Zheng et al. 2019). While CCD4 is considered responsible for the production of β-cyclocitral, ROS also contribute to its formation through nonenzymatic cleavage mechanisms (Ramel et al. 2012; Dickinson et al. 2019).

Plant CCDs are the main enzymes reported to synthesize DIAPOs. Both CCD8 and OsZAS can cleave 9-*cis*-configured β-apo-10′-carotenal and apo-10′-zeaxanthinal, respectively, to produce a same C9 DIAPO, 10,14′-diapocarotene-10,14′-dial (Jia et al. 2018; Wang et al. 2019). Members of CCD1 subfamily cleave a variety of linear, monocyclic, and bicyclic carotenoid and apocarotenoid substrates at multiple positions to produce various DIAPOs *in vitro* (Ilg et al. 2014). In another example, Diox1 cleaves β-apo-8′-carotenal to produce 8,15-diapocarotene-8,15-dial, as analyzed through GC–MS (Ruch et al. 2005). Although most of these DIAPOs are thought to be the side products or intermediates of carotenoids degradation, research on the biosynthesis of bixin, crocin, anchorene, and isoanchorene highlights the significance of DIAPOs as pigments and specific regulators of plant growth in plants.

3.3.2 Bixin Biosynthesis in *Bixa orellana*

Bixin, a dicarboxylic monomethyl ester apocarotenoid derived solely from the seeds of *Bixa orellana*, is one of the oldest pigments utilized by humans (Figure 3.2a). Its biosynthesis begins with the conversion of lycopene into bixin aldehyde (6,6′-diapolycopene-6,6′-dial), facilitated by the enzyme *B. orellana* lycopene cleavage dioxygenase (BoLCD) (Bouvier et al. 2003a). Bouvier and co-workers synthesized a 400-bp DNA probe to successfully screen for the *BoLCD* gene responsible for this conversion, revealing that *BoLCD* exhibits high sequence similarity (28% identity) to *Arabidopsis AtCCD1. In vitro* enzyme activity assays confirmed that BoLCD effectively converts lycopene into bixin aldehyde (Bouvier et al. 2003a).

Rodríguez-Ávila and co-workers further contributed to the understanding of bixin biosynthesis by isolating a new *CCD* gene, *BoCCD1*, and conducting evolutionary and expression analyses, which indicated its role in bixin production (Rodríguez-Ávila et al. 2011). Soares and co-workers identified that *BoCCD* was significantly induced during seed development, suggesting its involvement in carotenoid cleavage leading to bixin biosynthesis in the seeds of *B. orellana* (Soares et al. 2011). Notably, *BoCCD* displayed strong similarity with several plant CCD4 proteins.

Based on seed transcriptome data of *B. orellana*, Cárdenas-Conejo and co-workers identified nine CCD proteins, including four members from CCD1 subfamily (BoCCD1-1, BoCCD1-2, BoCCD1-3, and BoCCD1-4) and five members from CCD4 subfamily (BoCCD4-1, BoCCD4-2, BoCCD4-3, BoCCD4-4, and BoCCD4-5). Surprisingly, both the BoLCD and the BoCCD1 proteins described previously were not closely related to the CCD proteins found in this transcriptome. Subcellular localization predictions and gene expression changes during bixin accumulation suggested that BoCCD4-1, BoCCD4-2, BoCCD4-3, BoCCD1-3, and BoCCD1-4 might be involved in converting bixin aldehyde production (Cárdenas-Conejo et al. 2015). Later, combining *in situ* RT-qPCR in seed tissue with *in vitro* enzymatic assay analyses, it was shown that only BoCCD4-3 and BoCCD1-1 cleave lycopene at C5–C6 and C5′–C6′ to produce bixin aldehyde (Figure 3.2a) (Carballo-Uicab et al. 2019). However, using *in vivo* and *in vitro* approaches, Us-Camas and co-workers determined that the other identified BoCCD enzymes (BoCCD1-3, BoCCD1-4, BoCCD4-1, BoCCD4-2, and BoCCD4-4) also participate in the biosynthesis of bixin aldehyde from

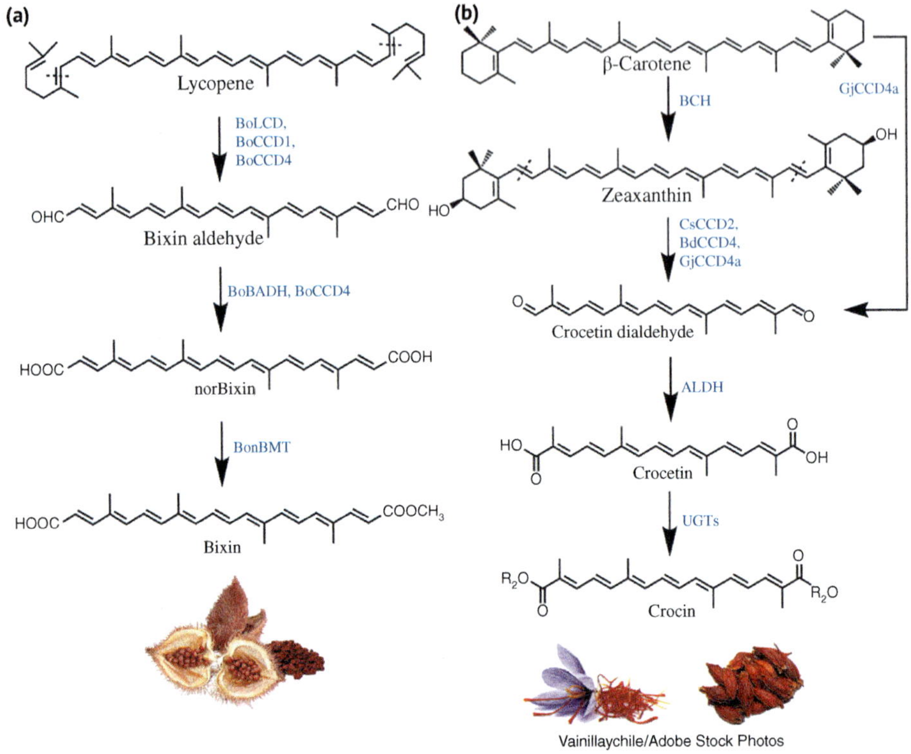

Figure 3.2 The biosynthesis routes of bixin and crocin in plants. (a) The biosynthesis route of bixin in *B. orellana*. (b) The biosynthesis route of crocin in saffron and *G. jasminoides*. The enzymes are marked in blue.

lycopene. Moreover, *in vivo* assays of BoCCD1-4, BoCCD4-1, BoCCD4-2, and BoCCD4-4 were also involved in the conversion of the oxidation of bixin aldehyde into norbixin, the second product of the bixin biosynthesis pathway (Us-Camaset al. 2022). In another study, Frusciante and co-workers reassessed the activity of six Bixa CCDs (BoCCD1-1, BoCCD1-3, BoCCD1-4, BoCCD4-2, BoCCD4-3, and BoCCD4-4) and found that none of them produced substantial amounts of bixin dialdehyde in *Escherichia coli*. Unexpectedly, BoCCD4-3 cleaved different carotenoids to yield the C20 DIAPO crocetin dialdehyde both *in vitro* and in tobacco (Fruscianteet al. 2022). Further clarification is still needed to finalize the bixin biosynthesis pathway in the future.

3.3.3 Crocin Biosynthesis in Saffron and Other Plants

Saffron accumulates significant amounts of the apocarotenoids crocetin and crocins within its stigmas, serving essential pigments for pollination and components of medicine (Tong et al. 2015). Crocin dialdehyde serves as an essential intermediate product in the proposed crocin biosynthetic pathway (Figure 3.2b) (Bouvier et al. 2003b). Utilizing deep transcriptome analysis of developing saffron stigmas, Ahrazem and co-workers identified an enzyme distinct from CCD1 and CCD4: saffron CsCCD2, which cleaves zeaxanthin at the C7–C8 and C7′–C8′ positions to yield crocetin dialdehyde via the C30 intermediate 3-OH-β-apo-8′-carotenal in *E. coli*. Furthermore, CCD2 was shown to cleave 3-OH-β-apo-10′-carotenal (C27) and 3-OH-β-apo-12′-carotenal (C25), targeting the C7–C8 double bond to produce a C17- and a C15-dialdehyde, respectively. While an endogenous aldehyde dehydrogenase in maize endosperm may be implicated in crocetin production, crocetin dialdehyde remains undetected (Frusciante et al. 2014). Notably, the synthetic pathways for crocin were introduced into tomato plants to induce CsCCD2L expression, resulting in the generation of crocetin dialdehyde and 2,6,6-trimethyl-4-hydroxy-1-carboxaldehyde-1-cyclohexene, thereby creating notable economic value (Ahrazem et al. 2022).

Homologs of CsCCD2 in *Buddleja davidii*, namely BdCCD4.1 and BdCCD4.3, exhibited C7–C8 and C7′–C8′ activity against zeaxanthin, producing crocetin dialdehyde both *in vitro* and *in vivo* (Ahrazem et al. 2017). In another study, when expressed in *E. coli*, a *CCD4* of *Gardenia jasminoides GjCCD4a* yielded detectable levels of crocetin dialdehyde using substrates such as lycopene, β-carotene, and zeaxanthin (Xu et al. 2020). Additionally, Zheng and co-workers further confirmed this through an engineered citrus callus system that GjCCD4a cleaved β-carotene to generate crocetin dialdehyde via a C30 β-apocarotenoid intermediate. *GjCCD4a* also produced a novel dialdehyde, C17-dialdehyde, by cleaving β-apo-10′-carotenal at the C7–C8 site. Importantly, GjCCD4a's enzyme activity in producing crocetin dialdehyde surpassed that of CsCCD2L (Zheng et al. 2022).

3.3.4 Anchorene and Isoanchorene

Recent research by Jia and co-workers identified a DIAPO, anchorene, playing essential roles in the regulation of root development (Jia et al. 2019). Anchorene efficiently promotes the formation of anchor roots (ANRs) which emerge from the collet just below the root–hypocotyl junction at as low a concentration as 5 μM. Anchorene was later identified in *Arabidopsis* as a carotenoid-derived metabolite by LC–MS, suggesting that it acts as an endogenous bioactive regulator in plants. The dependency of ANR formation on carotenoids was further underscored by the significant reduction in ANRs observed in carotenoid-deficient mutants, specifically ispH1 and psy, and carotenoid biosynthesis inhibitors NF- and CPTA-treated plants, while, anchorene application effectively rescued the reduction in ANR formation seen in the psy mutant and under the NF or CPTA conditions. As a natural plant metabolite, anchorene is predicted to be produced by cleaving the C11—C12 and C11′—C12′ double bonds in all carotenoids, starting from β-carotene in the carotenoid biosynthesis pathway. Feeding experiments demonstrated that all-*trans*-apo-12′-apocarotenoids, likely precursors of anchorene, lead to enhanced ANR formation. RNAseq analysis and auxin marker lines showed that anchorene regulates root development mainly through promoting auxin biosynthesis or transportation.

Isoanchorene, an analog of anchorene, can be produced by cleaving the C7—C8 and C15—C15′ double bonds in all carotenoids downstream of neurosporene (Jia et al. 2021). Despite differing only in the position of one methyl group, isoanchorene has distinct effects when applied to *Arabidopsis*. Research found that isoanchorene significantly inhibited primary root growth in a concentration-dependent manner. The combined application of anchorene and isoanchorene demonstrated additive effects on root development. Further evidence demonstrated that isoanchorene is also recognized as an endogenous plant metabolite that regulates root development through the modulation of auxin homeostasis.

3.4 Anchorene Regulates Root Development Through Modulating Auxin Homeostasis

Compared to the endogenous auxin indole-3-acetic acid (IAA) and auxin analog naphthaleneacetic acid (NAA), anchorene exhibits pleiotropic auxin-like properties, influencing various aspects of root formation (Ke et al. 2024). These effects include the emergence of adventitious roots, improved root gravitropism, increased hair density, and inhibition of primary root length. Notably, the ability of AR to induce these phenotypic changes relies on a functional auxin signaling pathway. In addition, RNA sequencing analysis further supports that anchorene regulates root development primarily by modulating auxin homeostasis (Jia et al. 2019).

Auxin homeostasis in plants is primarily regulated through its biosynthesis, transport, storage, and inactivation pathways (Fukui and Hayashi 2018). Among these, the indole-3-pyruvate (IPA) pathway is the well-characterized route for auxin biosynthesis. In this pathway, tryptophan is first catalyzed by TRYPTOPHAN AMINOTRANSFERASE OF ARABIDOPSIS 1/TRYPTOPHAN AMINOTRANSFERASE RELATED (TAA1/TAR) to produce IPA, which is then converted into IAA by YUCCA (YUC) enzymes (Stepanova et al. 2008; Dai et al. 2013). Pharmacological studies suggest that the auxin-like activities of anchorene are mediated through the TAA–YUC-dependent IPA biosynthetic pathway (Ke et al. 2024). When the IPA biosynthetic pathway was partially inhibited using specific chemical inhibitors of TAAs and YUCs, anchorene effectively restored auxin-deficient phenotypes. However, these findings also largely ruled out the possibility that anchorene directly targets TAA1 or YUCs to influence auxin accumulation.

The inactivation of IAA is primarily governed by the GH3–ILR1–DAO pathway. In this process, GRETCHEN HAGEN 3 (GH3) acyl amido synthetases conjugate IAA with L-amino acids, forming IAA–amino acid conjugates, which serve as the storage forms of auxin (Westfall et al. 2012; Hayashi et al. 2021). These conjugates are then hydrolyzed back into free IAA by IAA-Leu-Resistant1/ILR1-like (ILR1/ILL) amidohydrolases. Alternatively, DIOXYGENASE FOR AUXIN OXIDATION 1 (DAO1) irreversibly oxidizes IAA–amino acid conjugates into oxIAA–amino acid conjugates, which are further converted into inactive oxIAA by ILR1 hydrolase (Westfall et al. 2012; Porco et al. 2016; Zhang et al. 2016; Zhang and Peer 2017; Hayashi et al. 2021). Metabolite analyses related to auxin biosynthesis and catabolism revealed that anchorene treatment increased active IAA levels while reducing inactivated forms such as IAA-Glu and oxIAA (Ke et al. 2024). This suggests that anchorene directly influences auxin homeostasis by inhibiting the conjugation of IAA with amino acids. Notably, *Arabidopsis* mutants lacking GH3 proteins (*Δ8gh3*) showed no auxin-like phenotypes or induction of auxin-responsive genes in response to anchorene treatment, underscoring the critical role of GH3 proteins in mediating anchorene's biological functions. Furthermore, *in vitro* assays and bio-layer interferometry (BLI) assays demonstrated that anchorene directly binds to and inhibits GH3.17 activity (Linnewiel et al. 2009). These findings support the hypothesis that anchorene suppresses GH3.17 activity, likely by interfering with its binding to the endogenous substrate, IAA (Figure 3.3). However, the precise molecular mechanism by which anchorene inhibits GH3 proteins remains to be elucidated.

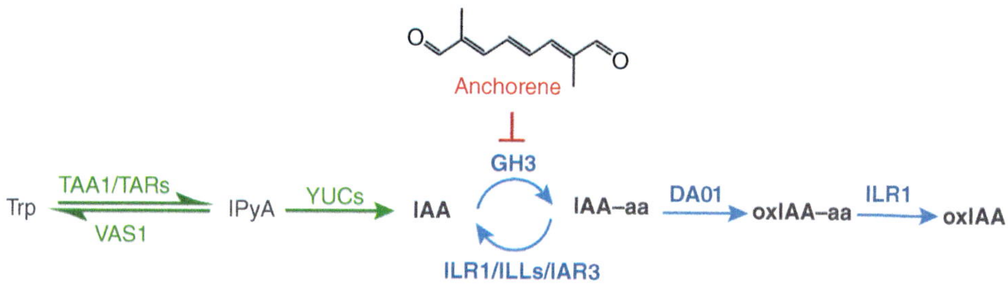

Figure 3.3 Anchorene regulates auxin homeostasis by suppressing GH3. Auxin biosynthesis enzymes are marked by green, while auxin catabolism enzymes are marked by blue.

3.5 Conclusion and Perspectives

Carotenoids are essential lipophilic isoprenoids with critical roles in human health and plant physiology, serving as precursors to vitamin A, antioxidants, and bioactive metabolites. Their oxidative cleavage generates apocarotenoids and DIAPOs. Carotenoid-derived DIAPOs, characterized by their dialdehyde structure, represent a fascinating class of metabolites with diverse biological functions in plants, ranging from growth regulation to stress adaptation. The discovery of anchorene as a regulator of root development through auxin homeostasis underscores the importance of DIAPOs in plant physiology. Similarly, the biosynthesis of crocin and bixin highlights their economic and ecological significance. However, the identification of new bioactive DIAPOs, the precise mechanisms underlying the biological activities of DIAPOs, particularly their interactions with key enzymes like GH3, remain to be fully elucidated. Future research will unravel the complex roles of DIAPOs and harness their potential for agricultural and industrial applications.

Author Contributions

K-PJ and CG conceived the work and wrote the chapter. YM and WL revised the chapter. All authors approved the submitted version.

Acknowledgments

We apologize to colleagues whose works were not referred due to space limitations. This work was supported by the National Natural Science Foundation of China (32470277 and 32170271).

Conflict of Interest

The authors declare no conflict of interest.

References

Ahrazem, O., Diretto, G., Argandoña, J. et al. (2017). Evolutionarily distinct carotenoid cleavage dioxygenases are responsible for crocetin production in *Buddleja davidii*. *Journal of Experimental Botany* 68, 4663–4677.

Ahrazem, O., Diretto, G., Rambla, J.L. et al. (2022). Engineering high levels of saffron apocarotenoids in tomato. *Horticulture Research* 23, uhac074.

Ahrazem, O., Gómez-Gómez, L., Rodrigo, M.J. et al. (2016). Carotenoid cleavage oxygenases from microbes and photosynthetic organisms: features and functions. *International Journal of Molecular Sciences* 17, 1781.

Al-Babili, S. and Bouwmeester, H.J. (2015). Strigolactones, a novel carotenoid-derived plant hormone. *Annual Review of Plant Biology* 66, 161–186.

Alder, A., Holdermann, I., Beyer, P. et al. (2008). Carotenoid oxygenases involved in plant branching catalyse a highly specific conserved apocarotenoid cleavage reaction. *Biochemical Journal* 416, 289–296.

Bouvier, F., Dogbo, O. and Camara, B. (2003a). Biosynthesis of the food and cosmetic plant pigment bixin (annatto). *Science* 300, 2089–2091.

Bouvier, F., Suire, C., Mutterer, J. et al. (2003b). Oxidative remodeling of chromoplast carotenoids: identification of the carotenoid dioxygenase *CsCCD* and *CsZCD* genes involved in crocus secondary metabolite biogenesis. *Plant Cell* 15, 47–62.

Carballo-Uicab, V.M., Cárdenas-Conejo, Y., Vallejo-Cardona, A.A. et al. (2019). Isolation and functional characterization of two dioxygenases putatively involved in bixin biosynthesis in annatto (*Bixa orellana* L.). *PeerJ* 21, e7064.

Cárdenas-Conejo, Y., Carballo-Uicab, V., Lieberman, M. et al. (2015). *De novo* transcriptome sequencing in *Bixa orellana* to identify genes involved in methylerythritol phosphate, carotenoid and bixin biosynthesis. *BMC Genomics* 16, 877.

Dai, X., Mashiguchi, K., Chen, Q. et al. (2013). The biochemical mechanism of auxin biosynthesis by an *Arabidopsis* YUCCA flavin-containing monooxygenase. *Journal of Biological Chemistry* 288, 1448–1457.

Dickinson, A.J., Lehner, K., Mi, J. et al. (2019). Beta-cyclocitral is a conserved root growth regulator. *Proceedings of the National Academy of Sciences of the United States of America* 116, 10563–10567.

Dickinson, A.J., Zhang, J., Luciano, M. et al. (2021). A plant lipocalin promotes retinal-mediated oscillatory lateral root initiation. *Science* 373, 1532–1536.

Frusciante, S., Demurtas, O.C., Sulli, M. et al. (2022). Heterologous expression of *Bixa orellana* cleavage dioxygenase 4-3 drives crocin but not bixin biosynthesis. *Plant Physiology* 188, 1469–1482.

Frusciante, S., Diretto, G., Bruno, M. et al. (2014). Novel carotenoid cleavage dioxygenase catalyzes the first dedicated step in saffron crocin biosynthesis. *Proceedings of the National Academy of Sciences of the United States of America* 111, 12246–12251.

Fukui, K. and Hayashi, K.I. (2018). Manipulation and sensing of auxin metabolism, transport and signaling. *Plant and Cell Physiology* 59, 1500–1510.

Giuliano, G., Al-Babili, S. and von Lintig, J. (2003). Carotenoid oxygenases: cleave it or leave it. *Trends in Plant Science* 8, 145–149.

Harrison, E.H. (2005). Mechanisms of digestion and absorption of dietary vitamin A. *Annual Review of Nutrition* 25, 87–103.

Hashimoto, H., Uragami, C. and Cogdell, R.J. (2016). Carotenoids and photosynthesis. In: *Subcellular Biochemistry:* Vol. 79: *Carotenoids in Nature* (ed. C. Stange), 111–139. Springer.

Hayashi, K.I., Arai, K., Aoi, Y. et al. (2021). The main oxidative inactivation pathway of the plant hormone auxin. *Nature Communications* 12, 6752.

Ilg, A., Beyer, P. and Al-Babili, S. (2009). Characterization of the rice carotenoid cleavage dioxygenase 1 reveals a novel route for geranial biosynthesis. *The FEBS Journal* 276, 736–747.

Ilg, A., Bruno, M., Beyer, P. et al. (2014). Tomato carotenoid cleavage dioxygenases 1A and 1B: relaxed double bond specificity leads to a plenitude of dialdehydes, mono-apocarotenoids and isoprenoid volatiles. *FEBS Open Bio* 4, 584–593.

Jia, K.P., Baz, L. and Al-Babili, S. (2018). From carotenoids to strigolactones. *Journal of Experimental Botany* 69, 2189–2204.

Jia, K.P., Dickinson, A.J., Mi, J. et al. (2019). Anchorene is a carotenoid-derived regulatory metabolite required for anchor root formation in *Arabidopsis. Science Advances* 5, eaaw6787.

Jia, K.P., Mi, J., Ablazov, A. et al. (2021). Iso-anchorene is an endogenous metabolite that inhibits primary root growth in Arabidopsis. *The Plant Journal* 107, 54–66.

Ke, D., Guo, J., Li, K. et al. (2022). Carotenoid-derived bioactive metabolites shape plant root architecture to adapt to the rhizospheric environments. *Frontiers in Plant Science* 13, 986414.

Ke, D., Xie, Y., Li, H. et al. (2024). Anchorene, a carotenoid-derived growth regulator, modulates auxin homeostasis by suppressing GH3-mediated auxin conjugation. *Journal of Integrative Plant Biology* 66, 2490–2504.

Linnewiel, K., Ernst, H., Caris-Veyrat, C. et al. (2009). Structure activity relationship of carotenoid derivatives in activation of the electrophile/antioxidant response element transcription system. *Free Radical Biology and Medicine* 47, 659–667.

Linnewiel-Hermoni, K., Motro, Y., Miller, Y. et al. (2014). Carotenoid derivatives inhibit nuclear factor kappa B activity in bone and cancer cells by targeting key thiol groups. *Free Radical Biology and Medicine* 75, 105–120.

Mi, J., Jia, K.P., Balakrishna, A. et al. (2019). A highly sensitive SPE derivatization-UHPLC-MS approach for quantitative profiling of carotenoid-derived dialdehydes from vegetables. *Journal of Agricultural and Food Chemistry* 67, 5899–5907.

Mi, J., Jia, K.P., Balakrishna, A. et al. (2020). A method for extraction and LC-MS-based identification of carotenoid-derived dialdehydes in plants. In: *Methods in Molecular Biology: Vol. 2083: Plant and Food Carotenoids* (eds. M. Rodriguez-Concepcion and R. Welsch), 177–188. Humana.

Moise, A.R., Al-Babili, S. and Wurtzel, E.T. (2014). Mechanistic aspects of carotenoid biosynthesis. *Chemical Reviews* 114, 164–193.

Moreno, J.C., Mi, J., Alagoz, Y. et al. (2021). Plant apocarotenoids: from retrograde signaling to interspecific communication. *The Plant Journal* 105, 351–375.

Nisar, N., Li, L., Lu, S. et al. (2015). Carotenoid metabolism in plants. *Molecular Plant* 8, 68–82.

Polívka, T. and Frank, H.A. (2010). Molecular factors controlling photosynthetic light harvesting by carotenoids. *Accounts of Chemical Research* 43, 1125–1134.

Porco, S., Pěnčík, A., Rashed, A. et al. (2016). Dioxygenase-encoding *AtDAO1* gene controls IAA oxidation and homeostasis in *Arabidopsis. Proceedings of the National Academy of Sciences of the United States of America* 113, 11016–11021.

Porta, H. and Rocha-Sosa, M. (2002). Plant lipoxygenases. Physiological and molecular features. *Plant Physiology* 130, 15–21.

Ramel, F., Birtic, S., Ginies, C. et al. (2012). Carotenoid oxidation products are stress signals that mediate gene responses to singlet oxygen in plants. *Proceedings of the National Academy of Sciences of the United States of America* 109, 5535–5540.

Rodríguez-Ávila, N.L., Narváez-Zapata, J.A., Ramírez-Benítez, J.E. et al. (2011). Identification and expression pattern of a new carotenoid cleavage dioxygenase gene member from *Bixa orellana*. *Journal of Experimental Botany* 62, 5385–5395.

Rodriguez-Concepcion, M., Avalos, J., Bonet, M.L. et al. (2018). A global perspective on carotenoids: metabolism, biotechnology, and benefits for nutrition and health. *Progress in Lipid Research* 70, 62–93.

Rosas-Saavedra, C. and Stange, C. (2016). Biosynthesis of carotenoids in plants: enzymes and color. *Subcellular Biochemistry* 79, 35–69.

Ruch, S., Beyer, P., Ernst, H. et al. (2005). Retinal biosynthesis in eubacteria: *in vitro* characterization of a novel carotenoid oxygenase from *Synechocystis* sp. PCC 6803. *Molecular Microbiology* 55, 1015–1024.

Scherzinger, D., Ruch, S., Kloer, D.P. et al. (2006). Retinal is formed from apo-carotenoids in *Nostoc* sp. PCC7120: *in vitro* characterization of an apo-carotenoid oxygenase. *Biochemical Journal* 398, 361–369.

Schwartz, S.H., Tan, B.C., Gage, D.A. et al. (1997). Specific oxidative cleavage of carotenoids by VP14 of maize. *Science* 276, 1872–1874.

Soares, V.L., Rodrigues, S.M., de Oliveira, T.M. et al. (2011). Unraveling new genes associated with seed development and metabolism in *Bixa orellana* L. by expressed sequence tag (EST) analysis. *Molecular Biology Reports* 38, 1329–1340.

Stepanova, A.N., Robertson-Hoyt, J., Yun, J. et al. (2008). *TAA1*-mediated auxin biosynthesis is essential for hormone crosstalk and plant development. *Cell* 133, 177–191.

Sui, X., Kiser, P.D., Lintig, J. et al. (2013). Structural basis of carotenoid cleavage: from bacteria to mammals. *Archives of Biochemistry and Biophysics* 539, 203–213.

Tong, Y., Yan, Y., Zhu, X. et al. (2015). Simultaneous quantification of crocetin esters and picrocrocin changes in Chinese saffron by high-performance liquid chromatography-diode array detector during 15 years of storage. *Pharmacognosy Magazine* 11, 540–545.

Us-Camas, R., Aguilar-Espinosa, M., Rodríguez-Campos, J. et al. (2022). Identifying *Bixa orellana* L. new carotenoid cleavage dioxygenases 1 and 4 potentially involved in bixin biosynthesis. *Frontiers in Plant Science* 13, 829089.

Voutilainen, S., Nurmi, T., Mursu, J. et al. (2006). Carotenoids and cardiovascular health. *The American Journal of Clinical Nutrition* 83, 1265–1271.

Walter, M.H. and Strack, D. (2011). Carotenoids and their cleavage products: biosynthesis and functions. *Natural Product Reports* 28, 663–692.

Wang, J.Y., Haider, I., Jamil, M. et al. (2019). The apocarotenoid metabolite zaxinone regulates growth and strigolactone biosynthesis in rice. *Nature Communications* 10, 810.

Wang, J.Y., Lin, P.Y. and Al-Babili, S. (2021). On the biosynthesis and evolution of apocarotenoid plant growth regulators. *Seminars in Cell & Developmental Biology* 109, 3–11.

Westfall, C.S., Zubieta, C., Herrmann, J. et al. (2012). Structural basis for prereceptor modulation of plant hormones by GH3 proteins. *Science* 336, 1708–1711.

Xu, Z., Pu, X., Gao, R. et al. (2020). Tandem gene duplications drive divergent evolution of caffeine and crocin biosynthetic pathways in plants. *BMC Biology* 18, 63.

Zhang, J., Lin, J.E., Harris, C. et al. (2016). DAO1 catalyzes temporal and tissue-specific oxidative inactivation of auxin in *Arabidopsis thaliana*. *Proceedings of the National Academy of Sciences of the United States of America* 113, 11010–11015.

Zhang, B., Liu, C., Wang, Y. et al. (2015). Disruption of a *CAROTENOID CLEAVAGE DIOXYGENASE 4* gene converts flower colour from white to yellow in *Brassica* species. *New Phytologist* 206, 1513–1526.

Zhang, J. and Peer, W.A. (2017). Auxin homeostasis: the DAO of catabolism. *Journal of Experimental Botany* 68, 3145–3154.

Zheng, X., Mi, J., Balakrishna, A. et al. (2022). *Gardenia* carotenoid cleavage dioxygenase 4a is an efficient tool for biotechnological production of crocins in green and non-green plant tissues. *Plant Biotechnology Journal* 20, 2202–2216.

Zheng, X., Zhu, K., Sun, Q. et al. (2019). Natural variation in *CCD4* promoter underpins species-specific evolution of red coloration in citrus peel. *Molecular Plant* 12, 1294–1307.

4

Apocarotenoid Metabolism and Redox Signaling in Plants

*Muhammad Aamir Manzoor[1], Irfan Ali Sabir[2], Iftikhar Hussain Shah[2], Muhammad Usman[2], Muhammad Azam[2], Ghulam Murtaza[3], Alam Sher[4], Muhammad Waheed Riaz[4], Zishan Ahmad[5], and Cheng Song[6]**

[1] *National Key Laboratory for Tropical Crop Breeding, School of Breeding and Multiplication (Sanya Institute of Breeding and Multiplication), Hainan University, Sanya, Hainan, China*
[2] *Department of Plant Science, School of Agriculture and Biology, Shanghai Jiao Tong University, Shanghai, China*
[3] *School of Agriculture, Yunnan University, Kunming, Yunnan, China*
[4] *Department of Agronomy, Shandong Agricultural University, Taian, Shandong, China*
[5] *Bamboo Research Institute, School of Life Sciences, Nanjing Forestry University, Nanjing, Jiangsu, China*
[6] *College of Biological and Pharmaceutical Engineering, West Anhui University, Lu'an, Anhui, China*

*Corresponding author: lanniao812329218@163.com

4.1 Introduction

Apocarotenoids represent a novel category of metabolites formed by the oxidative cleavage of carotenoids, which are essential molecules integral to several physiological functions in plants. Carotenoids, including lutein, β-carotene, and zeaxanthin, are recognized for their tasks in photoprotection, photosynthesis, and the synthesis of pigments (Moreno et al. 2021; Wang et al. 2021). The enzymes known as carotenoid cleavage dioxygenases (CCDs), which catalyze the oxidative breakdown of carotenoid molecules, are mostly responsible for converting these carotenoids into apocarotenoids (Wang et al. 2019; Hussain et al. 2022). In addition to producing different apocarotenoids, transformation modifies the plant's regulatory networks to adapt to environmental biotic and abiotic stresses (Zheng et al. 2021).

Beyond their function as metabolic by-products, apocarotenoids are important signaling molecules that have a key role in plant growth and stress responses (Figure 4.1). For instance, it has been shown that several apocarotenoids, like β-ionone, increase plant resilience to disease by modifying the expression of genes linked to defense and tolerance to stressful conditions (Felemban et al. 2023). In addition, apocarotenoids are involved in hormonal pathways, especially since these are the precursors to well-known plant hormones like strigolactones and abscisic acid (ABA), which are essential for regulating a variety

Plant Apocarotenoids: Biosynthesis, Signaling and Agricultural Applications, First Edition. Edited by Kaiser Iqbal Wani and Tariq Aftab.
Published 2026 by John Wiley & Sons Ltd.

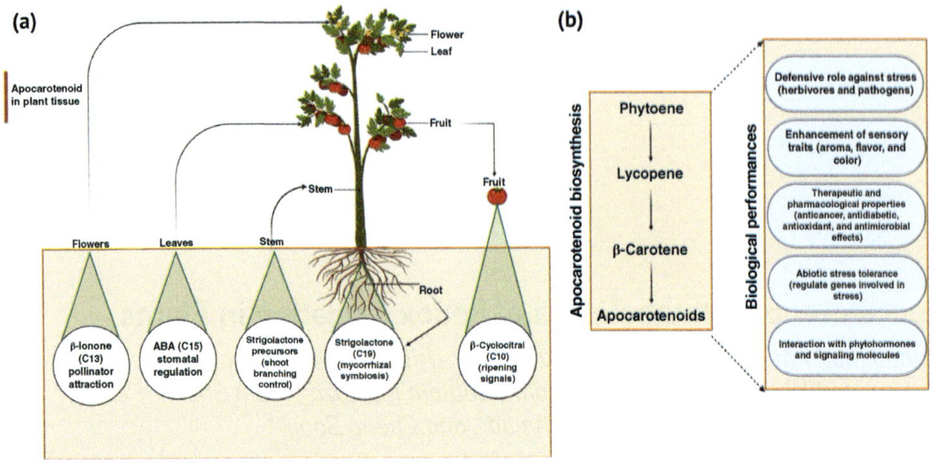

Figure 4.1 (a) Apocarotenoids exhibiting differential distribution across plant tissues, with specialized functions in each organ. The figure highlights key apocarotenoid compounds and their tissue-specific roles in plant physiology. (b) Apocarotenoid biosynthesis pathway and its biological functions.

of plant responses, including those related to drought adaptation and root architecture (Beltran and Stange 2016; Stauder et al. 2018). Their functions in regulating carotenoid synthesis and general plant health have been revealed by recent studies into apocarotenoid metabolism. Because apocarotenoids may change dramatically in response to environmental stresses, they aid in plant adaptation by improving redox signaling and modifying the profiles of reactive oxygen species (ROS) (Vieira et al. 2024). Apocarotenoids are essential components of retrograde signaling, the transmission from plastids to the nucleus that controls the plant's responsiveness to external changes, because of their capacity to affect gene expression linked to growth and developmental processes (Moreno et al. 2021).

The intricate functions of apocarotenoid metabolism will probably reveal new methods for enhancing plant resilience and modifying farming methods as research into this topic develops. In addition to improving our understanding of plant biology, knowing the biosynthetic routes and roles of these metabolites may inspire new ideas for crop management and improvement in the face of climate change (Ahrazem et al. 2019; Wang et al. 2021). Apocarotenoids are essential metabolites derived from carotenoid cleavage, playing vital functions in plant metabolism, development, and environmental adaptation. Their biosynthesis and activities demonstrate a complex interaction between main and secondary metabolisms, highlighting their potential to improve plant resilience and adaptation methods.

4.2 Redox Signaling in Plants

4.2.1 ROS Production and Scavenging

ROS, which include species like singlet oxygen ($^1O_2$), superoxide anion ($O_2^{\bullet -}$), hydrogen peroxide (H_2O_2), and hydroxyl radicals ($\bullet OH$), are chemically reactive molecules that include oxygen. These compounds are produced as by-products of the regular metabolic activities that take place in plants, especially during respiration and photosynthesis. For instance, singlet oxygen may be created during the light processes of photosynthesis when excited chlorophyll transfers energy to molecular oxygen. Likewise, another crucial location for the production of superoxide anions is the electron transport chain in mitochondria (Moreno et al. 2020).

ROS have important functions in cell communication under physiological conditions, favorably influencing the opening of stomata, growth control, and adaptations to biotic stress (Hussain et al. 2022). However, oxidative stress brought on by an excessive buildup of ROS may harm lipids, proteins, and DNA, upsetting cellular equilibrium (Wang et al. 2019). In response, plants have developed intricate antioxidant defenses to eliminate ROS, utilizing both enzyme-based and non-enzyme-based components. Central to this system are superoxide dismutase (SOD), which changes superoxide radicals into H_2O_2; catalase (CAT), which decomposes H_2O_2 into harmless water and oxygen; and peroxidases (POXs), which further assist in detoxifying H_2O_2 (Moreno et al. 2021). Glutathione (GSH) and ascorbate (vitamin C) are examples of nonenzymatic antioxidants that are essential components of the antioxidant defense system of plants. GSH contributes with sustaining the redox balance inside cells and is necessary for the optimal functioning of many antioxidant enzymes, whereas vitamin C may directly scavenge ROS and occurs in both oxidized and reduced forms (Fiorilli et al. 2019). Plant vitality depends on this dynamic equilibrium between ROS generation and removal, which permits response to changing environmental circumstances while reducing oxidative damage (Nisar et al. 2015).

4.2.2 Redox Sensors and Signaling Pathways

Redox sensors are essential for identifying changes in the cellular redox state, which are often brought on by variations in ROS. These sensors may be proteins with redox-sensitive cysteine residues that change their functional state when oxidized or reduced. Thioredoxins (TRXs) and glutaredoxins (GRXs) are well-known types of redox sensors. By taking part in the reversible redox alteration of cysteine residues, TRXs modify a variety of target proteins, potentially affecting protein–protein interactions or enzyme performance (Cao et al. 2015).

Through transcription factors, TRXs also play a role in controlling gene expression. For example, being redox-sensitive and triggered by a reduction reaction, the transcription factor NPR1, a key regulator of salicylic acid signaling implicated in systemic acquired resistance, may participate in downstream signaling pathways that enhance the expression of defense genes (Harrison and Quadro 2018). This indicates how redox signaling controls larger stress signaling networks in addition to its significance in the body's response to biotic stress. Other proteins, including protein kinases and phosphatases, may function as redox sensors in addition to TRXs and GRXs, using changes in the redox environment to alter signaling pathways. It is well known that ROS may activate mitogen-activated protein kinases (MAPKs), which spread extracellular signals and provide a link between the production of ROS and stress-responsive signaling cascades (Hussain et al. 2022). Furthermore, new research indicates that redox-sensitive transcription factors, like the NAC and WRKY families, have a role in the expression of genes that encode for stress response proteins. These transcription factors make a direct connection between redox signals and their expression (Amengual et al. 2011; Vieira et al. 2024). Therefore, redox sensors are responsible for the formation of a sophisticated network that enables plants to adaptably handle changes in their environment. This network allows plants to coordinate both defensive and growth responses in response to stress.

4.2.3 Role in Stress Responses

Redox signaling is crucial in coordinating plant responses to environmental stressors, including drought, salt, severe temperatures, and pathogens (Figure 4.2). Upon exposure to abiotic stresses, plants often exhibit elevated levels of ROS due to increased generation and diminished scavenging ability. The increase in ROS levels functions as both a stress indicator and a crucial signaling molecule that initiates several defensive responses (Wang et al. 2021).

During drought stress, the buildup of H_2O_2 may induce stomatal closure via signaling pathways including ABA, a vital hormone in the control of water loss (Figure 4.2). The ROS produced in guard cells may induce alterations in ion channels, leading to the closing of stomata, the plant's principal mechanism for transpiration and gas exchange, therefore conserving water (Moreno et al. 2021). Furthermore, ROS may stimulate the expression of genes responsible for osmotic adjustment and protective measures, including the generation of compatible solutes such as glycine betaine and proline, which stabilize cellular structures and preserve turgor pressure (Hussain et al. 2022). From the perspective of salt stress, research has shown that ROS act as secondary messengers that orchestrate ion homeostasis. ROS achieve this by modulating the expression of transporters and channels that are essential for maintaining sodium and potassium balance (Felemban et al. 2019). Another example of the intricacy of plant stress responses is provided by the interaction between ROS and phytohormones. For example, the interaction of ROS with hormones such as JA and ABA is essential for the activation of plant defense mechanisms against pathogens and the promotion of systemic acquired resistance (Zheng et al. 2021). The integration of ROS signaling with hormone pathways highlights the complex function of redox

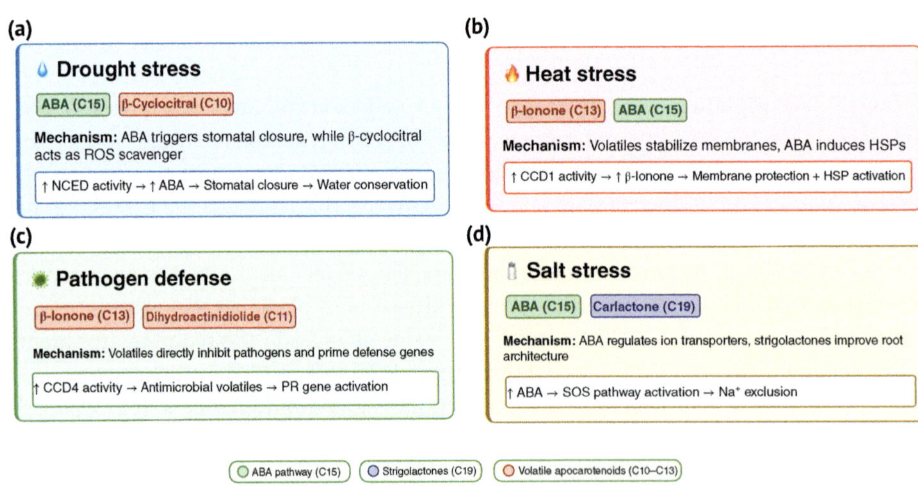

Figure 4.2 Apocarotenoid-mediated stress tolerance mechanisms in plants: (a) drought stress, (b) heat stress, (c) pathogen defense, and (d) salt stress; ABA (C15) and volatile derivatives (C10–C13) regulate drought, heat, salt, and pathogen responses via hormonal and ROS-scavenging pathways.

signaling as a key regulatory mechanism in coordinating physiological responses to environmental stress. Redox signaling, supported by the generation and management of ROS, functions as a vital mechanism via which plants detect and react to various environmental stressors. The balance among ROS generation, sensor activation, along with stress signaling is a vital adaptation mechanism that improves plant resilience and fitness in variable habitats (Lennicke and Cochemé 2021).

4.3 Interplay Between Apocarotenoid Metabolism and Redox Signaling

4.3.1 Stress-induced ROS and Apocarotenoid Production

For plants to adjust to environmental stresses, the interaction between apocarotenoid metabolism and ROS is essential (Figure 4.3). ROS concentrations usually rise in response to abiotic stressors like dryness, salt, or very high or low temperatures. This increase in ROS affects the formation of different apocarotenoids that are produced from carotenoids in addition to serving as a signal for triggering stress responses (D'Alessandro et al. 2019). For example, ABA and β-cyclocitral are two examples of apocarotenoids that play important roles in the process of modulating plant responses to stress. The synthesis of carotenoid CCDs, which are responsible for catalyzing the cleavage of carotenoids into apocarotenoid metabolites, is increased to a greater extent when stress is prolonged (Costabile et al. 2016). For instance, when there is a drought, increasing levels of ROS lead to greater expression of CCDs, which in turn leads to heightened levels of ABA, which helps stomatal closure and reduces water loss (Chater et al. 2015). Additionally, it has been shown that hypoxic and drought stresses may stimulate the production of certain apocarotenoids, which have the potential to mitigate the deleterious consequences of an excessive buildup of ROS (Hussain et al. 2022). The relationship demonstrates a feedback mechanism in which ROS signals generated by stress promote the synthesis of apocarotenoids linked to stress, increasing plant resistance to unfavorable circumstances (Abramczyk and Surmacki 2023).

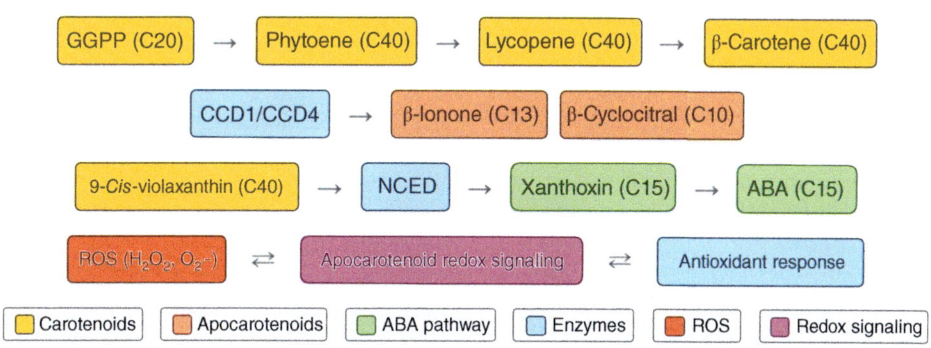

Figure 4.3 Apocarotenoid metabolism and redox signaling in plant. Apocarotenoids are derived from carotenoid cleavage by CCD enzymes (e.g., CCD1, NCED). Volatile apocarotenoids (β-ionone, β-cyclocitral) and ABA mediate stress responses. ROS and apocarotenoids interact in redox signaling, regulating oxidative stress and defense pathways.

4.3.2 Apocarotenoids Modulating Redox Homeostasis

Apocarotenoids function as signaling molecules that may alter the redox state of cells, especially under conditions of stress, and are essential for preserving redox homeostasis inside plant cells. By altering the glutathione/oxidized glutathione (GSH/GSSG) ratio, for instance, β-cyclocitral has been shown to impact redox signals and strengthen defenses against oxidative stress (Wani et al. 2017). Apocarotenoids may increase cellular antioxidant capability by aiding in the scavenging of excess ROS. Increased concentrations of these substances may encourage the action of antioxidant enzymes, strengthening the plant's defensive mechanisms under stress. Apocarotenoids affect a number of metabolic and signaling processes by directly taking part in redox reactions and regulating the cellular redox environment (Bandara et al. 2021). In addition, some apocarotenoids have the ability to serve as modulators of redox-sensitive transcription factors, which in turn triggers transcriptional reprogramming that coordinates responses to stress tolerance. During times of stress, for example, redox-sensitive transcription factors like Nrf2 may be activated by ROS in combination with apocarotenoid signaling. This activation results in the production of genes that boost antioxidant capacity and recovery pathways (Zhang et al. 2014; Juknat et al. 2019).

4.3.3 ABA and Stomatal Closure

An important apocarotenoid that mediates plant responses to drought stress is ABA. Under water-deficient situations, ABA levels rise dramatically, mostly in reaction to increased ROS levels. This rise causes stomatal closure, a vital physiological reaction that lowers transpiration water loss (Pridgeon and Hetherington 2021). Research has shown that by activating guard cells, inducible ROS generation promotes ABA signaling, which in turn results in increased expression of ion channels that regulate stomatal activity (Mehta et al. 2016). Apocarotenoid metabolism and redox control are closely linked by the coordination of ROS and ABA signaling, highlighting the crucial interaction between these components in the management of water stress.

4.3.4 β-Cyclocitral in Retrograde Signaling

Another significant apocarotenoid that comes from carotenoids is β-cyclocitral, which acts as a retrograde signal to affect the expression of genes related to stress response. It has been shown to take part in redox signaling pathways that facilitate communication between the nucleus and chloroplasts, enabling coordinated responses to stressful situations (D'Alessandro et al. 2019). This interaction is essential for responding to environmental stresses, since β-cyclocitral may enhance antioxidant defenses and modify metabolic pathways linked to stress tolerance (Abramczyk and Surmacki 2023).

4.3.5 Strigolactones in Stress Mitigation via Redox Homeostasis

Strigolactones, a category of apocarotenoids, serve as signaling molecules that modulate plant responses to abiotic stressors via redox homeostasis. These chemicals may regulate oxidative stress responses by affecting the generation of ROS and the functionality of antioxidant mechanisms. Strigolactones are associated with enhanced nutrient absorption and root growth, which are crucial under nutrient scarcity and drought conditions, hence improving plant resilience (Costabile et al. 2016; Hussain et al. 2022). This underscores the complex function of strigolactones in merging redox signaling with developmental mechanisms and stress alleviation techniques. The interaction between apocarotenoid metabolism and redox signaling is essential for plants to properly adapt to environmental stressors. Stress-induced ROS generation regulates the synthesis of apocarotenoids, which subsequently affect redox homeostasis and improve stress tolerance (Mansoor et al. 2024).

4.4 Conclusion and Future Perspectives

In summary, the interaction between apocarotenoid metabolism and redox signaling constitutes a complicated and essential connection in plant biology, especially under stress scenarios. The generation of ROS under environmental stressors functions as a signaling mechanism that triggers the synthesis of apocarotenoids and affects biological responses that improve stress tolerance. Crucial apocarotenoids, such as ABA and strigolactones, exemplify this connection by regulating stomatal closure and enhancing resource distribution under instances of abiotic stress. Moreover, apocarotenoids facilitate the preservation of redox equilibrium, offering preventive mechanisms that avert oxidative damage. Future studies should aim to clarify the molecular mechanisms regulating the interaction between apocarotenoids and ROS signaling, especially using sophisticated multi-omics methodologies that include genomic, transcriptomic, proteomic, and metabolomic data. Furthermore, investigating the function of apocarotenoids with rhizosphere bacteria may provide novel approaches for enhancing plant health and stress resilience. These possibilities may boost the development of robust crops to combat climate problems, hence contributing to sustainable agriculture practices. Comprehending the complex interplay of phytohormones, redox reactions, and stress adaptation mechanisms will be crucial for advancing future research and agricultural practices, as plants endeavor to thrive in progressively tough situations.

References

Abramczyk, H. and Surmacki, J.M. (2023). Control of mitochondrial electron transport chain flux and apoptosis by retinoic acid: Raman imaging in vitro human bronchial and lung cancerous cells. *Cancers* 15(18), 4535.

Ahrazem, O., Diretto, G., Argandoña Picazo, J. et al. (2019). The specialized roles in carotenogenesis and apocarotenogenesis of the phytoene synthase gene family in saffron. *Frontiers in Plant Science* 10, 249.

Amengual, J., Gouranton, E., van Helden, Y.G. et al. (2011). Beta-carotene reduces body adiposity of mice via BCMO1. *PLoS ONE* 6(6), e20644.

Bandara, S., Thomas, L.D., Ramkumar, S. et al. (2021). The structural and biochemical basis of apocarotenoid processing by β-carotene oxygenase-2. *ACS Chemical Biology* 16(3), 480–490.

Beltran, J.C.M. and Stange, C. (2016). Apocarotenoids: a new carotenoid-derived pathway. In: *Subcellular Biochemistry: Vol. 79: Carotenoids in Nature: Biosynthesis, Regulation and Function* (ed. C. Syange), 239–272. Springer.

Cao, H., Wang, J., Dong, X. et al. (2015). Carotenoid accumulation affects redox status, starch metabolism, and flavonoid/anthocyanin accumulation in citrus. *BMC Plant Biology* 15, 27.

Chater, C., Peng, K., Movahedi, M. et al. (2015). Elevated CO_2-induced responses in stomata require ABA and ABA signaling. *Current Biology* 25(20), 2709–2716.

Costabile, B.K., Kim, Y.-K., Iqbal, J. et al. (2016). β-Apo-10′-carotenoids modulate placental microsomal triglyceride transfer protein expression and function to optimize transport of intact β-carotene to the embryo. *Journal of Biological Chemistry* 291(35), 18525–18535.

D'Alessandro, S., Mizokami, Y., Légeret, B. et al. (2019). The apocarotenoid β-cyclocitric acid elicits drought tolerance in plants. *iScience* 19, 461–473.

Felemban, A., Braguy, J., Zurbriggen, M.D. et al. (2019). Apocarotenoids involved in plant development and stress response. *Frontiers in Plant Science* 10, 1168.

Felemban, A., Moreno, J.C., Mi, J. et al. (2023). β-Ionone regulates *Arabidopsis thaliana* transcriptome and increases its resistance against *Botrytis cinerea*. *bioRxiv* 2023.2005. 2002.539130.

Fiorilli, V., Wang, J.Y., Bonfante, P. et al. (2019). Apocarotenoids: old and new mediators of the arbuscular mycorrhizal symbiosis. *Frontiers in Plant Science* 10, 1186.

Harrison, E.H. and Quadro, L. (2018). Apocarotenoids: emerging roles in mammals. *Annual Review of Nutrition* 38(1), 153–172.

Hussain, K., Kumar, A., Fayaz, M. et al. (2022). CstMYB14 links ROS signaling, apocarotenoid metabolism, and stress response in *Crocus sativus* L. *Physiologia Plantarum* 174(3), e13712.

Juknat, A., Gao, F., Coppola, G. et al. (2019). miRNA expression profiles and molecular networks in resting and LPS-activated BV-2 microglia—effect of cannabinoids. *PLoS ONE* 14(2), e0212039.

Lennicke, C. and Cochemé, H.M. (2021). Redox metabolism: ROS as specific molecular regulators of cell signaling and function. *Molecular Cell* 81(18), 3691–3707.

Mansoor, S., Mir, M.A., Karunathilake, E. et al. (2024). Strigolactones as promising biomolecule for oxidative stress management: a comprehensive review. *Plant Physiology and Biochemistry* 206, 108282.

Mehta, M., Rajmani, R.S. and Singh, A. (2016). *Mycobacterium tuberculosis* WhiB3 responds to vacuolar pH-induced changes in mycothiol redox potential to modulate phagosomal maturation and virulence. *Journal of Biological Chemistry* 291(6), 2888–2903.

Moreno, J.C., Mi, J., Agrawal, S. et al. (2020). Expression of a carotenogenic gene allows faster biomass production by redesigning plant architecture and improving photosynthetic efficiency in tobacco. *The Plant Journal* 103(6), 1967–1984.

Moreno, J.C., Mi, J., Alagoz, Y. et al. (2021). Plant apocarotenoids: from retrograde signaling to interspecific communication. *The Plant Journal* 105(2), 351–375.

Nisar, N., Li, L., Lu, S. et al. (2015). Carotenoid metabolism in plants. *Molecular Plant* 8(1), 68–82.

Pridgeon, A.J. and Hetherington, A.M. (2021). ABA signalling and metabolism are not essential for dark-induced stomatal closure but affect response speed. *Scientific Reports* 11(1), 5751.

Stauder, R., Welsch, R., Camagna, M. et al. (2018). Strigolactone levels in dicot roots are determined by an ancestral symbiosis-regulated clade of the *PHYTOENE SYNTHASE* gene family. *Frontiers in Plant Science* 9, 255.

Vieira, E.A., Gaspar, M., Caldeira, C.F. et al. (2024). Desiccation tolerance in the resurrection plant *Barbacenia graminifolia* involves changes in redox metabolism and carotenoid oxidation. *Frontiers in Plant Science* 15, 1344820.

Wang, J.Y., Haider, I., Jamil, M. et al. (2019). The apocarotenoid metabolite zaxinone regulates growth and strigolactone biosynthesis in rice. *Nature Communications* 10(1), 810.

Wang, J.Y., Lin, P.-Y. and Al-Babili, S. (2021). On the biosynthesis and evolution of apocarotenoid plant growth regulators. *Seminars in Cell & Developmental Biology* 109, 3–11.

Wani, Z.A., Kumar, A., Sultan, P. et al. (2017). *Mortierella alpina* CS10E4, an oleaginous fungal endophyte of *Crocus sativus* L. enhances apocarotenoid biosynthesis and stress tolerance in the host plant. *Scientific Reports* 7(1), 8598.

Zhang, B., Wang, X.-Q., Chen, H.-Y. et al. (2014). Involvement of the Nrf2 pathway in the regulation of pterostilbene-induced apoptosis in HeLa cells via ER stress. *Journal of Pharmacological Sciences* 126(3), 216–229.

Zheng, X., Yang, Y. and Al-Babili, S. (2021). Exploring the diversity and regulation of apocarotenoid metabolic pathways in plants. *Frontiers in Plant Science* 12, 787049.

Section II

Key Apocarotenoids and Their Roles

5

Strigolactones: Discovery, Biological Activity, Biosynthesis, and Applications

*Peng Zhong[1,2], Jianhua Jia[1,2], Jiali Liu[1,2], and Xiaohui Yan[1,2]**

[1] *State Key Laboratory of Component-based Chinese Medicine, Tianjin University of Traditional Chinese Medicine, Tianjin, China*
[2] *Haihe Laboratory of Modern Chinese Medicine, Tianjin, China*

*Corresponding author: yanxh@tjutcm.edu.cn

5.1 Introduction

Strigolactones (SLs) are a class of β-carotene-derived plant hormones that regulate physiological metabolism and play a role in the response to various biotic and abiotic stresses. SLs were initially characterized as host recognition signals for root-parasitic plants, such as *Striga* and *Orobanche* (Cook et al. 1966). More than 40 SLs have been identified over the last five decades. All natural SLs contain a methylbutenolide moiety (D-ring) that connects to the other part via an enol ether bond. SLs can promote plant growth and development through mechanisms such as stimulating the germination of rare seeds, shaping root structures, enhancing ideal branching, controlling parasitic weeds, stimulating nutrient absorption, alleviating drought, and promoting mycorrhizal symbiosis, thus overcoming bottlenecks in artificial cultivation (Gomez-Roldan et al. 2008; Umehara et al. 2008; Lopez-Obando et al. 2015; Waters et al. 2017; Kodama et al. 2022). Beyond their agricultural applications, SLs also exhibit interesting roles in regulating apoptosis and inflammation in animal cells (Johnson et al. 1981; Zheng et al. 2018; Yang et al. 2022). In this chapter, we review the discovery, biological activity, biosynthesis, and agricultural applications of SLs. Our objective was to provide a theoretical foundation for the generation and application of natural and unnatural SLs.

Plant Apocarotenoids: Biosynthesis, Signaling and Agricultural Applications, First Edition. Edited by Kaiser Iqbal Wani and Tariq Aftab.
Published 2026 by John Wiley & Sons Ltd.

5.2 Discovery of SLs

Since the isolation of strigol (**1**) and (+)-strigyl acetate (**2**) from the root exudation of cotton (*Gossypium hirsutum*) in 1966, more than 30 canonical and 10 noncanonical SLs have been identified so far (Cook et al. 1966, 1972). SLs are generally categorized into canonical and noncanonical subfamilies. Canonical SLs consist of three fused rings (A-, B-, and C-rings) connected to the D-ring. According to the configuration of the C-ring, canonical SLs are further divided into two subgroups: strigol type and orobanchol (OB) type (Figure 5.1). Akiyama et al. (2005) isolated 5-deoxystrigol (5DS, **3**) from *Lotus japonicus*.

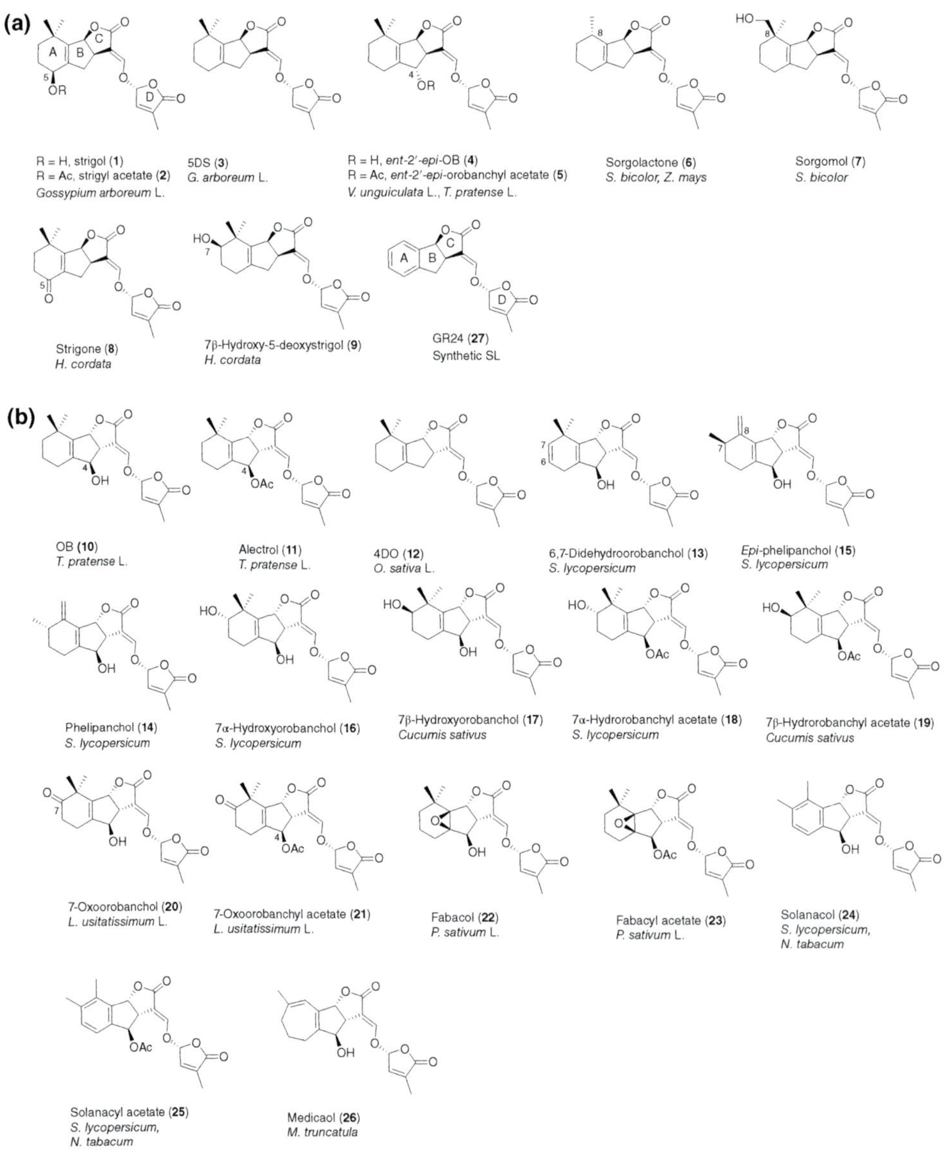

Figure 5.1 Canonical and noncanonical SLs: (a) strigol-type SLs and (b) OB-type SLs.

Ueno et al. (2011) isolated hydroxylated *ent*-2′-*epi*-OB (**4**) and C4-hydroxyacetylated *ent*-2′-*epi*-orobanchyl acetate (**5**) from cowpea (*Vigna unguiculata* L.) and red clover (*Trifolium pratense* L.). Hauck et al. (1992) isolated sorgolactone (**6**), a sorghum ketone with a single methyl group at C8, from maize (*Zea mays*) and sorghum (*Sorghum bicolor*). Xie et al. (2008) isolated and identified sorgomol (**7**), which features a hydroxymethyl group at C8, from sorghum. Kisugi et al. (2013a,b) isolated strigone (**8**) and 7β-hydroxy-5DS (**9**) from *Houttuynia cordata*. Yokota et al. (1998) isolated OB (**10**) and alectrol (**11**) from the root secretion of red clover (*T. pratense* L.) as germination stimulants of *Orobanche minor*, an obligate root holophrastic plant. Cardoso et al. (2014) isolated 4-deoxyorobanchol (4DO, **12**) from rice (*Oryza sativa*) root. Wakabayashi et al. (2022a) isolated three novel canonical SLs from tomato: 6,7-didehydroorobanchol (**13**), phelipanchol (**14**), and *epi*-phelipanchol (**15**). Kohlen et al. (2013) identified 7α-hydroxyorobanchol (**16**) and 7β-hydroxyorobanchol (**17**) in the root secretions of *Solanum lycopersicum*. Khetkam et al. (2014) isolated 7α-hydroxyorobanchyl acetate (**18**) and 7β-hydroxyorobanchyl acetate (**19**) from cucumber (*Cucumis sativus*). Xie et al. (2009b) isolated 7-oxoorobanchol (**20**) and 7-oxoorobanchyl acetate (**21**) from flax (*Linum usitatissimum* L.). It was hypothesized that pea (*Pisum sativum* L.) might produce fabacol (**22**), in which the double bond of the A-ring is oxidized to a cyclopropyl ether, which has not been identified so far. Instead, fabacyl acetate (**23**), an acetylation derivative of **22**, was isolated from this plant (Xie et al. 2009a). Xie et al. (2007, 2013) identified solanacol (**24**) and solanacyl acetate (**25**), which feature an aromatic A-ring, from tobacco (*Nicotiana tabacum*). Tokunaga et al. (2015) isolated medicaol (**26**) from *Medicago truncatula*, with a unique seven-membered cycloheptadiene in the A-ring. Due to the scarcity of natural SLs, dozens of unnatural SLs, exemplified by GR24 (**27**), have been synthesized to investigate the biological activities of this important family of plant hormones (Bromhead et al. 2014). These synthetic SLs hold great promise as agrichemicals for crop protection (Uraguchi et al. 2018; Kaniganti et al. 2022).

Noncanonical SLs lack the typical B- and C-rings of canonical SLs and are structurally more unstable than the latter ones (Figure 5.2) (Yoneyama et al. 2018b). Some noncanonical SLs have been identified as biosynthetic intermediates. Alder et al. (2012) found that carlactone (CL, **28**) is formed from β-carotene by the action of three enzymes, β-carotene isomerase (all-*trans*-/9-*cis*-β-carotene isomerase DWARF27, D27) and carotenoid cleavage dioxygenases (CCD7 and CCD8). CL also serves as a key biosynthetic intermediate for all SLs (Seto et al. 2014). CYP711A1 (MORE AXILLARY GROWTH1, MAX1) catalyzes oxidation of the C19 methyl group of CL to produce carlactonoate (CLA, **29**), which is subsequently converted by CLA methyltransferase (CLAMT) to afford methyl carlactonoate (MeCLA, **30**) (Abe et al. 2014; Mashiguchi et al. 2022). Brewer et al. (2016) then identified a side branch oxidoreductase (LBO [lateral branching oxidoreductase]) in *Arabidopsis thaliana* that oxidizes MeCLA to 1′-HO-MeCLA (**31**). Baz et al. (2018) showed that 3-hydroxycarlactone (**32**) is also an intermediate in SL biosynthesis. In *L. japonicus*, LjMAX1 converts CLA into 18-hydroxycarlactonoate (**33**) (Mori et al. 2020b). Heliolactone (**34**), isolated from sunflower (*Helianthus annuus* L.), has been shown to effectively promote the germination of root-parasitic weed seeds (Ueno et al. 2014). Recently, Li et al. (2024) identified 4-oxo-methyl carlactonoate (4-oxo-MeCLA, **35**) in the root secretions of rice. Kim et al. (2014) isolated avenaol (**36**), which contains an all-*cis* cyclopropane structure, from the allelopathic plant black oat (*Avena strigosa*). Charnikhova et al. (2017, 2018) isolated zealactone (**37**) and zeapyranolactone (**38**) from maize (*Z. mays*). Lotuslactone (**39**)

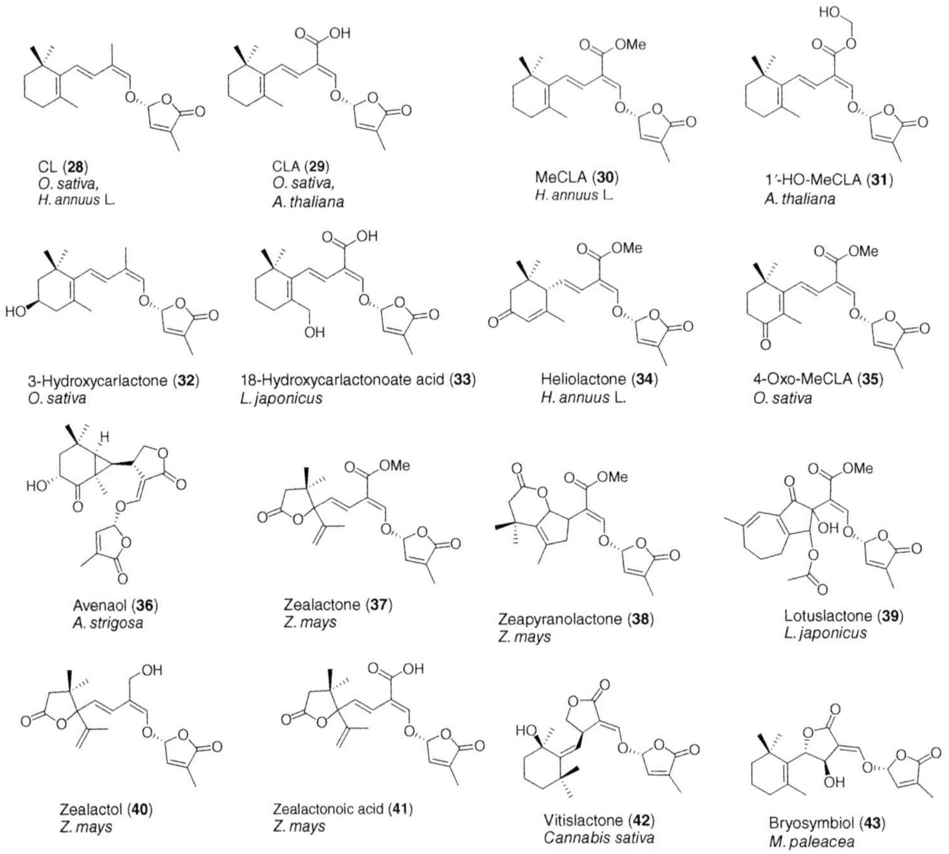

Figure 5.2 Chemical structures of noncanonical SLs.

was isolated from *L. japonicus* with a unique seven-membered cycloheptadiene ring (Xie et al. 2019). Li et al. (2023) reported the isolation of zealactol (**40**) and zealactoic acid (**41**) from *Z. mays*. In the same year, vitislactone (**42**) was isolated from the exudates of grapevine rootstocks in response to nitrogen starvation (Lailheugue et al. 2023). Kodama et al. (2022) isolated bryosymbiol (**43**) from the bryophyte *Marchantia paleacea*. They observed that bryosymbiol formation was enhanced in arbuscular mycorrhizal (AM) symbiosis environments and that AM symbiosis was impaired in the absence of bryosymbiol.

5.3 Biological Activities of SLs

5.3.1 Regulating Plant Root Growth and Development

SLs have been shown to promote the elongation of seminal/primary roots and adventitious roots (ARs), while repressing lateral root (LR) formation (Figure 5.3). *Arabidopsis* has often been used as a model plant in these studies by applying exogenous SL analogs or analyzing the phenotypes of SL-deficient mutants (Ruyter-Spira et al. 2011). A few *MAX* genes participate in SL biosynthesis and signaling in *Arabidopsis*. Compared to the wild type,

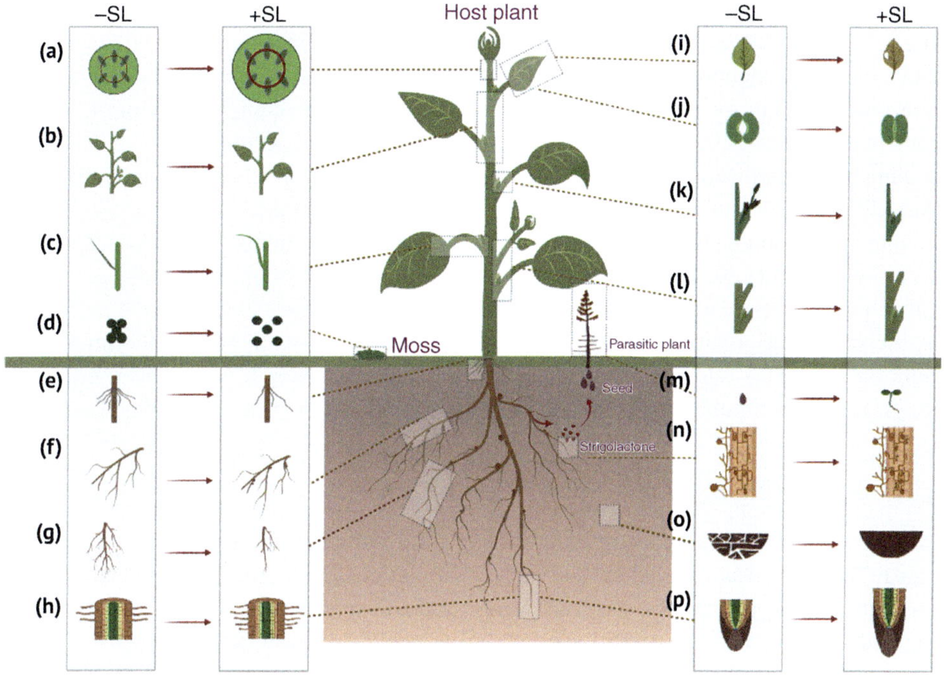

Figure 5.3 Activities of SLs in regulating plant growth and development: (a) promote stem elongation, (b) inhibit shoot branching, (c) regulate the leaf angle, (d) regulate moss population, (e) inhibit AR formation, (f) increase the nodule numbers, (g) inhibit secondary LR formation, (h) lengthen root hair, (i) promote leaf senescence, (j) close stoma, (k) inhibit axillary bud growth, (l) promote internode length, (m) promote parasitic seed germination, (n) increase mycelial branching, (o) adaptation to drought, and (p) increase primary root length.

Arabidopsis mutated in either *MAX3* (*AtCCD7*), *MAX4* (*AtCCD8*), or cytochrome P450 *MAX1* (*AtCYP711A1*) displayed an increased number of shoot branches, and the mutant phenotype could be rescued by the exogenous application of GR24 (Benková et al. 2003; Kapulnik et al. 2011). Mayzlish-Gati et al. (2012) found that under low phosphorus stress, SLs induce root hair development in *Arabidopsis*. LR formation is another key determinant of root system architecture. In both tomato and *Arabidopsis*, SL biosynthesis and signaling mutants exhibit obviously increased LR density, whereas exogenous GR24 application to these mutants has the reverse effects (Koltai 2011). At low concentration (0.1 μM), GR24 application reduced the root diameter and increased the fine-root density, whereas at high concentration (10 μM), it reduced the LR length and enhanced the LR density in grapevine (*Vitis vinifera* L.). Effect of GR24 on nodule number depends on the concentration. *Medicago truncatula* treated with a lower concentration (0.1 μM) of GR24 had a slightly increased nodule number, whereas higher concentrations (2–5 μM) of GR24 strongly reduce the number of nodules. In the SL biosynthesis (*d10*) and signaling (*d3*) mutants of *O. sativa*, AR elongation was significantly decreased, but their LR number was not noticeably different from that of the wild type (Arite et al. 2012).

SLs also participate in the regulation of hypocotyl elongation. Max2 is a leucine-rich repeat protein that positively regulates photomorphogenesis in *Arabidopsis*. It has been reported that *max2* mutant seedlings are highly sensitive to blue, red, and far-red lights. Jia et al. (2014) found that, at low concentrations, SLs inhibit hypocotyl elongation via *MAX2*-mediated signaling pathway; at high concentrations, SLs function through a *MAX2*-independent mechanism. Karrikins (KARs), butanolide compounds from burning vegetation, are structurally similar to SLs, and stimulate seed germination and enhance seedling photomorphogenesis. In *Arabidopsis*, the karrikin insensitive2 protein (KAI2) and DWARF14 (AtD14) are necessary for the response to KARs and SLs. Interestingly, KAI2 can mediate responses to both KARs and SLs, whereas AtD14 mediates only SL responses (Waters et al. 2013). Neither 5DS nor 4DO participated in KAI2-dependent seed germination or hypocotyl elongation, but these two compounds both were active in AtD14-mediated pypocotyl elongation and secondary shoot growth (Scaffidi et al. 2014). Furthermore, the enantiomer of 5DS (*ent*-5DS) was active against KAI2 in growth and gene expression assays. The main reason for this selectivity is the configuration of the butanolide D-ring.

5.3.2 Stimulating Germination of Root-parasitic Weeds

Cook et al. (1966) initially found that strigol could induce germination of *Striga* seeds (Figure 5.3). OB was later identified as a germination stimulant (Yokota et al. 1998). Correlation analysis between SL content and rates of *Striga* parasitism in African rice revealed that rice varieties with lower SL levels exhibited lower *Striga* parasitism rates (Jamil et al. 2011). The tomato *SL-ORT1* mutant was unable to produce SLs and exhibited high resistance to *Phelipanche* and *Orobanche* spp. When *SL-ORT1* was grown together with susceptible tomato cv. M-82, it was parasitized by *Phelipanche aegyptiaca*. Therefore, it was deduced that the resistance of the tomato *SL-ORT1* mutant to parasitic weeds is due to SL deficiency (Dor et al. 2011).

SLs exhibit broad seed germination stimulant activity in various root-parasitic plants. Nomura et al. (2013) evaluated the seed germination stimulant activities of 36 SLs on two *Striga* species, *Striga hermonthica* and *Striga gesnerioides*. All the tested SLs showed activities toward the seeds of *S. hermonthica*. However, only a few SLs were active against *S. gesnerioides*. Different sensitivities may be critical for the survival of root-parasitic plants. *Orobanche minor* is a root-holoparasitic plant that is sensitive to various SLs. In contrast, some *Orobanche* species, including *Orobanche cumana*, *Orobanche hederae*, and *Orobanche densiflora*, exhibit high selectivity for specific SLs. Kim et al. (2010) investigated the germination stimulant activities of 11 SLs on *O. minor* seeds and found that the lipophilicity and stability of the SLs played crucial roles in their stimulatory efficacy.

The characterization of SL signaling pathways in parasitic plants laid a foundation to address the question of the sensing of SLs by parasite seeds (Waters et al. 2017). In obligate parasites such as witchweeds (*Striga*, *Alectra* spp.) and broomrapes (*Orobanche*, *Phelipanche* spp.), SL perception is mediated by KAI2d proteins, similar to the mechanism of D14. The *KAI2d* gene is present only in the parasitic weed genomes. The encoded KAI2d proteins are predicted to have enlarged ligand-binding pockets that differ from those of AtKAI2. Binding of SLs to KAI2d results in activation of the KAI2d protein and its association with SCF (MAX2) and a SAMX1-type protein (SMXL). SMXL is then polyubiquitinated by SCF (MAX2) and degraded by the 26S proteasome to initiate downstream responses (Nelson et al. 2012; Waters et al. 2012; Nomura et al. 2024).

5.3.3 Promoting AM Fungal Symbiosis

Arbuscular mycorrhizal fungi (AMFs) are obligate symbiotic fungi that colonize plant roots. They possess characteristic finely branched hyphal structures called arbuscules. AMFs form mutualistic associations with plants and provide various benefits, such as improving mineral nutrient acquisition and water uptake, and protecting plants against biotic and abiotic stresses (Besserer et al. 2006). The synergistic interaction between AMFs and SLs has a significant effect on plant development. SLs modulate AMFs colonization within plant roots (Akiyama et al. 2005; Yoneyama et al. 2007). AMFs can produce and secrete diffusible signals (Myc factors) that promote symbiotic signaling, resulting in oscillations in nuclear-associated calcium. Detailed studies of these molecules revealed their identities as chitooligosaccharides and lipo-chitooligosaccharides (Maillet et al. 2011; Genre et al. 2013).

AMFs have been shown to stimulate the production and release of SLs in plants (Figure 5.3) (Bouwmeester et al. 2007). Compared with the wild type, the *O. sativa d3* mutant exhibited a significantly reduced AMFs colonization rate, whereas the *d14* mutant showed an increased colonization rate. Further studies have shown that neither D14 nor its paralog (D14L) is essential for AMFs symbiosis with *O. sativa*, whereas D3 is indispensable (Yoshida et al. 2012). In the SL-deficient mutant of pea, AMFs colonization was significantly reduced compared with that in the wild-type plant. Exogenous application of GR24 to the mutant restored AMFs colonization (Foo 2013). However, the application of GR24 to four AMF nonhost plants, *Arabidopsis*, *Spinacia oleracea* L., *Lupinus polyphyllus* Lindl., and *Fagopyrum esculentum* Moench, did not establish AMFs colonization. Therefore, SL is not the only factor that mediates communication between plant and AMFs (Illane et al. 2011).

The tomato *SL-ORT1* mutant exhibited lower levels of AMFs (*Glomus intraradices*) colonization. Analysis of the root extracts from this mutant revealed only minute amounts of solanacol and didehydro-OB (Kapulnik et al. 2011). *SL-ORT1* shoots exhibited increased lateral shoot branching, whereas exogenous application of GR24 to the mutant restored the wild-type phenotype. Besserer et al. (2006) revealed that SLs stimulated spore germination of two AMFs at concentrations as low as 10^{-13} M. Treatment of *Gigaspora rosea*, a member of the AMFs, with GR24 led to a rapid (within minutes) increase in NADH concentration, NADP dehydrogenase activity, and ATP content in the fungal cell. Several days after GR24 treatment, upregulation of the genes involved in mitochondrial metabolism and hyphal growth and enhancement of fungal mitotic activity was observed (Besserer et al. 2008). These results clearly demonstrate the role of SLs in switching AMFs toward a pre-symbiotic state. López-Ráez et al. (2011) found that once symbiosis between AMFs and tomato is established, SL production is significantly reduced. Similar findings were observed in *S. bicolor* (Lendzemo et al. 2007). Colonization with AMFs has been shown to reduce *Orobanche* parasitism in sunflower (*H. annuus*). This effect is attributed to the decreased SL levels and the inhibitory effect of AMFs on *Orobanche* seed germination (Louarn et al. 2012).

5.3.4 Influencing the Nodulation Process of Rhizobia

SLs have been shown to be involved in plant nodulation (Nod). Application of GR24 (10 and 100 nM) to alfalfa (*Medicago sativa*) resulted in clear Nod by *Sinorhizobium meliloti*

(Soto et al. 2010). It has been reported that the SL biosynthetic mutant of pea has low Nod, whereas the SL response mutant has increased Nod. Application of GR24 to the wild type or SL-deficient mutant (*rms1*) of pea showed increased nodule number. Further grafting experiments revealed that nodule number and SL levels in the roots of *rms1* mutant were not affected by grafting to wild-type scions. These results indicate that SLs in the root regulate nodule number (Foo et al. 2011).

The Nod factors are essential for intracellular rhizobium infection. NODULATION SIGNALING PATHWAY1 (NSP1) and NSP2 are GRAS-type transcription factors involved in Nod factor signaling. The *nsp1* mutant of *M. truncatula* completely lost the ability to synthesize SLs, whereas the *nsp2* mutant produced a higher amount of OB than the wild-type plant. The expression of *D27* in the *nsp1*/*nsp2* mutant of *M. truncatula* is significantly reduced. Notably, both NSP1 and NSP2 have dual functions in regulating gene expression after rhizobium-induced signaling and SL biosynthesis under nonsymbiotic conditions (Liu et al. 2011; Foo 2013).

5.3.5 Regulating Shoot Branching

Shoot branching is a crucial process for plant growth and yield. It is regulated by a series of factors such as hormones, genotype, environment, and development. Auxin and cytokinin have long been identified as the repressors and promoters of bud outgrowth, respectively. SLs were later found to inhibit shoot branching in the enhanced shoot branching mutants, including the *rms* mutant of pea, *max* mutant of *Arabidopsis*, and *dwarf* mutant of rice. Gomez-Roldan et al. (2008) observed that exogenous GR24 application in an SL-deficient mutant (*CCD8*) of *P. sativum* restored the wild-type branching phenotype. Tomatoes expressing an *SlCCD7* antisense construct exhibited a marked increase in branching and no obvious changes in root carotenoids (Vogel et al. 2010). Similarly, mutations in *CCD7* or *CCD8* in *Petunia hybrida* resulted in the enhanced production of a branching promoter and loss of branching inhibitor (Drummond et al. 2009). Han et al. (2021) reported that application of different concentrations of GR24 to Mongolian oak (*Quercus mongolica*) seedlings reduced the number and biomass of lateral branches.

Auxin is mainly produced in the shoot apex and young leaves, and plays an important role in regulating plant bud outgrowth. SLs are also involved in auxin-mediated bud inhibition. Buds from SL-deficient mutants are resistant to apically supplied auxin (Crawford et al. 2010). Moreover, the expression of *CCD7* and *CCD8* genes was upregulated by auxin. It has been proposed that SL biosynthesis is upregulated by auxin in the polar auxin transport stream (PATS), and the produced SLs move acropetally into buds and inhibit bud growth (Domagalska and Leyser 2011). SLs do not interfere with the endocytosis or internalization of PIN, but specifically uncouple the effect of auxin on endocytosis and trafficking processes (Müller et al. 2015). The transcription factor BRANCHED1 (BRC1) has been proposed to integrate plant hormone-mediated regulatory signals. Seale et al. (2017) showed that SL regulates bud activation through parallel effects on *BRC1* transcription and the shoot auxin network.

5.3.6 Enhancing Tolerance to Abiotic Stresses

During their lifetime, plants are frequently challenged by various abiotic stress factors, including extreme temperatures, drought, salinity, sunlight, and heavy metals. To cope

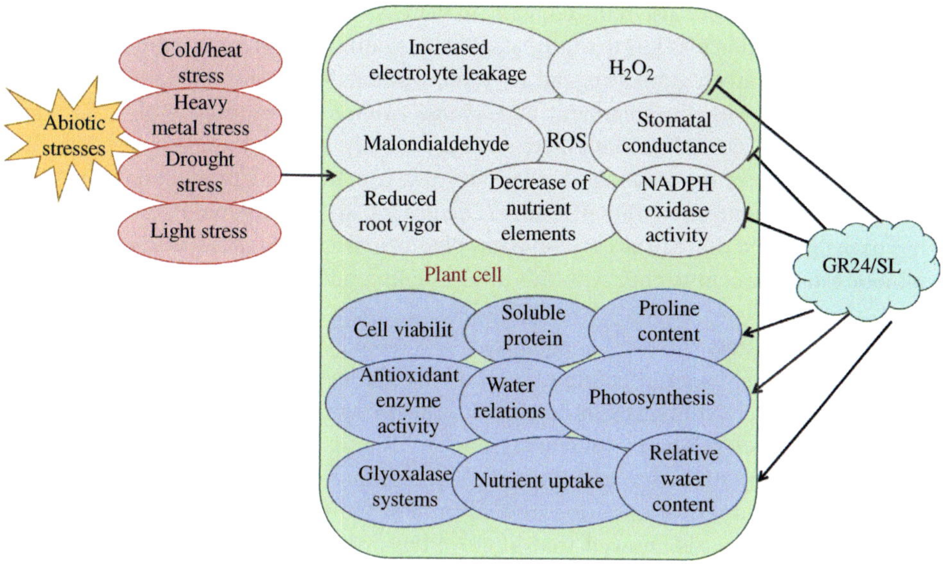

Figure 5.4 Effect of SLs in regulating plant abiotic stresses.

with these environmental stresses, plants have evolved diverse strategies to sense diverse environmental parameters and initiate signaling pathways. Genetic studies have provided clear evidence that SL biosynthesis is regulated by various abiotic stresses (Figure 5.4). SLs have drawn great attention because of their crucial roles in plant adaptation to various abiotic stresses (Qi et al. 2024).

5.3.6.1 Salt Stress

Soil salinization is one of the most critical agricultural and environmental challenges worldwide (Ruiz-Lozano et al. 2012). Excess soluble salts in the soil are harmful to plants by hindering seed germination, growth and development, and flowering and fruiting (Hong et al. 2009). Exogenous application of SLs can enable plants to alleviate salt stress by leveraging soil microbes, especially AMFs, which help plants to improve water uptake, enhance nutrient absorption, and upregulate osmotic regulators (Sadhana 2014). Compared to the wild-type plant, the *Arabidopsis* SL biosynthesis mutants (*max3* and *max4*) and signaling mutant (*max2*) exhibited reduced tolerance to salt stress during the seed germination and mature stages (Ha et al. 2014). Exogenous application of GR24 significantly reduced H_2O_2 and malondialdehyde levels in rice seedlings and *Salvia miltiorrhiza* L., mitigating salt-induced cellular damage (Ling et al. 2020; Sharifi and Bidabadi 2020). Compared to the control group (without GR24), the SL-treated group alleviated the adverse effects on growth status, leaf photosynthesis, and physiological/biochemical indices in the rice seedlings. Salt stress has been reported to negatively affect multiple growth parameters in sunflower, including photosynthetic pigment levels, the net CO_2 assimilation rate, and stomatal conductance. Treatment with GR24 effectively reversed these adverse effects and increased Na^+, K^+, and Ca^{2+} levels in both the shoots and roots (Sarwar and Shahbaz 2020; Zulfiqar et al. 2021). Similarly, Iftikhar et al. (2023) revealed that salt stress significantly reduced the biomass and chlorophyll content in maize. The K^+ and Ca^{2+} concentrations in the shoots and roots were also reduced, with an increase in Na^+ concentration in these tissues. Foliar application of GR24 to salt-treated maize significantly increased the chlorophyll content, kernel number, grain size, and K^+ level, while reducing the Na^+ level. Ren et al. (2018) found that with the treatment of *Sesbania cannabina* seedlings with exogenous abscisic acid, the H_2O_2 and SL levels increased sequentially, which ultimately enhanced the salt stress tolerance of the seedlings. Under salt stress conditions, SL production, ABA content, and AMFs colonization rates in lettuce (*Lactuca sativa* cv. "Romana") were correlated. It has been proposed that salt stress stimulates plants to increase SL production to promote symbiotic establishment to cope with this unfavorable condition (Aroca et al. 2013).

5.3.6.2 Drought Stress

Drought is a critical factor affecting plant growth and development (Banks et al. 2019). According to a recent study, SLs could enhance the drought tolerance of maize seedlings by improving water relations, increasing photosynthetic pigments and gas exchange, and enhancing the activities of antioxidant enzymes (Sattar et al. 2022). In *Festuca arundinacea*, both axillary bud initiation and outgrowth are strongly inhibited by drought stress. qRT-PCR analysis showed that after 14 days of drought stress, the expression of genes responsible for axillary bud development was downregulated, whereas genes involved in axillary bud dormancy and SL biosynthesis were upregulated. Therefore, SL has been proposed to be involved in regulating tiller production under drought condition (Zhuang et al. 2017). The expression of *FaD14* at 7 and 14 days was significantly upregulated under drought stress, confirming the important role of SL under drought stress (Bennett et al. 2016). The barley (*Hordeum vulgare* L.) *hvd14.d* mutant, which has a missense mutation in the SL receptor HvD14, is highly sensitive to drought stress (Marzec et al. 2016). Further studies on this mutant demonstrated that the drought-sensitive phenotype might be attributed to disturbed ABA metabolism and signaling pathways (Marzec et al. 2020).

AMFs can alleviate drought stress in plants. Ruiz-Lozano et al. (2016) showed that under drought stress, AM fungus (*Rhizophagus irregularis*) induces SL biosynthesis and improves drought tolerance in tomato and lettuce. Treatment of winter wheat (*Triticum aestivum* L.) and *V. vinifera* L. with GR24 decreased electrolyte leakage, stomatal opening, and reactive oxygen species (ROS), whereas the water content, chlorophyll content, and photosynthesis rate were enhanced (Sedaghat et al. 2017, 2020, 2021; Min et al. 2019). Moreover, it was found that foliar application of SL or MeJA improved the drought tolerance of *Dracocephalum kotschyi* Boiss and enhanced essential oil production in this herbal plant (Bidabadi and Sharifi 2021).

5.3.6.3 Temperature Stress

High temperature severely affects the growth and fitness of plants. Elevated temperatures reduce the seed germination rate, decrease photosynthetic efficiency, and restrict plant growth (Reed et al. 2022). Extreme heat exposure can lead to plant mortality. Hu and coworkers observed that the application of GR24 to tall fescue could upregulate cell cycle-related genes and downregulate auxin transport-related genes in the tips of crown root under heat stress. During this process, SL and auxin interact in a feedback manner to control root elongation (Hu et al. 2018). Omoarelojie et al. (2020) demonstrated that GR24 could mediate tolerance to heat stress during seed germination in *Lupinus angustifolius*. *Lupinus angustifolius* seeds pretreated with GR24 had higher germination indices and proline content, with lower lipid peroxidation levels. Application of GR24 to *Arabidopsis* can rescue germination inhibition by high temperatures in SL biosynthetic mutants but not in signaling mutants. Further investigation has revealed that SL alleviates heat stress by modulating GA and ABA levels (Toh et al. 2012; Lechat et al. 2015).

Low-temperature stress also poses negative impact on crop production. SL has been shown to participate in low-temperature tolerance in peas and *Arabidopsis*. Dark chilling had no obvious influence on photosynthesis in wild-type plants. However, photosynthesis was significantly decreased in the *max* mutant of *Arabidopsis* and the *rms4* mutant of pea. It was thus proposed that SL alleviates chilling-induced photosynthesis inhibition and enhances chilling tolerance in plants (Cooper et al. 2018). Exogenous application of GR24 to rape (*Brassica rapa* L.) seedlings significantly alleviated low-temperature stress. GR24 treatment increased cell viability, photosynthesis, and soluble protein content in the seedlings. H_2O_2 plays a central role in SL-mediated cold tolerance (Zhang et al. 2020).

5.3.6.4 Light Stress

Light is an important environmental factor for plants. When light intensity is unsuitable for plant growth, it causes abiotic stress and physiological damage in plants. Seedlings of the *max2* mutant of *Arabidopsis* possess longer hypocotyls and petioles in light than those of the wild type. These results suggest that SLs might repress primordia formation by axillary meristem (Stirnberg et al. 2002). Bharti et al. (2015) observed light-induced increase of SL levels in the seedling roots and leaves of sunflower. Similarly, Koltai et al. (2011) reported that above a certain threshold, light intensity was a positive regulator of SL biosynthesis in tomato roots. Sensing the ratio of red to far-red (R/FR) light positively regulates AMFs symbiosis in tomato and *L. japonicus*. Under low R/FR light conditions, the jasmonic acid (JA) and SL levels were reduced, and AM colonization was suppressed. When these two plants are grown in light with high R/FR ratios, the JA and SL amounts and AM colonization are higher than those in plants grown at low R/FR ratio (Nagata et al. 2015). The exogenous application of GR24 to tomato seedlings can effectively alleviate growth inhibition and increase yield under low-light stress. It has been proposed that SLs enhance plant tolerance to low-light stress by promoting light energy utilization, reducing ROS damage, and enhancing photosynthesis (Lu et al. 2019).

5.3.6.5 Heavy Metal Stress

Heavy metal contamination of water and soil is an important problem for plant growth. Exposure to heavy metals can lead to various physiological and biochemical changes (Bhoi et al. 2021). Excess cadmium (Cd) in the nutrient solution caused obvious deleterious effects in switchgrass (*Panicum virgatum*) seedlings, such as reduced biomass, water content, root length, root surface area, number of root tips, uptake of trace elements, photosynthetic pigments, and gas exchange. In addition, the contents of Cd and MDA, and the activities of superoxide (SOD), peroxide (POD), and catalase (CAT) were markedly increased. The treatment of Cd-stressed *P. virgatum* seedlings with GR24 (1 μM) significantly alleviated these adverse effects (Tai et al. 2017). Similarly, exogenous application of 1-μM GR24 also mitigated Cd-induced growth inhibition in *H. vulgare* L. Notably, GR24 was shown to increase NO, glutathione, and ascorbic acid levels to eliminate Cd-induced oxidative damage (Qiu et al. 2021). The wild type and *d10* and *d17* mutants exhibited distinct phenotypic differences under arsenic (As) stress. The transcription levels of antioxidant enzyme genes were higher in the wild-type roots than those in the mutant roots. Conversely, the expression of metal transporter genes was upregulated in the mutants relative to that in the wild type. These transport proteins play crucial roles in facilitating the uptake of bioavailable metal cations from the soil, maintaining metal ion homeostasis, and ensuring normal plant growth and development. Wild type and SL-deficient mutants (*d10* and *d7*) of rice have different responses to arsenic toxicity. The wild type has lower arsenate uptake, GSH biosynthesis, and antioxidant defense responses than the mutants, suggesting that SLs are involved in arsenic stress mitigation (Mostofa et al. 2021).

5.3.7 Pharmacological Activities of SLs

Many plant hormones such as abscisic acid, salicylic acid, and cytokinins exhibit various pharmacological activities (Klessig et al. 2016; Lievens et al. 2017). SLs and their analogs have been found to possess anti-inflammatory and antioxidant effects. SIRT1 is an NAD^+-dependent deacetylase that plays a crucial role in enhancing insulin signaling, glucose uptake, GLUT4 translocation, and mitochondrial biogenesis. GR24 can influence glucose metabolism by upregulating and activating SIRT1. Additionally, GR24 induces the expression of phase-II detoxifying enzymes in hepatic and macrophage cells, suggesting the medical potential of SLs in mitigating inflammation-mediated chronic disorders (Modi et al. 2017). GR24 also exhibits anti-inflammatory activity both *in vitro* (in RAW264.7) and *in vivo* (in zebrafish larvae) (Zheng et al. 2018). SLs were found to selectively inhibit cancer cell proliferation and induce apoptosis without affecting normal cells (Pollock et al. 2012). Yang et al. (2022) investigated the cytotoxicity of a panel of GR24 analogs against colorectal cancer cells. Notably, one GR24 analog exhibited potent cytotoxicity against colorectal cancer cells but not against normal human cells. This analog inhibited tumor growth in an HCT116 xenograft mouse model with low toxicity. Two SL analogs TIT3 and TIT7 exhibited dose- and time-dependent inhibition of HepG2 cell viability. Furthermore, TIT3 and TIT7 demonstrated significant antiproliferative effects on HepG2 hepatocellular carcinoma cells, while exerting minimal effects on baby hamster kidney (BHK) cells (Hasan et al. 2018). Pollock et al. (2014) and Mayzlish-Gati et al. (2015) examined the cytotoxicity of the SL analogs EG5, EG9c, ST357, ST362, and MEB55 in various cancer cell lines. All of these compounds can inhibit cancer cell growth, induce cellular stress responses, cause cell cycle arrest, and promote apoptosis. Further studies have demonstrated that ST362 and MEB55 induce DNA double-strand breaks (DSBs), thereby activating the DNA damage response in osteosarcoma cells. However, no DNA DSBs or apoptosis was detected in noncancerous cells, indicating the potential clinical applications of these compounds (Croglio et al. 2016). Two SL analogs, IND and EGO10, have been shown to inhibit glioma cancer cell growth by inducing apoptosis and G1 cell cycle arrest. These two compounds are promising antiglioma agents (Antika et al. 2022). SL analogs have also been reported to induce the expression of heme oxygenase (HO-1) and NAD(P)H dehydrogenase [quinone] 1 (NQO1), and inhibit the replication of human cytomegalovirus (HCMV) and angiogenesis (Tumer et al. 2018; Carillo et al. 2019; Biolatti et al. 2020). These results indicate that SLs and their analogs are promising drug candidates.

5.4 Biosynthesis of SLs

Early studies on carotene-deficient mutants or with carotene biosynthetic inhibitors have suggested that SLs are biosynthesized via the carotenoid pathway. SLs are derived from *trans*-β-carotene and their biosynthesis involves isomerization, bond cleavage, and multiple oxygenation steps to form the canonical and noncanonical SLs (Figure 5.5) (Matusova et al. 2005; Alder et al. 2012).

5.4.1 The Biosynthetic Pathway of SLs

The enzymes involved in SL biosynthesis include D27, carotenoid cleavage dioxygenases CCD7 and CCD8, cytochrome P450 oxygenases, methyltransferases, and LBO (Brewer et al. 2016). Analysis of *Arabidopsis max1–max4* mutants showed that all four *MAX* genes are related to SLs. MAX1, MAX3, and MAX4 participate in SL biosynthesis, whereas MAX2 is responsible for SL sensing and signal transduction (Stirnberg et al. 2002, 2007; Booker et al. 2005). Subsequent studies have shown that MAX1 belongs to the cytochrome P450 family, whereas MAX3 (CCD7) and MAX4 (CCD8) belong to the carotenoid cleavage dioxygenase family (Sorefan et al. 2003; Booker et al. 2004; Lazar and Goodman 2006).

D27 catalyzes the reversible isomerization of all-*trans*-β-carotene to 9-*cis*-β-carotene. Rice tillering dwarf mutants with high tiller numbers were obtained by β-irradiation and have been extensively used to study SLs' function and biosynthesis (Iwata and Omura 1977; Ishikawa et al. 2005). The rice *d27* mutant exhibited a highly tillered and dwarf phenotype. *OsD27* encodes an iron-containing protein that is mainly expressed in juvenile leaves, leaf buds, inflorescence initiation zones, LRs, and crown roots (Lailheugue et al. 2023). Mutational studies have shown that D27 is essential for SLs' production in *O. sativa* (Lailheugue et al. 2023). All the characterized D27 enzymes, including those from

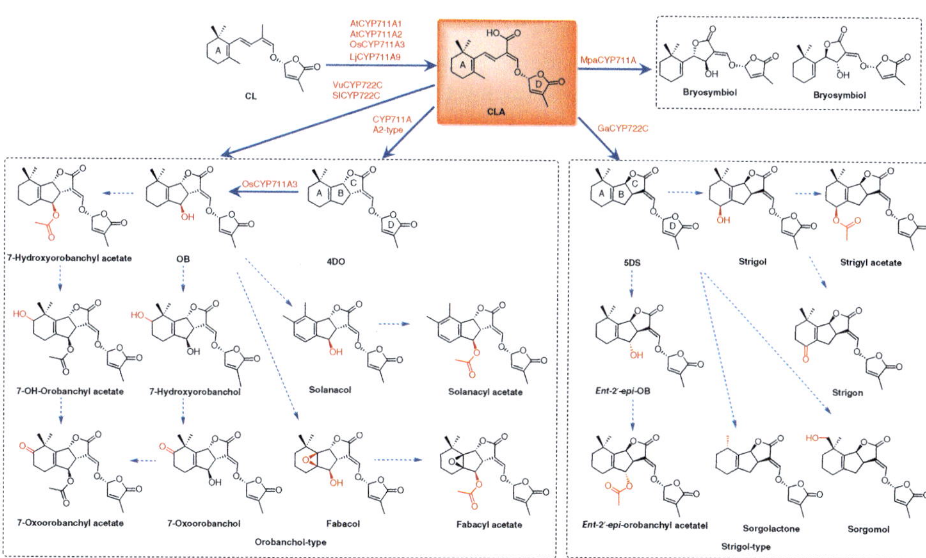

Figure 5.5 The downstream biosynthetic pathway of SLs.

Arabidopsis, *Dunaliella bardawil*, *T. aestivum*, *Psathyrostachys juncea*, and *Crocus sativus*, all show high specificity for the C9–C10 double bond during the isomerization (Bruno and Al-Babili 2016; Davidi and Pick 2017; Abuauf et al. 2018; López-Jiménez et al. 2022; Ren et al. 2024). In addition to plants, D27 is also present in some microorganisms. For example, CaD27 from *Cyanobacterium aponinum* has been characterized as an all-*trans*/*cis*- and *cis*/*cis*-β-carotene isomerase. However, CaD27 exhibits cis/cis conversion preference (Alvarez et al. 2024).

After isomerization, CCD7 catalyzes the cleavage of 9-*cis*-β-carotene to produce 9-*cis*-β-apo-10′-carotenal and β-ionone. The *Arabidopsis AtMAX3* gene encodes this dioxygenase. Subsequent studies identified *PsRMS5* from *P. sativum* and *PhDAD3* from *P. hybrida* as genes encoding CCD7 (Morris et al. 2001; Drummond et al. 2009). In *O. sativa*, CCD7 is encoded by *OsD17/HTD1* gene (Wang et al. 2020). In *Dactylis glomerata*, *DgCCD7A* and *DgCCD7D* have been identified as the genes encoding CCD7 (Xu et al. 2024). *CsCCD7* and *CsCCD8* from *Crocus sativus* were the first to be identified in a non-grass monocotyledonous plant (Rubio-Moraga et al. 2014). *In vitro* experiments demonstrated that recombinant AtCCD7 could cleave the C9–C10 double bond of β-carotene and lycopene (Schwartz et al. 2004).

CCD8 acts after CCD7 to convert to form CL and heptanal (Sorefan et al. 2003; Seto et al. 2014). 9-*Cis*-β-apo-10′-carotenal is preferred to the corresponding all-*trans*-isomer (Alder et al. 2008). The *AtMAX4* gene from *Arabidopsis*, the *PsRMS1* gene from *P. sativum*, the *PhDAD1* gene from *P. hybrida*, and the *DWARF10* (*D10*) gene in *O. sativa* were all found to encode CCD8 proteins. Their mutagenesis resulted in the abolishment of SL biosynthesis (Snowden et al. 2005; Arite et al. 2007). Conversion to CL is proposed to include several reactions: the isomerization of the C12–C13 double bond, sequential oxygenation of the C14 atom, the formation of a cyclic endoperoxide at C14, and migration of the C10—C11 and C14—C15 bonds, to finally yield CL.

CL is the central metabolite of the SL biosynthetic pathway and is a common precursor of canonical and noncanonical SLs (Al-Babili and Bouwmeester 2015; Yang et al. 2019; Alvi et al. 2022). In *Arabidopsis*, the cytochrome P450 MAX1 (CYP711A1) successively oxidizes CL to generate 19-hydroxy-CL, 19-oxo-CL, and carlactonoic acid (CLA) (Yoneyama et al. 2018a). All active CYP711A enzymes identified to date can convert CL to CLA (Zhang et al. 2014; Kawada et al. 2023). According to the final product, enzymes of the CYP711A family are further categorized into three types: the A1, A2, and A3 types. The A1-type CYP711A enzyme produces CLA as its final product. The A2-type CYP711A enzymes also oxidize the C18 of CLA to facilitate the B–C-ring closure of 4DO, with Os900 from rice as an example (Sigalas et al. 2023). A3-type CYP711A enzymes catalyze the hydroxylation of 4DO to afford canonical SL OB (Chesterfield et al. 2020; Wakabayashi et al. 2022b). In addition to CYP722C and CYP728B, two subfamilies of enzymes belonging to the CYP85 clan of plant CYP450s are also involved in SL biosynthesis. Enzymes of group-I CYP722C convert CLA to OB, and those of group-II CYP722C convert CLA to 5DS. The CYP728B protein from *sorghum* can catalyze the oxidation of 5DS to yield sorgomol (Jia et al. 2017; Mori et al. 2020a; Wakabayashi et al. 2020).

The enzyme catalyzing CLA methylation was recently identified in *Arabidopsis*. CLAMT, encoded by At4g36470, can methylate CLA to produce MeCLA using *S*-adenosyl-L-methionine (SAM) as a methyl doner (Wakabayashi et al. 2019; Mashiguchi et al. 2022). LBO is an enzyme of the 2-oxoglutarate and Fe^{II}-dependent dioxygenase superfamily that oxidizes MeCLA to form hydroxymethyl CLA (1′-HO-MeCLA) (Yoneyama et al. 2020).

5.4.2 Heterologous Biosynthesis of SLs

SLs are produced in plants with extremely low levels and poor stability (Lin et al. 2009). Chemical synthesis is the prevailing approach for obtaining SLs and their analogs. With advances in elucidating SL biosynthetic pathways and synthetic biology techniques, biosynthesis of SLs in microbial hosts is now possible. Wu et al. (2021) achieved heterologous production of SLs using a mixed culture of *Escherichia coli* and *Saccharomyces cerevisiae*. The SL biosynthetic pathway was divided into two parts. The upper part from β-carotene to CL was achieved in *E. coli* by expressing the *D27*, *CCD7*, and *CCD8* genes, whereas the lower part was constituted in *S. cerevisiae* to take advantage of its superior ability to express plant CYP450s. The recombinant *E. coli* and *S. cerevisiae* were then co-cultured to produce 5DS (6.65 μg/L), 4DO (3.46 μg/L), and OB (19.36 μg/L).

5.5 Conclusion and Future Perspectives

In conclusion, numerous natural SLs have been characterized over the last six decades. They have many essential roles in controlling plant development, enhancing tolerance to abiotic stresses, and acting as germination stimulants for destructive parasitic weeds. Our understanding of biosynthesis, signaling, and interactions with other plant hormones is rapidly accumulating. SL analogs have been used as agrochemicals to control *Striga* and *Orobanche* by inducing suicidal seed germination. Investigation of SL biosynthetic pathways and signaling would facilitate crop breeding to address grand challenges in agriculture.

Due to the high structural complexity of SLs, their chemical synthesis is not feasible for large-scale production (Morris and McErlean 2016; Zorrilla et al. 2022). Biosynthesis of SLs in microbial hosts holds great potential for producing these compounds in high yields and generating superior SL derivatives by manipulating the vast reservoir of enzymes in nature.

Acknowledgments

This work is supported by the National Key R&D Program of China (2020YFA0907900) and the Special Project for Technological Innovation in New Productive Forces of Modern Chinese Medicines (24ZXZKSY00010).

References

Abe, S., Sado, A., Tanaka, K. et al. (2014). Carlactone is converted to carlactonoic acid by MAX1 in *Arabidopsis* and its methyl ester can directly interact with AtD14 in vitro. *Proceedings of the National Academy of Sciences of the United States of America* 111(50), 18084–18089.

Abuauf, H., Haider, I., Jia, K.P. et al. (2018). The Arabidopsis *DWARF27* gene encodes an all-*trans*-/9-*cis*-β-carotene isomerase and is induced by auxin, abscisic acid and phosphate deficiency. *Plant Science* 277, 33–42.

Akiyama, K., Matsuzaki, K.-i. and Hayashi, H. (2005). Plant sesquiterpenes induce hyphal branching in arbuscular mycorrhizal fungi. *Nature* 435(7043), 824–827.

Al-Babili, S. and Bouwmeester, H.J. (2015). Strigolactones, a novel carotenoid-derived plant hormone. *Annual Review of Plant Biology* 66, 161–186.

Alder, A., Holdermann, I., Beyer, P. et al. (2008). Carotenoid oxygenases involved in plant branching catalyse a highly specific conserved apocarotenoid cleavage reaction. *Biochemical Journal* 416, 289–296.

Alder, A., Jamil, M., Marzorati, M. et al. (2012). The path from β-carotene to carlactone, a strigolactone-like plant hormone. *Science* 335(6074), 1348–1351.

Alvarez, D., Yang, Y., Saito, Y. et al. (2024). Characterization of a β-carotene isomerase from the cyanobacterium *Cyanobacteria aponinum. Philosophical Transactions of the Royal Society B* 379, 20230360.

Alvi, A.F., Sehar, Z., Fatma, M. et al. (2022). Strigolactone: an emerging growth regulator for developing resilience in plants. *Plants* 11(19), 2604.

Antika, G., Cinar, Z., Seçen, E. et al. (2022). Strigolactone analogs: two new potential bioactiphores for glioblastoma. *ACS Chemical Neuroscience* 13(5), 572–580.

Arite, T., Iwata, H., Ohshima, K. et al. (2007). *DWARF10*, an *RMS1/MAX4/DAD1* ortholog, controls lateral bud outgrowth in rice. *Plant Journal* 51(6), 1019–1029.

Arite, T., Kameoka, H. and Kyozuka, J. (2012). Strigolactone positively controls crown root elongation in rice. *Journal of Plant Growth Regulation* 31(2), 165–172.

Aroca, R., Ruiz-Lozano, J.M., Zamarreño, A.M. et al. (2013). Arbuscular mycorrhizal symbiosis influences strigolactone production under salinity and alleviates salt stress in lettuce plants. *Journal of Plant Physiology* 170(1), 47–55.

Banks, J.M., Percival, G.C. and Rose, G. (2019). Variations in seasonal drought tolerance rankings. *Trees* 33(4), 1063–1072.

Baz, L., Mori, N., Mi, J. et al. (2018). 3-Hydroxycarlactone, a novel product of the strigolactone biosynthesis core pathway. *Molecular Plant* 11(10), 1312–1314.

Benková, E., Michniewicz, M., Sauer, M. et al. (2003). Local, efflux-dependent auxin gradients as a common module for plant organ formation. *Cell* 115(5), 591–602.

Bennett, T., Liang, Y., Seale, M. et al. (2016). Strigolactone regulates shoot development through a core signalling pathway. *Biology Open* 5(12), 1806–1820.

Besserer, A., Bécard, G., Jauneau, A. et al. (2008). GR24, a synthetic analog of strigolactones, stimulates the mitosis and growth of the arbuscular mycorrhizal fungus *Gigaspora rosea* by boosting its energy metabolism. *Plant Physiology* 148(1), 402–413.

Besserer, A., Puech-Pagès, V., Kiefer, P. et al. (2006). Strigolactones stimulate arbuscular mycorrhizal fungi by activating mitochondria. *PLoS Biology* 4(7), e226.

Bharti, N., Tripathi, S. and Bhatla, S.C. (2015). Photomodulation of strigolactone biosynthesis and accumulation during sunflower seedling growth. *Plant Signaling & Behavior* 10(8), e1049792.

Bhoi, A., Yadu, B., Chandra, J. et al. (2021). Contribution of strigolactone in plant physiology, hormonal interaction and abiotic stresses. *Planta* 254(2), 28.

Bidabadi, S.S. and Sharifi, P. (2021). Strigolactone and methyl jasmonate-induced antioxidant defense and the composition alterations of different active compounds in *Dracocephalum kotschyi Boiss* under drought stress. *Journal of Plant Growth Regulation* 40(2), 878–889.

Biolatti, M., Blangetti, M., D'Arrigo, G. et al. (2020). Strigolactone analogs are promising antiviral agents for the treatment of human cytomegalovirus infection. *Microorganisms* 8(5), 703.

Booker, J., Auldridge, M., Wills, S. et al. (2004). MAX3/CCD7 is a carotenoid cleavage dioxygenase required for the synthesis of a novel plant signaling molecule. *Current Biology* 14(14), 1232–1238.

Booker, J., Sieberer, T., Wright, W. et al. (2005). *MAX1* encodes a cytochrome P450 family member that acts downstream of *MAX3/4* to produce a carotenoid-derived branch-inhibiting hormone. *Developmental Cell* 8(3), 443–449.

Bouwmeester, H.J., Roux, C., Lopez-Raez, J.A. et al. (2007). Rhizosphere communication of plants, parasitic plants and AM fungi. *Trends in Plant Science* 12(5), 224–230.

Brewer, P.B., Yoneyama, K., Filardo, F. et al. (2016). *LATERAL BRANCHING OXIDOREDUCTASE* acts in the final stages of strigolactone biosynthesis in *Arabidopsis*. *Proceedings of the National Academy of Sciences of the United States of America* 113(22), 6301–6306.

Bromhead, L.J., Visser, J. and McErlean, C.S.P. (2014). Enantioselective synthesis of the strigolactone mimic (+)-GR24. *The Journal of Organic Chemistry* 79(3), 1516–1520.

Bruno, M. and Al-Babili, S. (2016). On the substrate specificity of the rice strigolactone biosynthesis enzyme *DWARF27*. *Planta* 243(6), 1429–1440.

Cardoso, C., Charnikhova, T., Jamil, M. et al. (2014). Differential activity of *Striga hermonthica* seed germination stimulants and *Gigaspora rosea* hyphal branching factors in rice and their contribution to underground communication. *PLoS ONE* 9(8), e104201.

Carillo, P., Martínez-Poveda, B., Medina, M.Á. et al. (2019). The strigolactone analog GR-24 inhibits angiogenesis *in vivo* and *in vitro* by a mechanism involving cytoskeletal reorganization and VEGFR2 signalling. *Biochemical Pharmacology* 168, 366–383.

Charnikhova, T.V., Gaus, K., Lumbroso, A. et al. (2017). Zealactones. Novel natural strigolactones from maize. *Phytochemistry* 137, 123–131.

Charnikhova, T.V., Gaus, K., Lumbroso, A. et al. (2018). Zeapyranolactone—a novel strigolactone from maize. *Phytochemistry Letters* 24, 172–178.

Chesterfield, R.J., Vickers, C.E. and Beveridge, C.A. (2020). Translation of strigolactones from plant hormone to agriculture: achievements, future perspectives, and challenges. *Trends in Plant Science* 25(11), 1087–1106.

Cook, C.E., Whichard, L.P., Turner, B. et al. (1966). Germination of witchweed (*Striga lutea* Lour.): isolation and properties of a potent stimulant. *Science* 154(3753), 1189–1190.

Cook, C.E., Whichard, L.P., Wall, M. et al. (1972). Germination stimulants. II. structure of strigol, a potent seed germination stimulant for witchweed (*Striga lutea*). *Journal of the American Chemical Society* 94(17), 6198–6199.

Cooper, J.W., Hu, Y., Beyyoudh, L. et al. (2018). Strigolactones positively regulate chilling tolerance in pea and in *Arabidopsis*. *Plant, Cell & Environment* 41(6), 1298–1310.

Crawford, S., Shinohara, N., Sieberer, T. et al. (2010). Strigolactones enhance competition between shoot branches by dampening auxin transport. *Development* 137(17), 2905–2913.

Croglio, M.P., Haake, J.M., Ryan, C.P. et al. (2016). Analogs of the novel phytohormone, strigolactone, trigger apoptosis and synergize with PARP inhibitors by inducing DNA damage and inhibiting DNA repair. *Oncotarget* 7(12), 13984–14001.

Davidi, L. and Pick, U. (2017). Novel 9-*cis*/*all-trans* β-carotene isomerases from plastidic oil bodies in *Dunaliella bardawil* catalyze the conversion of *all-trans* to 9-*cis* β-carotene. *Plant Cell Reports* 36(6), 807–814.

Domagalska, M.A. and Leyser, O. (2011). Signal integration in the control of shoot branching. *Nature Reviews Molecular Cell Biology* 12(4), 211–221.

Dor, E., Yoneyama, K., Wininger, S. et al. (2011). Strigolactone deficiency confers resistance in tomato line *SL-ORT1* to the parasitic weeds *Phelipanche* and *Orobanche* spp. *Phytopathology* 101(2), 213–222.

Drummond, R.S., Martínez-Sánchez, N.M., Janssen, B.J. et al. (2009). *Petunia hybrida CAROTENOID CLEAVAGE DIOXYGENASE7* is involved in the production of negative and positive branching signals in petunia. *Plant Physiology* 151(4), 1867–1877.

Foo, E. (2013). Auxin influences strigolactones in pea mycorrhizal symbiosis. *Journal of Plant Physiology* 170(5), 523–528.

Foo, E. and Davies, N.W. (2011). Strigolactones promote nodulation in pea. *Planta* 234(5), 1073–1081.

Genre, A., Chabaud, M., Balzergue, C. et al. (2013). Short-chain chitin oligomers from arbuscular mycorrhizal fungi trigger nuclear Ca^{2+} spiking in *Medicago truncatula* roots and their production is enhanced by strigolactone. *New Phytologist* 198(1), 190–202.

Gomez-Roldan, V., Fermas, S., Brewer, P.B. et al. (2008). Strigolactone inhibition of shoot branching. *Nature* 455(7210), 189–194.

Ha, C.V., Leyva-González, M.A., Osakabe, Y. et al. (2014). Positive regulatory role of strigolactone in plant responses to drought and salt stress. *Proceedings of the National Academy of Sciences of the United States of America* 111(2), 851–856.

Han, X.Y., Hao, X., Qiu, S. et al. (2021). Strigolactone regulates plant architecture by inhibiting lateral branch growth in *Quercus mongolica* seedlings. *Scandinavian Journal of Forest Research* 36(5), 333–343.

Hasan, M.N., Choudhry, H., Razvi, S.S. et al. (2018). Synthetic strigolactone analogues reveal anti-cancer activities on hepatocellular carcinoma cells. *Bioorganic & Medicinal Chemistry Letters* 28(6), 1077–1083.

Hauck, C., Müller, S. and Schildknecht, H. (1992). A germination stimulant for parasitic flowering plants from *Sorghum bicolor*, a genuine host plant. *Journal of Plant Physiology* 139(4), 474–478.

Hong, C.Y., Chao, Y.Y., Yang, M.Y. et al. (2009). NaCl-induced expression of glutathione reductase in roots of rice *Oryza sativa* L. seedlings is mediated through hydrogen peroxide but not abscisic acid. *Plant and Soil* 320(1–2), 103–115.

Hu, Q., Zhang, S. and Huang, B. (2018). Strigolactones and interaction with auxin regulating root elongation in tall fescue under different temperature regimes. *Plant Science* 271, 34–39.

Iftikhar, I., Shahbaz, M. and Wahid, M.A. (2023). Potential role of foliage applied strigolactone (GR24) on photosynthetic pigments, gas exchange attributes, mineral nutrients and yield components of *Zea mays* L. under saline regimes. *Journal of Crop Health* 75(3), 577–591.

Illane, A., García-Garrido, J.M., Sampedro, I. et al. (2011). Strigolactones seem not to be involved in the nonsusceptibility of arbuscular mycorrhizal (AM) nonhost plants to AM fungi. *Botany* 89(4), 285–288.

Ishikawa, S., Maekawa, M., Arite, T. et al. (2005). Suppression of tiller bud activity in tillering dwarf mutants of rice. *Plant and Cell Physiology* 46(1), 79–86.

Iwata, N. and Omura, T. (1977). Linkage studies in rice *Oryza sativa* L.: on some mutants derived from chronic gamma irradiation. *Journal of the Faculty of Agriculture, Kyushu University* 21, 117–127.

Jamil, M., Charnikhova, T., Cardoso, C. et al. (2011). Quantification of the relationship between strigolactones and *Striga hermonthica* infection in rice under varying levels of nitrogen and phosphorus. *Weed Research* 51(4), 373–385.

Jia, K.-P., Baz, L. and Al-Babili, S. (2017). From carotenoids to strigolactones. *Journal of Experimental Botany* 69(9), 2189–2204.

Jia, K.P., Luo, Q., He, S.B. et al. (2014). Strigolactone-regulated hypocotyl elongation is dependent on cryptochrome and phytochrome signaling pathways in *Arabidopsis*. *Molecular Plant* 7(3), 528–540.

Johnson, A.W., Gowada, G., Hassanali, A. et al. (1981). The preparation of synthetic analogues of strigol. *Journal of the Chemical Society, Perkin Transactions 1* 1981, 1734–1743.

Kaniganti, S., Bhattacharya, J., Petla, B.P. et al. (2022). Strigolactone, a neglected plant hormone, with a great potential for crop improvement: crosstalk with other plant hormones. *Environmental and Experimental Botany* 204, 105072.

Kapulnik, Y., Delaux, P.M., Resnick, N. et al. (2011). Strigolactones affect lateral root formation and root-hair elongation in *Arabidopsis*. *Planta* 233(1), 209–216.

Kawada, K., Saito, T., Onoda, S. et al. (2023). Synthesis of carlactone derivatives to develop a novel inhibitor of strigolactone biosynthesis. *ACS Omega* 8(15), 13855–13862.

Khetkam, P., Xie, X., Kisugi, T. et al. (2014). 7α- and 7β-Hydroxyorobanchyl acetate as germination stimulants for root parasitic weeds produced by cucumber. *Journal of Pesticide Science* 39(3), 121–126.

Kim, H.I., Kisugi, T., Khetkam, P. et al. (2014). Avenaol, a germination stimulant for root parasitic plants from *Avena strigosa*. *Phytochemistry* 103, 85–88.

Kim, H.I., Xie, X.N., Kim, H.S. et al. (2010). Structure–activity relationship of naturally occurring strigolactones in *Orobanche minor* seed germination stimulation. *Journal of Pesticide Science* 35(3), 344–347.

Kisugi, T., Nakatani, Y., Xie, X. et al. (2013a). Identification of 7β-hydroxy-5-deoxystrigol from root exudates of dokudami. *The Japanese Society for Chemical Regulation of Plants* 48, 58.

Kisugi, T., Xie, X., Kim, H.I. et al. (2013b). Strigone, isolation and identification as a natural strigolactone from *Houttuynia cordata*. *Phytochemistry* 87, 60–64.

Klessig, D.F., Tian, M. and Choi, H.W. (2016). Multiple targets of salicylic acid and its derivatives in plants and animals. *Frontiers in Immunology* 7, 206.

Kodama, K., Rich, M.K., Yoda, A. et al. (2022). An ancestral function of strigolactones as symbiotic rhizosphere signals. *Nature Communications* 13(1), 3974.

Kohlen, W., Charnikhova, T., Bours, R. et al. (2013). Tomato strigolactones: a more detailed look. *Plant Signaling & Behavior* 8(1), e22785.

Koltai, H. (2011). Strigolactones are regulators of root development. *New Phytologist* 190(3), 545–549.

Koltai, H., Cohen, M., Chesin, O. et al. (2011). Light is a positive regulator of strigolactone levels in tomato roots. *Journal of Plant Physiology* 168(16), 1993–1996.

Lailheugue, V., Merlin, I., Boutet, S. et al. (2023). Vitislactone, a non-canonical strigolactone exudated by grapevine rootstocks in response to nitrogen starvation. *Phytochemistry* 215, 113837.

Lazar, G. and Goodman, H.M. (2006). *MAX1*, a regulator of the flavonoid pathway, controls vegetative axillary bud outgrowth in *Arabidopsis*. *Proceedings of the National Academy of Sciences of the United States of America* 103(2), 472–476.

Lechat, M.M., Brun, G., Montiel, G. et al. (2015). Seed response to strigolactone is controlled by abscisic acid-independent DNA methylation in the obligate root parasitic plant *Phelipanche ramosa* L. Pomel. *Journal of Experimental Botany* 66(11), 3129–3140.

Lendzemo, V.W., Kuyper, T.W., Matusova, R. et al. (2007). Colonization by arbuscular mycorrhizal fungi of sorghum leads to reduced germination and subsequent attachment and emergence of *Striga hermonthica*. *Plant Signal & Behavior* 2(1), 58–62.

Li, C., Dong, L., Durairaj, J. et al. (2023). Maize resistance to witchweed through changes in strigolactone biosynthesis. *Science* 379(6627), 94–99.

Li, C., Haider, I., Wang, J.Y. et al. (2024). OsCYP706C2 diverts rice strigolactone biosynthesis to a noncanonical pathway branch. *Science Advances* 10(35), eadq3942.

Lievens, L., Pollier, J., Goossens, A. et al. (2017). Abscisic acid as pathogen effector and immune regulator. *Frontiers in Plant Science* 8, 587.

Lin, H., Wang, R.X., Qian, Q. et al. (2009). DWARF27, an iron-containing protein required for the biosynthesis of strigolactones, regulates rice tiller bud outgrowth. *Plant Cell* 21(5), 1512–1525.

Ling, F., Su, Q., Jiang, H. et al. (2020). Effects of strigolactone on photosynthetic and physiological characteristics in salt-stressed rice seedlings. *Scientific Reports* 10(1), 6183.

Liu, W., Kohlen, W., Lillo, A. et al. (2011). Strigolactone biosynthesis requires the symbiotic GRAS-type transcription factors NSP1 and NSP2. *Plant Cell* 23(10), 3853–3865.

López-Jiménez, A.J., Morote, L., Niza, E. et al. (2022). Subfunctionalization of D27 isomerase genes in saffron. *International Journal of Molecular Sciences* 23(18), 10543.

Lopez-Obando, M., Ligerot, Y., Bonhomme, S. et al. (2015). Strigolactone biosynthesis and signaling in plant development. *Development* 142(21), 3615–3619.

López-Ráez, J.A., Charnikhova, T., Fernández, I. et al. (2011). Arbuscular mycorrhizal symbiosis decreases strigolactone production in tomato. *Journal of Plant Physiology* 168(3), 294–297.

Louarn, J., Carbonne, F., Delavault, P. et al. (2012). Reduced germination of *Orobanche cumana* seeds in the presence of arbuscular mycorrhizal fungi or their exudates. *PLoS ONE* 7(11), e49273.

Lu, T., Yu, H., Li, Q. et al. (2019). Improving plant growth and alleviating photosynthetic inhibition and oxidative stress from low-light stress with exogenous GR24 in tomato (*Solanum lycopersicum* L.) seedlings. *Frontiers in Plant Science* 10, 490.

Maillet, F., Poinsot, V., André, O. et al. (2011). Fungal lipochitooligosaccharide symbiotic signals in arbuscular mycorrhiza. *Nature* 469(7328), 58–63.

Marzec, M., Daszkowska-Golec, A., Collin, A. et al. (2020). Barley strigolactone signalling mutant *hvd14.d* reveals the role of strigolactones in abscisic acid-dependent response to drought. *Plant, Cell & Environment* 43(9), 2239–2253.

Marzec, M., Gruszka, D., Tylec, P. et al. (2016). Identification and functional analysis of the *HvD14* gene involved in strigolactone signaling in *Hordeum vulgare*. *Physiologia Plantarum* 158, 341–355.

Mashiguchi, K., Seto, Y., Onozuka, Y. et al. (2022). A carlactonoic acid methyltransferase that contributes to the inhibition of shoot branching in *Arabidopsis*. *Proceedings of the National Academy of Science of the United States of America* 119(14), e2111565119.

Matusova, R., Rani, K., Verstappen, F.W.A. et al. (2005). The strigolactone germination stimulants of the plant-parasitic *Striga* and *Orobanche* spp. are derived from the carotenoid pathway. *Plant Physiology* 139(2), 920–934.

Mayzlish-Gati, E., De-Cuyper, C., Goormachtig, S. et al. (2012). Strigolactones are involved in root response to low phosphate conditions in *Arabidopsis. Plant Physiology* 160(3), 1329–1341.

Mayzlish-Gati, E., Laufer, D., Grivas, C.F. et al. (2015). Strigolactone analogs act as new anti-cancer agents in inhibition of breast cancer in xenograft model. *Cancer Biology & Therapy* 16(11), 1682–1688.

Min, Z., Li, R., Chen, L. et al. (2019). Alleviation of drought stress in grapevine by foliar-applied strigolactones. *Plant Physiology and Biochemistry* 135, 99–110.

Modi, S., Yaluri, N., Kokkola, T. et al. (2017). Plant-derived compounds strigolactone GR24 and pinosylvin activate SIRT1 and enhance glucose uptake in rat skeletal muscle cells. *Scientific Reports* 7(1), 17606.

Mori, N., Nomura, T. and Akiyama, K. (2020a). Identification of two oxygenase genes involved in the respective biosynthetic pathways of canonical and non-canonical strigolactones in *Lotus japonicus. Planta* 251(2), 40.

Mori, N., Sado, A., Xie, X. et al. (2020b). Chemical identification of 18-hydroxycarlactonoic acid as an LjMAX1 product and *in planta* conversion of its methyl ester to canonical and non-canonical strigolactones in *Lotus japonicus. Phytochemistry* 174, 112349.

Morris, J.C. and McErlean, C.S.P. (2016). Synthesis of highly *enantio*-enriched stereoisomers of hydroxy-GR24. *Organic & Biomolecular Chemistry* 14(4), 1236–1238.

Morris, S.E., Turnbull, C.G.N., Murfet, I.C. et al. (2001). Mutational analysis of branching in pea. Evidence that *Rms1* and *Rms5* regulate the same novel signal. *Plant Physiology* 126(3), 1205–1213.

Mostofa, M.G., Rahman, M.M. and Nguyen, K.H. (2021). Strigolactones regulate arsenate uptake, vacuolar-sequestration and antioxidant defense responses to resist arsenic toxicity in rice roots. *Journal of Hazardous Materials* 415, 125589.

Müller, D., Waldie, T., Miyawaki, K. et al. (2015). Cytokinin is required for escape but not release from auxin mediated apical dominance. *The Plant Journal* 82(5), 874–886.

Nagata, M., Yamamoto, N., Shigeyama, T. et al. (2015). Red/far red light controls arbuscular mycorrhizal colonization via jasmonic acid and strigolactone signaling. *Plant & Cell Physiology* 56(11), 2100–2109.

Nelson, D.C., Flematti, G.R., Ghisalberti, E.L. et al. (2012). Regulation of seed germination and seedling growth by chemical signals from burning vegetation. *Annual Review of Plant Biology* 63, 107–130.

Nomura, S., Nakashima, H., Mizutani, M. et al. (2013). Structural requirements of strigolactones for germination induction and inhibition of *Striga gesnerioides* seeds. *Plant Cell Reports* 32(6), 829–838.

Nomura, T., Seto, Y. and Kyozuka, J. (2024). Unveiling the complexity of strigolactones: exploring structural diversity, biosynthesis pathways, and signaling mechanisms. *Journal of Experimental Botany* 75(4), 1134–1147.

Omoarelojie, L.O., Kulkarni, M.G., Finnie, J.F. et al. (2020). Synthetic strigolactone (*rac*-GR24) alleviates the adverse effects of heat stress on seed germination and photosystem II function in lupine seedlings. *Plant Physiology and Biochemistry* 155, 965–979.

Pollock, C.B., Koltai, H., Kapulnik, Y. et al. (2012). Strigolactones: a novel class of phytohormones that inhibit the growth and survival of breast cancer cells and breast cancer stem-like enriched mammosphere cells. *Breast Cancer Research and Treatment* 134(3), 1041–1055.

Pollock, C.B., McDonough, S., Wang, V.S. et al. (2014). Strigolactone analogues induce apoptosis through activation of p38 and the stress response pathway in cancer cell lines and in conditionally reprogrammed primary prostate cancer cells. *Oncotarget* 5(6), 1683–1698.

Qi, J., Mao, Y., Cui, J. et al. (2024). The role of strigolactone in resistance to environmental stress in plants. *Physiologia Plantarum* 176(4), e14419.

Qiu, C.W., Zhang, C., Wang, N.H. et al. (2021). Strigolactone GR24 improves cadmium tolerance by regulating cadmium uptake, nitric oxide signaling and antioxidant metabolism in barley (*Hordeum vulgare* L.). *Environmental Pollution* 273, 116486.

Reed, R.C., Bradford, K.J. and Khanday, I. (2022). Seed germination and vigor: ensuring crop sustainability in a changing climate. *Heredity* 128(6), 450–459.

Ren, X.M., Ai, Q., Li, Z. et al. (2024). The *Psathyrostachys juncea DWARF27* gene encodes an all-*trans*-/9-*cis*-β-carotene isomerase in the control of plant branches in *Arabidopsis thaliana* by strigolactones. *G3 Genes Genomes Genetics* 14(9), jkae147.

Ren, C.G., Kong, C.C. and Xie, Z.H. (2018). Role of abscisic acid in strigolactone-induced salt stress tolerance in arbuscular mycorrhizal *Sesbania cannabina* seedlings. *BMC Plant Biology* 18(1), 74.

Rubio-Moraga, A., Ahrazem, O., Pérez-Clemente, R.M. et al. (2014). Apical dominance in saffron and the involvement of the branching enzymes CCD7 and CCD8 in the control of bud sprouting. *BMC Plant Biology* 14, 171.

Ruiz-Lozano, J.M., Aroca, R., Zamarreño Á, M. et al. (2016). Arbuscular mycorrhizal symbiosis induces strigolactone biosynthesis under drought and improves drought tolerance in lettuce and tomato. *Plant, Cell & Environment* 39(2), 441–452.

Ruiz-Lozano, J.M., Porcel, R., Azcón, C. et al. (2012). Regulation by arbuscular mycorrhizae of the integrated physiological, response to salinity in plants: new challenges in physiological and molecular studies. *Journal of Experimental Botany* 63(11), 4033–4044.

Ruyter-Spira, C., Kohlen, W., Charnikhova, T. et al. (2011). Physiological effects of the synthetic strigolactone analog GR24 on root system architecture in Arabidopsis: another belowground role for strigolactones? *Plant Physiology* 155(2), 721–734.

Sadhana, B. (2014). Arbuscular mycorrhizal fungi (AMF) as a biofertilizer—a review. *International Journal of Current Microbiology and Applied Sciences* 3(4), 384–400.

Sarwar, Y. and Shahbaz, M. (2020) Modulation in growth, photosynthetic pigments, gas exchange attributes and inorganic ions in sunflower (*Helianthus annuus* L.) by strigolactones (GR24) achene priming under saline conditions. *Pakistan Journal of Botany* 52(1), 23–31.

Sattar, A., Ul-Allah, S., Ijaz, M. et al. (2022). Exogenous application of strigolactone alleviates drought stress in maize seedlings by regulating the physiological and antioxidants defense mechanisms. *Cereal Research Communications* 50(2), 263–272.

Scaffidi, A., Waters, M.T., Sun, Y.K. et al. (2014). Strigolactone hormones and their stereoisomers signal through two related receptor proteins to induce different physiological responses in *Arabidopsis. Plant Physiology* 165(3), 1221–1232.

Schwartz, S.H., Qin, X. and Loewen, M.C. (2004). The biochemical characterization of two carotenoid cleavage enzymes from *Arabidopsis* indicates that a carotenoid-derived compound inhibits lateral branching. *Journal of Biological Chemistry* 279(45), 46940–46945.

Seale, M., Bennett, T. and Leyser, O. (2017). *BRC1* expression regulates bud activation potential but is not necessary or sufficient for bud growth inhibition in *Arabidopsis. Development* 144(9), 1661–1673.

Sedaghat, M., Emam, Y., Mokhtassi-Bidgoli, A. et al. (2021). The potential of the synthetic strigolactone analogue GR24 for the maintenance of photosynthesis and yield in winter wheat under drought: investigations on the mechanisms of action and delivery modes. *Plants* 10(6), 1223.

Sedaghat, M., Sarvestani, Z.T., Emam, Y. et al. (2020). Foliar-applied GR24 and salicylic acid enhanced wheat drought tolerance. *Russian Journal of Plant Physiology* 67(4), 733–739.

Sedaghat, M., Tahmasebi-Sarvestani, Z., Emam, Y. et al. (2017). Physiological and antioxidant responses of winter wheat cultivars to strigolactone and salicylic acid in drought. *Plant Physiology and Biochemistry* 119, 59–69.

Seto, Y., Sado, A., Asami, K. et al. (2014). Carlactone is an endogenous biosynthetic precursor for strigolactones. *Proceedings of the National Academy of Sciences of the United States of America* 111(4), 1640–1645.

Sharifi, P. and Bidabadi, S.S. (2020). Strigolactone could enhances gas-exchange through augmented antioxidant defense system in *Salvia nemorosa* L. plants subjected to saline conditions stress. *Industrial Crops and Products* 151, 112460.

Sigalas, P.P., Buchner, P., Thomas, S.G. et al. (2023). Nutritional and tissue-specific regulation of cytochrome P450 CYP711A *MAX1* homologues and strigolactone biosynthesis in wheat. *Journal of Experimental Botany* 74(6), 1890–1910.

Snowden, K.C., Simkin, A.J., Janssen, B.J. et al. (2005). The *decreased apical dominance1/ Petunia hybrida CAROTENOID CLEAVAGE DIOXYGENASE8* gene affects branch production and plays a role in leaf senescence, root growth, and flower development. *The Plant Cell* 17(3), 746–759.

Sorefan, K., Booker, J., Haurogné, K. et al. (2003). *MAX4* and *RMS1* are orthologous dioxygenase-like genes that regulate shoot branching in *Arabidopsis* and pea. *Genes & Development* 17(12), 1469–1474.

Soto, M.J., Fernández-Aparicio, M., Castellanos-Morales, V. et al. (2010). First indications for the involvement of strigolactones on nodule formation in alfalfa (*Medicago sativa*). *Soil Biology & Biochemistry* 42(2), 383–385.

Stirnberg, P., Furner, I.J. and Leyser, H.M.O. (2007). MAX2 participates in an SCF complex which acts locally at the node to suppress shoot branching. *Plant Journal* 50(1), 80–94.

Stirnberg, P., van de Sande, K. and Leyser, H.M.O. (2002). *MAX1* and *MAX2* control shoot lateral branching in *Arabidopsis. Development* 129(5), 1131–1141.

Tai, Z., Yin, X., Fang, Z. et al. (2017). Exogenous GR24 alleviates cadmium toxicity by reducing cadmium uptake in switchgrass (*Panicum virgatum*) seedlings. *International Journal of Environmental Research and Public Health* 14(8), 852.

Toh, S., Kamiya, Y., Kawakami, N. et al. (2012). Thermoinhibition uncovers a role for strigolactones in *Arabidopsis* seed germination. *Plant and Cell Physiology* 53(1), 107–117.

Tokunaga, T., Hayashi, H. and Akiyama, K. (2015). Medicaol, a strigolactone identified as a putative didehydro-orobanchol isomer, from *Medicago truncatula. Phytochemistry* 111, 91–97.

Tumer, T.B., Yılmaz, B., Ozleyen, A. et al. (2018). GR24, a synthetic analog of strigolactones, alleviates inflammation and promotes Nrf2 cytoprotective response: in vitro and in silico evidences. *Computational Biology and Chemistry* 76, 179–190.

Ueno, K., Furumoto, T., Umeda, S. et al. (2014). Heliolactone, a non-sesquiterpene lactone germination stimulant for root parasitic weeds from sunflower. *Phytochemistry* 108, 122–128.

Ueno, K., Nomura, S., Muranaka, S. et al. (2011). *Ent*-2′-*epi*-orobanchol and its acetate, as germination stimulants for *Striga gesnerioides* seeds isolated from cowpea and red clover. *Journal of Agricultural and Food Chemistry* 59(19), 10485–10490.

Umehara, M., Hanada, A., Yoshida, S. et al. (2008). Inhibition of shoot branching by new terpenoid plant hormones. *Nature* 455(7210), 195–200.

Uraguchi, D., Kuwata, K., Hijikata, Y. et al. (2018). A femtomolar-range suicide germination stimulant for the parasitic plant *Striga hermonthica*. *Science* 362(6420), 1301–1305.

Vogel, J.T., Walter, M.H., Giavalisco, P. et al. (2010). SlCCD7 controls strigolactone biosynthesis, shoot branching and mycorrhiza-induced apocarotenoid formation in tomato. *The Plant Journal* 61(2), 300–311.

Wakabayashi, T., Hamana, M., Mori, A. et al. (2019). Direct conversion of carlactonoic acid to orobanchol by cytochrome P450 CYP722C in strigolactone biosynthesis. *Science Advances* 5(12), eaax9067.

Wakabayashi, T., Moriyama, D., Miyamoto, A. et al. (2022a). Identification of novel canonical strigolactones produced by tomato. *Frontiers in Plant Science* 13, 1064378.

Wakabayashi, T., Shida, K., Kitano, Y. et al. (2020). CYP722C from *Gossypium arboreum* catalyzes the conversion of carlactonoic acid to 5-deoxystrigol. *Planta* 251(5), 97.

Wakabayashi, T., Ueno, K. and Sugimoto, Y. (2022b). Structure elucidation and biosynthesis of orobanchol. *Frontiers in Plant Science* 13, 835160.

Wang, Y., Shang, L., Yu, H. et al. (2020). A strigolactone biosynthesis gene contributed to the green revolution in rice. *Molecular Plant* 13(60), 923–932.

Waters, M.T., Brewer, P.B., Bussell, J.D. et al. (2012). The *Arabidopsis* ortholog of rice DWARF27 acts upstream of MAX1 in the control of plant development by strigolactones. *Plant Physiology* 159(3), 1073–1085.

Waters, M.T., Gutjahr, C., Bennett, T. et al. (2017). Strigolactone signaling and evolution. *Annual Review of Plant Biology* 68, 291–322.

Waters, M.T. and Smith, S.M. (2013). KAI2- and MAX2-mediated responses to karrikins and strigolactones are largely independent of HY5 in *Arabidopsis* seedlings. *Molecular Plant* 6(1), 63–75.

Wu, S., Ma, X.Q., Zhou, A.Q. et al. (2021). Establishment of strigolactone-producing bacterium-yeast consortium. *Science Advances* 7(38), eahb4048.

Xie, X., Kusumoto, D., Takeuchi, Y. et al. (2007). 2′-Epi-orobanchol and solanacol, two unique strigolactones, germination stimulants for root parasitic weeds, produced by tobacco. *Journal of Agricultural and Food Chemistry* 55(20), 8067–8072.

Xie, X., Mori, N., Yoneyama, K. et al. (2019). Lotuslactone, a non-canonical strigolactone from *Lotus japonicus*. *Phytochemistry* 157, 200–205.

Xie, X., Yoneyama, K., Harada, Y. et al. (2009a). Fabacyl acetate, a germination stimulant for root parasitic plants from *Pisum sativum*. *Phytochemistry* 70(2), 211–215.

Xie, X., Yoneyama, K., Kisugi, T. et al. (2013). Confirming stereochemical structures of strigolactones produced by rice and tobacco. *Molecular Plant* 6(1), 153–163.

Xie, X., Yoneyama, K., Kurita, J. et al. (2009b). 7-Oxoorobanchyl acetate and 7-oxoorobanchol as germination stimulants for root parasitic plants from flax (*Linum usitatissimum*). *Bioscience, Biotechnology, and Biochemistry* 73(6), 1367–1370.

Xie, X., Yoneyama, K., Kusumoto, D. et al. (2008). Sorgomol, germination stimulant for root parasitic plants, produced by *Sorghum bicolor. Tetrahedron Letters* 49(13), 2066–2068.

Xu, X., Liang, Y., Feng, G. et al. (2024). A favorable natural variation in *CCD7* from orchardgrass confers enhanced tiller number. *The Plant Journal* 121(2), e17200.

Yang, S.T., Fan, J.B., Liu, T.T. et al. (2022). Development of strigolactones as novel autophagy/mitophagy inhibitors against colorectal cancer cells by blocking the autophagosome-lysosome fusion. *Journal of Medicinal Chemistry* 65(14), 9706–9717.

Yang, T., Lian, Y.K. and Wang, C.Y. (2019). Comparing and contrasting the multiple roles of butenolide plant growth regulators: strigolactones and karrikins in plant development and adaptation to abiotic stresses. *International Journal of Molecular Sciences* 20(24), 6270.

Yokota, T., Sakai, H., Okuno, K. et al. (1998). Alectrol and orobanchol, germination stimulants for *Orobanche minor*, from its host red clover. *Phytochemistry* 49(7), 1967–1973.

Yoneyama, K., Akiyama, K., Brewer, P.B. et al. (2020). Hydroxyl carlactone derivatives are predominant strigolactones in *Arabidopsis. Plant Direct* 4(5), e00219.

Yoneyama, K., Mori, N., Sato, T. et al. (2018a). Conversion of carlactone to carlactonoic acid is a conserved function of MAX1 homologs in strigolactone biosynthesis. *New Phytologist* 218(4), 1522–1533.

Yoneyama, K., Xie, X., Kusumoto, D. et al. (2007). Nitrogen deficiency as well as phosphorus deficiency in sorghum promotes the production and exudation of 5-deoxystrigol, the host recognition signal for arbuscular mycorrhizal fungi and root parasites. *Planta* 227(1), 125–132.

Yoneyama, K., Xie, X., Yoneyama, K. et al. (2018b). Which are the major players, canonical or non-canonical strigolactones? *Journal of Experimental Botany* 69(9), 2231–2239.

Yoshida, S., Kameoka, H., Tempo, M. et al. (2012). The D3 F-box protein is a key component in host strigolactone responses essential for arbuscular mycorrhizal symbiosis. *New Phytologist* 196(4), 1208–1216.

Zhang, Y.X., van Dijk, A.D.J., Scaffidi, A. et al. (2014). Rice cytochrome P450 MAX1 homologs catalyze distinct steps in strigolactone biosynthesis. *Nature Chemical Biology* 10(12), 1028–1033.

Zhang, X., Zhang, L., Sun, Y. et al. (2020). Hydrogen peroxide is involved in strigolactone induced low temperature stress tolerance in rape seedlings (*Brassica rapa* L.). *Plant Physiology and Biochemistry* 157, 402–415.

Zheng, J.X., Han, Y.S., Wang, J.C. et al. (2018). Strigolactones: a plant phytohormone as novel anti-inflammatory agents. *MedChemComm* 9(1), 181–188.

Zhuang, L.L., Wang, J. and Huang, B.R. (2017). Drought inhibition of tillering in *Festuca arundinacea* associated with axillary bud development and strigolactone signaling. *Environmental and Experimental Botany* 142, 15–23.

Zorrilla, J.G., Rial, C., Varela, R.M. et al. (2022). Strategies for the synthesis of canonical, non-canonical and analogues of strigolactones, and evaluation of their parasitic weed germination activity. *Phytochemistry Reviews* 21(5), 1627–1659.

Zulfiqar, H., Shahbaz, M., Ahsan, M. et al. (2021). Strigolactone (GR24) induced salinity tolerance in sunflower (*Helianthus annuus* L.) by ameliorating morpho-physiological and biochemical attributes under in vitro conditions. *Journal of Plant Growth Regulation* 40(5), 2079–2091.

6

Strigolactone Perception and Response Under Nitrogen and Phosphorus Deficiency

Ishrat Mehmood[1], Abdullah[2], Kaiser Iqbal Wani[3,4,], Muhammad Aamir Manzoor[3,4], and Tariq Aftab[2]*

[1] *Department of Biophysics, Panjab University, Chandigarh, India*
[2] *Department of Botany, Aligarh Muslim University, Aligarh, Uttar Pradesh, India*
[3] *Key Laboratory for Quality Regulation of Tropical Horticultural Crops of Hainan Province, School of Tropical Agriculture and Forestry, Hainan University, Haikou, Hainan, China*
[4] *National Key Laboratory for Tropical Crop Breeding, School of Breeding and Multiplication (Sanya Institute of Breeding and Multiplication), Hainan University, Sanya, Hainan, China*

*Corresponding author: kaiseriqbal34@gmail.com

6.1 Introduction

In the intricate web of life, plants are the silent players of survival, constantly adapting to the intricacies of their environment. Among the myriad challenges they face, nutrient scarcity—particularly deficiencies in nitrogen (N) and phosphorus (P)—stands as a formidable barrier to growth, reproduction, and resilience. These macronutrients are the lifeblood of cellular machinery: N fuels the synthesis of proteins, nucleic acids, and chlorophyll, while P anchors energy transfer (ATP), membrane integrity, and genetic material. When soils are deficient in these critical elements, plants do not simply wither; they deploy a sophisticated arsenal of biochemical and physiological strategies to recalibrate their metabolism, prioritize resource allocation, and communicate with microbial allies in the rhizosphere. At the heart of this adaptive symphony lies a class of signaling molecules known as strigolactones (SLs), a class of carotenoid-derived plant hormones that have emerged as pivotal regulators of plant development and stress adaptation (Wani et al. 2023). SLs were initially discovered for their role in stimulating seed germination of parasitic plants like *Striga*—a sinister adaptation that exploits host plants' distress signals (Cook et al. 1966). However, their evolutionary significance extends far beyond this parasitic relationship. Over the past two decades, SLs have been recognized as multifunctional hormones regulating shoot architecture, root development, and symbiotic interactions with arbuscular mycorrhizal fungi (AMFs), which assist plants in nutrient acquisition (Akiyama et al. 2005; Umehara et al. 2008). Under nutrient-limiting conditions, particularly N and P scarcities, SLs biosynthesis and

Plant Apocarotenoids: Biosynthesis, Signaling and Agricultural Applications, First Edition. Edited by Kaiser Iqbal Wani and Tariq Aftab.
Published 2026 by John Wiley & Sons Ltd.

signaling are dynamically regulated, enabling plants to optimize resource acquisition and sustain growth (Yoneyama et al. 2012; Marzec 2016). Plants increase SL production, a response that reflects their dual role as endogenous coordinators of growth and extracellular negotiators of microbial partnerships. For instance, in P-deficient soils, SLs exuded from roots attract AMFs, forming symbiotic networks that enhance phosphate uptake (López-Ráez et al. 2008). Yet, this survival tactic comes with a Faustian bargain: the same SLs can beckon parasitic weeds, turning nutrient stress into a crisis of infestation.

The perception and signaling of SLs are orchestrated by finely tuned molecular machinery. The α/β-hydrolase receptor D14 (DWARF14) interacts with SLs, triggering a conformational change that enables binding to the F-box protein MAX2 (More Axillary Growth 2). This complex recruits SMXL (SUPPRESSOR OF MAX2-LIKE) proteins for ubiquitination and proteasomal degradation, derepressing downstream transcriptional programs (Yao et al. 2016a). Under N or P deficiency, this pathway is hijacked to prioritize stress adaptation over growth. For example, SL signaling suppresses shoot branching—a "selfish" trait under ample nutrients—to conserve resources, while redirecting carbon to roots to explore nutrient-rich soil patches (Kohlen et al. 2011a). Recent studies reveal that nutrient stress also modulates the SL biosynthesis *carotenoid cleavage dioxygenase* (*CCD*) genes *CCD7* and *CCD8*, creating feedback loops that fine-tune hormone levels (Marzec et al. 2020). However, the interplay between N and P deficiencies in regulating SL dynamics remains enigmatic. While both nutrients induce SL production, their distinct metabolic roles suggest divergent signaling crosstalk. For instance, P deficiency predominantly enhances SL exudation to recruit AMFs, whereas N limitation may prioritize internal SL signaling to reshape root architecture (Ito et al. 2022).

Intriguingly, SLs do not act in isolation. Their signaling intersects with other hormonal pathways, such as auxin and cytokinin, to balance growth and stress responses. Under P deficiency, SLs antagonize auxin transport to inhibit lateral root formation while promoting elongated root hairs—a paradox that underscores their context-dependent roles (Ruyter-Spira et al. 2011a, b). Similarly, N scarcity alters cytokinin distribution, which may synergize with SLs to suppress shoot branching (Brewer et al. 2009). This hormonal crosstalk is further complicated by environmental variables like light quality and soil microbiota, painting a picture of SLs as molecular integrators of plant–environment dialogs. The agricultural implications of these discoveries are profound. As climate change and intensive farming exacerbate soil nutrient depletion, understanding SL biology offers avenues for breeding crops with enhanced nutrient efficiency or engineered SL exudation to optimize AM symbiosis without attracting parasites (Koltai 2011). For instance, rice varieties with modulated SL biosynthesis exhibit improved phosphate uptake under low-P conditions (Ito et al. 2022). Yet, challenges persist: the duality of SLs as symbiont signals and parasitic triggers demands precision in manipulating these pathways. Moreover, the plasticity of SL responses across species and environments highlights the need for context-specific solutions. This chapter will provide the recent advancements, weaving together molecular, physiological, and ecological perspectives.

6.2 Biosynthesis and Signaling Pathways of SLs in Plants

SLs, a class of carotenoid-derived plant hormones, play pivotal roles in regulating plant architecture, stress responses, and symbiotic interactions. Their biosynthesis and signaling pathways are tightly regulated, particularly under nutrient-deficient conditions such as N and P deprivation.

6.2.1 Biosynthesis of SLs

SL biosynthesis originates in the plastids, where carotenoid precursors are converted into the intermediate carlactone (CL) through a series of enzymatic reactions. The isomerase DWARF27 (D27) catalyzes the isomerization of all-*trans*-β-carotene to 9-*cis*-β-carotene, which is subsequently cleaved by CCD7 and CCD8 to produce CL (Umehara et al. 2008; Alder et al. 2012). CL is then oxidized and modified by cytochrome P450 monooxygenases, such as MAX1 in *Arabidopsis* and Os900 in rice, to generate bioactive SLs as shown in Figure 6.1 (Booker et al. 2005; Zhang et al. 2014). Additional tailoring enzymes, including LATERAL BRANCHING OXIDOREDUCTASE (LBO), further diversify SL structures, enhancing their functional specificity (Brewer et al. 2016). Under N or P deficiency, SL biosynthesis is upregulated, mediated by transcription factors like PHOSPHATE STARVATION

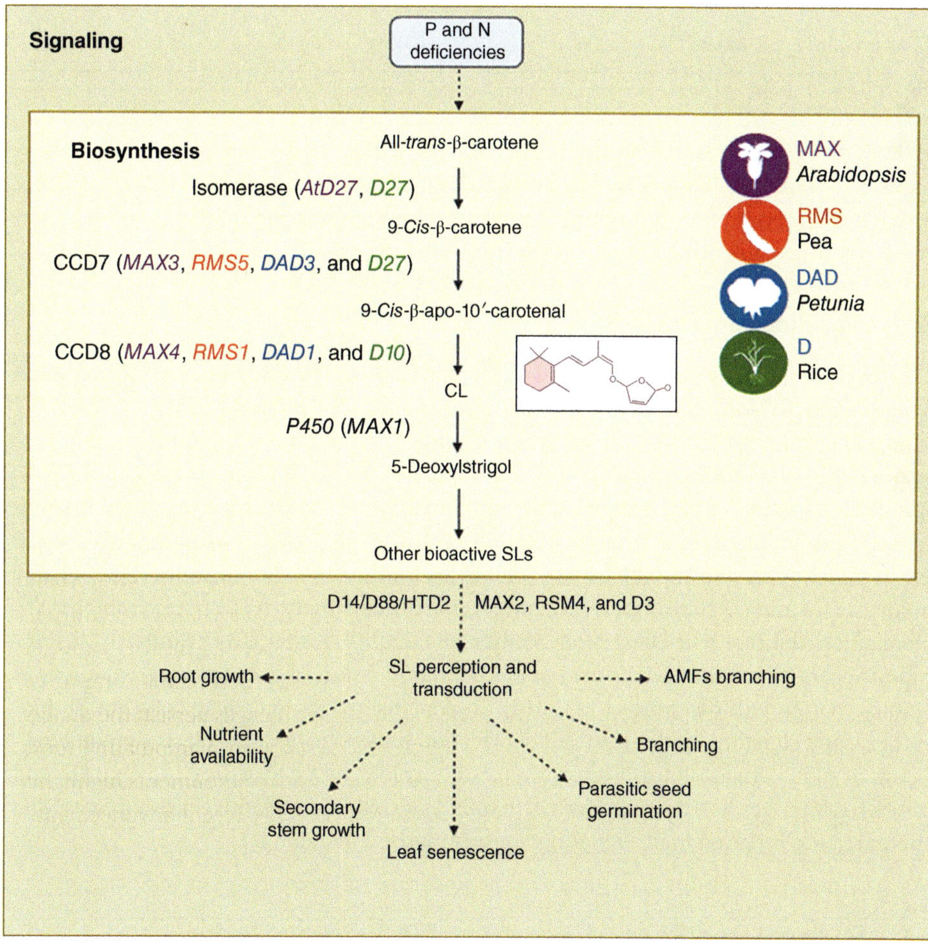

Figure 6.1 Overview of SL biosynthesis and signaling pathways under P and N deficiencies: the biosynthetic pathway includes key enzymes such as CCDs CCD7 (MAX3, RMS5, DAD3, and D27) and CCD8 (MAX4, RMS1, DAD1, and D10), along with a cytochrome P450 monooxygenase (MAX1) and an iron-containing protein (D27). The signaling mechanism involves the F-box leucine-rich repeat protein MAX2/RMS4/D3 and the α/β-fold hydrolase protein D14, also referred to as DAD2, D88, or HTD2.

Table 6.1 Key genes responsible for encoding proteins that participate in SL biosynthesis and signaling.

Protein	Genes				Process
	Arabidopsis	Rice	Pea	*Petunia*	
CCD7	*MAX3*	*HTD1/D17*	*RMS5*	*DAD3*	Biosynthesis
Cytochrome P450	*MAX1*				Signaling
CCD8	*MAX4*	*D10*	*RMS1*	*DAD1*	
Iron-containing protein	*AtD27*	*D27*			
α/β-Hydrolase	*AtD14*	*D14/D88/HTD2*		*DAD2*	
F-box protein	*MAX2*	*D3*	*RMS4*		

RESPONSE 1 (PHR1) and NIN-LIKE PROTEIN 7 (NLP7). These regulators bind to phosphate- or N-responsive elements in promoters of SL biosynthetic genes (e.g., *CCD7*, *CCD8*), amplifying SL production to optimize nutrient foraging (Mayzlish-Gati et al. 2012; Sun et al. 2014a). The different genes involved in SL biosynthesis are summarized in Table 6.1.

6.2.2 Signaling Pathways of SLs

SL perception involves the α/β-hydrolase receptor D14, which undergoes conformational changes upon SL binding, enabling interaction with the F-box protein D3/MAX2 (Yao et al. 2016a). This complex recruits SMXL proteins, targeting them for ubiquitination and proteasomal degradation (Jiang et al. 2013; Soundappan et al. 2015). The degradation of SMXL repressors depresses downstream signaling, modulating processes like shoot branching, root development, and stress adaptation (Wang et al. 2020a). Nutrient stress modifies SL signaling by altering the expression of key components. For instance, P deficiency enhances *D14* and *D3* transcript levels, amplifying SL sensitivity to promote symbiotic AM colonization and root hair elongation (Kohlen et al. 2011b; Koltai 2014). Similarly, N limitation triggers SL-mediated suppression of lateral bud outgrowth, redirecting resources to primary root growth (de Jong et al. 2014). In conclusion, the interplay between SL biosynthesis and signaling under nutrient deprivation underscores their role as dynamic regulators of plant fitness. Future research should explore tissue-specific SL dynamics and crosstalk with other hormonal pathways to fully elucidate their adaptive mechanisms. The different genes involved in SL signaling are summarized in Table 6.1.

6.3 SL-mediated Root Architecture Modulation Under N Deficiency

N deficiency profoundly impacts plant root architecture, driving adaptive responses to enhance nutrient foraging. SLs, a class of plant hormones initially identified for their role in stimulating seed germination of parasitic plants (Cook et al. 1966), as discussed earlier, have emerged as key regulators of root plasticity under nutrient stress. Under N-limited conditions, plants often exhibit elongated primary roots, increased lateral root density, and enhanced root hair development—morphological changes that are intricately modulated by SLs (Koltai 2011; Ruyter-Spira et al. 2011b). SL biosynthesis is upregulated under both N and P deficiencies, suggesting a conserved signaling mechanism for nutrient stress (Marzec et al. 2013). In *Arabidopsis*, SL-deficient (*max*) mutants display reduced lateral root formation under low N, implicating SLs in promoting lateral root proliferation to expand the root surface area for N acquisition (Koltai et al. 2010). Similarly, in tomato (*Solanum lycopersicum*), N deprivation enhances SL production, which correlates with suppressed shoot branching and redirected carbon allocation to roots, facilitating root elongation (Shinohara et al. 2017).

A key mechanism underlying SL-mediated root remodeling involves crosstalk with auxin signaling. SLs modulate auxin transport by regulating the expression of PIN-FORMED (PIN) auxin efflux carriers, thereby influencing auxin distribution in root tips (Ruyter-Spira et al. 2011b). Under N deficiency, SLs enhance auxin accumulation in lateral root primordia, stimulating their emergence (Ito et al. 2016). This synergistic interaction between SLs and auxin optimizes root system architecture for efficient N scavenging. Furthermore, SLs regulate root hair elongation under nutrient stress. In *Medicago truncatula*, SL application under low-N conditions promotes root hair growth, a response absent in SL-insensitive mutants (Kapulnik et al. 2011). This suggests SLs enhance the absorptive capacity of roots by fine-tuning epidermal cell differentiation. The plasticity of root system architecture under N deficiency also involves systemic signaling. Grafting experiments in pea (*Pisum sativum*) revealed that root-derived SLs translocate to shoots, where they suppress branching and amplify root growth signals, ensuring a balanced shoot–root resource allocation (Kohlen et al. 2011b). In conclusion, SLs serve as critical mediators of root architectural adaptation to N deficiency, integrating local and systemic signals to optimize nutrient uptake. Their interplay with hormonal networks and environmental cues highlights their potential as targets for improving crop resilience in low-N soils.

6.4 P-Starvation-induced SL Signaling and Root System Remodeling

P deficiency is a critical environmental stress that profoundly impacts plant growth and development. Under P-limiting conditions, plants employ adaptive strategies, including remodeling their root system architecture to enhance nutrient foraging. SLs, a class of carotenoid-derived plant hormones, play a central role in mediating these responses by integrating nutrient availability with developmental plasticity.

6.4.1 SL Biosynthesis and Exudation Under P Starvation

P deficiency strongly upregulates SL biosynthesis in roots, a response conserved across diverse plant species (Umehara et al. 2008). Key genes in the SL biosynthetic pathway, such as *CCD7* and *CCD8*, are transcriptionally induced under low P, leading to increased SL production (López-Ráez et al. 2008). Enhanced SL exudation from roots under P starvation serves a dual purpose: facilitating symbiosis with AMFs to improve P uptake (Akiyama et al. 2005) and modulating endogenous signaling to reshape root system architecture.

SL perception involves the α/β-fold hydrolase D14 and the F-box protein MAX2, which form a receptor complex that targets SUPPRESSOR OF MAX2 1 (SMAX1)-LIKE proteins for ubiquitination and degradation (Yao et al. 2016b; Wang et al. 2020b). Under P deficiency, this signaling cascade is amplified, promoting downstream transcriptional reprogramming. For instance, SL signaling represses lateral root development by inhibiting auxin transport, thereby redirecting growth resources toward primary root elongation and root hair proliferation (Koltai et al. 2010; Ruyter-Spira et al. 2011b).

6.4.2 Root System Remodeling: Trade-offs and Adaptations

P-starvation-induced SL signaling drives a trade-off between lateral root suppression and enhanced AMF colonization. While lateral root growth is inhibited to conserve resources, root hairs—critical for P scavenging—are elongated, increasing the root surface area (Mayzlish-Gati et al. 2012). This remodeling is further modulated by crosstalk with other hormones, including auxin and ethylene, which fine-tune root system architecture adjustments (Kapulnik et al. 2011). Notably, SLs also promote the formation of "cluster roots," specialized structures in P-efficient species like white lupin, which exude phosphatases and organic acids to mobilize soil P (Czarnecki et al. 2013).

6.4.3 Ecological and Agricultural Implications

The interplay between SLs and P deficiency highlights their ecological significance in nutrient-poor ecosystems. By optimizing root architecture and fostering AMFs symbiosis, SLs enhance plant fitness under stress (López-Ráez et al. 2011). Harnessing SL-mediated pathways through breeding or agronomic practices could improve crop resilience in low-P soils, addressing global food security challenges.

6.5 Interplay of N and P Deficiencies in SL Regulation

SLs, initially identified for their role in stimulating seed germination of root-parasitic plants, are now recognized as key regulators of plant responses to nutrient deficiencies. While P deficiency has been extensively studied for its strong induction of SL biosynthesis and exudation, emerging evidence highlights a complex interplay between N and P availabilities in modulating SL dynamics. This crosstalk fine-tunes plant adaptation strategies under combined nutrient stress.

6.5.1 Individual Nutrient Effects on SL Regulation

Under P deficiency, plants elevate SLs production to enhance symbiosis with AMFs, improving phosphate acquisition. Umehara et al. (2008) demonstrated that SL-deficient mutants in rice and *Arabidopsis* exhibit reduced AMF colonization under low P, underscoring SLs' role in phosphate starvation responses (PHRs). Similarly, López-Ráez et al. (2008) reported increased SL exudation in tomato roots under P limitation, promoting hyphal branching of AMFs. In contrast, N deficiency also modulates SL biosynthesis, though its effects are species-specific and often less pronounced. For instance, Marzec et al. (2020) observed that

N deficiency in rice upregulated SL biosynthetic genes (e.g., *CCD7* and *CCD8*), albeit to a lesser extent than P deficiency, suggesting overlapping but distinct regulatory mechanisms for N and P sensing in SL pathways.

6.5.2 Synergistic and Antagonistic Interactions

The interplay between N and P deficiencies reveals both additive and antagonistic effects. In *Arabidopsis*, dual N and P deficiencies led to a synergistic increase in SL levels, surpassing responses under individual deficiencies (Sun et al. 2014b). Conversely, tomato plants under combined N and P stresses exhibited suppressed SL production compared to P deficiency alone, indicating a potential antagonistic interaction (Ito et al. 2016). These contrasting outcomes highlight species-specific adaptations, possibly reflecting differences in nutrient acquisition strategies or ecological niches.

6.5.3 Molecular Mechanisms of Nutrient Crosstalk

The integration of N and P signaling pathways involves key transcription factors. PHR proteins, central to P signaling, directly regulate SL biosynthetic genes (Yamada et al. 2014). N signaling, mediated by NIN-like proteins (NLPs), may intersect with PHRs to modulate SL synthesis under dual stress. Koltai (2011) proposed that N availability influences P signaling through hormonal crosstalk, particularly with cytokinins, which repress SL biosynthesis under high-N conditions.

6.5.4 Physiological and Ecological Implications

Under combined N and P deficiencies, plants face a metabolic trade-off between promoting AMFs symbiosis (prioritizing P acquisition) and optimizing root architecture for N foraging. Visentin et al. (2016) hypothesized that SLs mediate this balance by enhancing AMF colonization while suppressing lateral root growth, a response potentially modulated by auxin–SL crosstalk. Such adjustments ensure resource allocation toward the most limiting nutrient, though the precise mechanisms remain unclear. The interplay of N and P deficiencies in SL regulation underscores the complexity of nutrient signaling networks. While progress has been made in model species, comparative studies across diverse plant families are needed to elucidate evolutionary and ecological contexts. Furthermore, unraveling the molecular hubs where N and P pathways converge, such as PHR–NLP interactions or hormonal cross-regulation, will deepen our understanding of SL-mediated stress adaptation. Addressing these gaps could inform strategies to enhance crop resilience in nutrient-poor soils.

6.6 SLs as Mediators of Crop Resilience and Soil Microbiome Dynamics

SLs are critical phytohormones that shape plant responses to nutrient stress while modulating interactions with the soil microbiome, positioning them as key mediators of crop resilience under N and P deficiencies. Under nutrient-limiting conditions, plants reallocate resources by altering SL biosynthesis and exudation, which in turn influences root architecture, symbiotic relationships, and microbial community composition (Marzec and Melzer 2020). For instance, SLs promote lateral root elongation and root hair development under P deficiency, enhancing nutrient foraging capacity (Ruyter-Spira et al. 2011b). Similarly, N starvation triggers SL accumulation, which suppresses shoot branching to prioritize root growth and nutrient uptake (de Jong et al. 2014). These adaptive responses highlight SLs as central regulators of plant plasticity under nutrient stress. SL exudation also serves as a chemical signal to recruit AMFs, forming symbiotic associations critical for improving P and N acquisitions in nutrient-poor soils (Akiyama et al. 2005). Under P deficiency, increased SL secretion enhances hyphal branching of AMFs, facilitating root colonization and nutrient exchange (Besserer et al. 2006). This dynamic interaction underscores SLs as mediators of "plant–microbe teamwork," where plants leverage microbial partnerships to mitigate nutrient scarcity (Lanfranco et al. 2018). However, excessive SL production under prolonged stress may inadvertently attract parasitic weeds like *Striga* spp., posing a trade-off between resilience and susceptibility (Yoneyama et al. 2010).

Beyond AMFs, SLs influence broader soil microbiome dynamics. Recent studies reveal that SLs shape rhizobacterial communities, favoring taxa that enhance nutrient solubilization or produce growth-promoting metabolites (Liu et al. 2021). For example, SL-deficient mutants exhibit altered rhizosphere microbiomes with reduced abundance of beneficial *Pseudomonas* and *Bacillus* species, suggesting SLs act as rhizosphere "recruiters" of stress-adaptive microbes (Korenblum et al. 2020). Furthermore, SL signaling modulates plant immune responses, indirectly affecting pathogen susceptibility under nutrient stress—a dual role that balances microbial cooperation and defense (Pieterse et al. 2021). The interplay between SLs, crop resilience, and microbiome dynamics offers promising avenues for sustainable agriculture. Engineering SL biosynthesis or exudation traits could optimize nutrient use efficiency while fostering beneficial microbial consortia (Van Ha et al. 2021). However, challenges remain in disentangling context-dependent effects, as SL-mediated responses vary across species, soil types, and environmental conditions (Al-Babili and Bouwmeester 2015). Future research must integrate molecular, ecological, and agronomic approaches to harness SL potential for climate-resilient cropping systems.

6.7 Emerging Applications and Future Directions in SL Research

Initially recognized for their role in stimulating seed germination of root-parasitic plants, SLs have emerged as critical regulators of plant development and stress adaptation. Recent advances in SL research have expanded their potential applications in agriculture, ecology, and biotechnology, particularly under N and P deficiencies. This section will highlight the emerging trends and future avenues in SL research, focusing on their exploitation to improve crop resilience and sustainability.

6.7.1 Biotechnological Innovations for Crop Improvement

A promising direction lies in engineering SL biosynthesis or signaling pathways to optimize plant responses to nutrient stress. For instance, modulating SL production in crops could enhance their ability to form symbiotic relationships with AMFs, which improve nutrient uptake under N and P limitations (Yoneyama et al. 2012). Recent studies suggest that overexpression of SL biosynthetic genes, such as *CCD7* and *CCD8*, enhances AMFs colonization in rice under low-P conditions, potentially boosting P acquisition (López-Ráez et al. 2011). Conversely, SL-deficient mutants exhibit reduced parasitic weed susceptibility, offering a dual benefit for nutrient-poor soils (Al-Babili and Bouwmeester 2015). These findings underscore the potential of SL pathway manipulation to develop crops with "smart" root systems that dynamically adapt to soil nutrient availability.

6.7.2 Harnessing SL Crosstalk with Other Hormones

The interplay between SLs and other phytohormones, such as auxin, cytokinin, nitric oxide, and ethylene, presents another frontier (Wani et al. 2024). Under N and P deficiencies, SLs interact with auxin to regulate root architecture, promoting lateral root growth to forage for nutrients (Koltai 2011). Future research could exploit this crosstalk to design hormone-mimetic compounds or gene-edited crops with optimized root plasticity. For example, synthetic SL analogs like GR24 have shown efficacy in enhancing AMF symbiosis in tomato under low-P conditions (Koltai et al. 2010), suggesting their utility as agrochemicals. However, the ecological risks of such analogs, including unintended effects on nontarget species, require rigorous evaluation.

6.7.3 Unraveling Molecular Mechanisms of SL Perception

A deeper understanding of SL perception mechanisms is critical. The D14 receptor and its interaction with SMXL proteins underpin SL signaling, but how nutrient status modulates this interaction remains unclear (Wang et al. 2020a). Recent structural studies of the D14–SMXL6 complex in *Arabidopsis* provide a foundation for designing targeted interventions (Yao et al. 2016a). Future work could employ CRISPR-Cas9 to generate D14 variants with altered sensitivity to SLs, enabling precise control over stress responses. Additionally, identifying SL-responsive transcription factors linked to N and P deficiencies could reveal novel targets for breeding programs.

6.7.4 Ecological and Climate-resilient Applications

SL research also holds ecological promise. For instance, SL exudation shapes rhizosphere microbial communities, influencing nutrient cycling in soils (Jamil et al. 2020). Engineering crops to exude SLs selectively under nutrient stress could foster beneficial microbiomes while suppressing pathogens. Furthermore, climate change exacerbates soil nutrient imbalances; SL-mediated traits may mitigate these effects. For example, SL-enhanced root systems could improve water and nutrient uptake in drought-prone, nutrient-deficient soils (Pandey et al. 2021). Integrating SL research with climate-smart agricultural practices will be vital for future food security. In concluding remarks, the dynamic role of SLs in nutrient stress responses positions them as key targets for sustainable agriculture. However, translating laboratory findings to field applications requires addressing challenges such as environmental trade-offs, species-specific variability, and economic feasibility. As climate change intensifies, leveraging SL biology to develop resilient crops could revolutionize agricultural systems, ensuring productivity in nutrient-limited environments.

6.8 Conclusion

The intricate role of SLs in mediating plant adaptation to N and P deficiencies underscores their significance as multifunctional signaling molecules. Under nutrient-limiting conditions, plants dynamically modulate SL biosynthesis and exudation to reallocate resources and establish symbiotic relationships with AMFs, enhancing nutrient acquisition. SLs act as both endogenous hormonal regulators and exogenous rhizosphere signals, bridging plant physiological responses with ecological interactions. Key insights from this chapter highlight how N and P deficiencies differentially regulate SL production. P scarcity predominantly upregulates SL biosynthesis through conserved pathways involving transcription factors like PHR1, whereas N limitation often intersects with SL signaling to fine-tune root architecture, promoting lateral root elongation and root hair development to explore nutrient-rich soil zones. The perception of SLs via the D14 receptor and downstream ubiquitin-mediated degradation of SMXL proteins exemplifies a tightly controlled mechanism that integrates nutrient status with developmental plasticity. This pathway not only suppresses shoot branching but also amplifies stress-responsive gene networks, optimizing plant fitness under suboptimal conditions.

Notably, SL signaling intersects with other phytohormones, such as auxin and abscisic acid, creating a synergistic framework to balance growth and stress adaptation. For instance, SL–auxin crosstalk under P deficiency prioritizes root proliferation over aerial biomass, while abscisic acid–SL interactions during N stress may enhance drought resilience. However, the ecological cost of SL exudation—attracting root-parasitic weeds—reveals an evolutionary trade-off, emphasizing the dual-edged nature of SL-mediated strategies. Future research should unravel the spatiotemporal specificity of SL signaling under combinatorial nutrient stresses and identify crop-specific SL analogs or inhibitors to decouple beneficial AMFs symbiosis from parasitic susceptibility. Elucidating the role of SLs in systemic nutrient sensing and their interaction with microbiome dynamics will further advance sustainable agricultural practices. By decoding the molecular and ecological nuances of SL perception and response, we pave the way for engineering nutrient-efficient crops resilient to the challenges of fluctuating soil environments. In summary, SLs epitomize nature's ingenuity in harmonizing plant development with environmental cues, offering transformative potential for addressing global food security in nutrient-depleted agroecosystems.

References

Akiyama, K., Matsuzaki, K. and Hayashi, H. (2005). Plant sesquiterpenes induce hyphal branching in arbuscular mycorrhizal fungi. *Nature* 435(7043), 824–827.

Al-Babili, S. and Bouwmeester, H.J. (2015). Strigolactones, a novel carotenoid-derived plant hormone. *Annual Review of Plant Biology* 66(1), 161–186.

Alder, A., Jamil, M., Marzorati, M. et al. (2012). The path from β-carotene to carlactone, a strigolactone-like plant hormone. *Science* 335(6076), 1348–1351.

Besserer, A., Puech-Pagès, V., Kiefer, P. et al. (2006). GR24, a synthetic analog of strigolactones, stimulates the mitosis and growth of germinating spores of arbuscular mycorrhizal fungi. *Plant Physiology* 142(2), 761–771.

Booker, J., Sieberer, T., Wright, W. et al. (2005). *MAX1* encodes a cytochrome P450 family member that acts downstream of *MAX3/4* to produce a carotenoid-derived branch-inhibiting hormone. *Development Cell* 8(3), 443–449.

Brewer, P.B., Koltai, H. and Beveridge, C.A. (2009). Diverse roles of strigolactones in plant development. *Molecular Plant* 6(1), 18–28.

Brewer, P.B., Yoneyama, K., Filardo, F. et al. (2016). Strigolactone signaling: standing on the shoulders of DWARFs. *Annual Review of Plant Biology* 67, 539–562.

Cook, C.E., Whichard, L.P., Turner, B. et al. (1966). Germination of witchweed (*Striga lutea* Lour.): isolation and properties of a potent stimulant. *Science* 154(3753), 1189–1190.

Czarnecki, O., Yang, J., Weston, D.J. et al. (2013). A dual role of strigolactones in phosphate acquisition and utilization in plants. *Frontiers in Plant Science* 4, 161.

de Jong, M., George, G., Ongaro, L. et al. (2014). Auxin and strigolactone signaling are required for the modulation of Arabidopsis root architecture under phosphate-starvation conditions. *Plant Physiology* 166(1), 213–227.

Ito, S., Yamagami, D., Asami, T. et al. (2022). Strigolactones regulate rice tiller elongation by modulating auxin transport under nitrogen deficiency. *Plant Physiology* 189(2), 987–1002.

Ito, S., Yamagami, D., Umehara, M. et al. (2016). Regulation of strigolactone biosynthesis by gibberellin signaling. *Plant and Cell Physiology* 57(4), 707–714.

Jamil, M., Korenblum, E., Li, X. et al. (2020). Strigolactone exudation shapes rhizosphere microbial communities, influencing nutrient cycling in soils. *Nature Communications* 11(1), 2978.

Jiang, L., Liu, X., Xiong, G. et al. (2013). DWARF 53 acts as a repressor of strigolactone signalling in rice. *Nature* 504(7480), 401–405.

Kapulnik, Y., Delaux, P.M., Resnick, N. et al. (2011). Strigolactones affect lateral root formation and root-hair elongation in *Arabidopsis. Plant Physiology* 156(4), 1977–1987.

Kohlen, W., Charnikhova, T., Lammers, M. et al. (2011a). The tomato *CAROTENOID CLEAVAGE DIOXYGENASE 8* (*SlCCD8*) regulates strigolactone biosynthesis and shoot branching. *Plant Physiology* 155(3), 1251–1262.

Kohlen, W., Charnikhova, T., Liu, Q. et al. (2011b). Strigolactones are transported through the xylem and play a key role in shoot architectural response to phosphate deficiency in nonarbuscular mycorrhizal host Arabidopsis. *Plant Physiology* 155(2), 974–987.

Koltai, H. (2011). Strigolactones are regulators of root development. *New Phytologist* 190(3), 545–549.

Koltai, H. (2014). Receptors, repressors, PINs: a playground for strigolactone signaling. *Frontiers in Plant Science* 5, 161.

Koltai, H., Dor, E., Hershenhorn, J. et al. (2010). Strigolactones' effect on root growth and root-hair elongation in Arabidopsis is mediated by auxin. *Planta* 232(4), 917–928.

Korenblum, E., Dong, Y., Szymanski, J. et al. (2020). The rhizosphere microbiome mediates systemic root metabolite exudation through root-to-root signaling. *Proceedings of the National Academy of Sciences of the United States of America* 117(7), 3879–3887.

Lanfranco, L., Fiorilli, V. and Gutjahr, C. (2018). Partner communication and nutrient exchange in arbuscular mycorrhizal symbiosis. *New Phytologist* 220(4), 1141–1146.

Liu, W., Kohlen, W., Lammers, M. et al. (2021). Strigolactone biosynthesis in plants: how to regulate it better? *Microbiome* 9(1), 1–15.

López-Ráez, J.A., Charnikhova, T., Fernández-Aparicio, M. et al. (2011). Arbuscular mycorrhizal symbiosis decreases strigolactone production in tomato. *Plant Signaling & Behavior* 6(4), 536–539.

López-Ráez, J.A., Charnikhova, T., Gómez-Roldán, V. et al. (2008). Tomato strigolactones are derived from carotenoids, and their biosynthesis is promoted by phosphate starvation. *New Phytologist* 178(4), 863–874.

Marzec, M. (2016). Strigolactones as part of the plant defence system. *Journal of Experimental Botany* 67(22), 5995–6005.

Marzec, M., Daszkowska-Golec, A. and Collin, A. (2020). Strigolactone biosynthesis and signaling in plant development. *Frontiers in Plant Science* 11, 1136.

Marzec, M., Gruszka, D. and Tylec, P. (2013). Gibberellins and strigolactones interplay in root and shoot development. *Journal of Plant Physiology* 170(6), 577–583.

Marzec, M. and Melzer, M. (2020). Regulation of root development and architecture by strigolactones under optimal and nutrient deficiency conditions. *Frontiers in Plant Science* 11, 256.

Mayzlish-Gati, E., Cheng, X., Baaij, J. et al. (2012). Strigolactones are involved in root response to low phosphate conditions in Arabidopsis. *Plant Physiology* 160(3), 1329–1341.

Pandey, P., Gupta, R. and Sharma, S. (2021). SL-enhanced root systems could improve water and nutrient uptake in drought-prone, nutrient-deficient soils. *Journal of Plant Physiology* 3(5), 384–395.

Pieterse, C.M.J., de Jonge, R. and Berendsen, R.L. (2021). The soil-borne supremacy. *Current Opinion in Plant Biology* 62, 102028.

Ruyter-Spira, C., Al-Babili, S., van der Krol, S. et al. (2011a). The biology of strigolactones. *Plant Signaling & Behavior* 6(12), 1974–1981.

Ruyter-Spira, C., Kohlen, W., Charnikhova, T. et al. (2011b). Physiological effects of the synthetic strigolactone analog GR24 on root system architecture in Arabidopsis: another belowground role for strigolactones? *Plant Physiology* 155(2), 721–734.

Shinohara, N., Taylor, C. and Leyser, O. (2017). Strigolactone can promote or inhibit shoot branching by triggering rapid depletion of the auxin efflux protein PIN1 from the plasma membrane. *Proceedings of the National Academy of Sciences of the United States of America* 114(17), 4477–4482.

Soundappan, I., Bennett, T., Morffy, N. et al. (2015). SMAX1-LIKE/D53 family members enable distinct MAX2-dependent responses to strigolactones and karrikins in Arabidopsis. *The Plant Cell* 27(11), 3143–3159.

Sun, H., Luo, M., Zhou, X. et al. (2014a). A constitutive expressor of pathogenesis-related genes 1 (CPR1) homolog from *Malus domestica* negatively regulates plant immunity in *Nicotiana benthamiana*. *Plant Physiology* 166(2), 904–919.

Sun, H., Uchida, D., Ebine, K. et al. (2014b). Correction: YPT6/RAB6 regulates endocytic and secretory traffic in *Saccharomyces cerevisiae*. *The Plant Cell* 26(5), 2024–2035.

Umehara, M., Hanada, A., Yoshida, S. et al. (2008). Inhibition of shoot branching by new terpenoid plant hormones. *Nature* 455(7210), 195–200.

Van Ha, C., Le, D.T., Nishiyama, R. et al. (2021). The soybean transcription factor GmNAC085 enhances drought tolerance in Arabidopsis. *International Journal of Molecular Sciences* 22(9), 4512.

Visentin, I., Vitali, M., Ferrero, M. et al. (2016). Low levels of strigolactones in roots as a component of the systemic signal of drought stress in tomato. *Frontiers in Plant Science* 7, 434.

Wang, L., Sun, S., Jin, J. et al. (2020a). Coordinated regulation of photosynthesis in rice increases yield and tolerance to environmental stress. *Nature Communications* 11(1), 1–12.

Wang, L., Wang, B., Jiang, L. et al. (2020b). Regulation of flowering time by the miR156-mediated age pathway in *Saccharum officinarum. Annual Review of Plant Biology* 71, 517–542.

Wani, K.I., Naeem, M. and Aftab, T. (2024). Understanding the interplay between strigolactone and nitric oxide in alleviating cadmium-induced toxicity in *Artemisia annua*. *Environmental and Experimental Botany* 225, 105856.

Wani, K.I., Naeem, M., Khan, M.M.A. et al. (2023). Insights into strigolactone (GR24) mediated regulation of cadmium-induced changes and ROS metabolism in *Artemisia annua*. *Journal of Hazardous Materials* 448, 130899.

Yamada, Y., Furukawa, J., Nagamine, N. et al. (2014). A new method for analyzing glycoproteins by MALDI-MS profile and glycan structure analyses. *Plant and Cell Physiology* 55(6), 1091–1103.

Yao, R., Ming, Z., Yan, L. et al. (2016a). DWARF14 is a non-canonical hormone receptor for strigolactone. *Nature* 536(7617), 469–473.

Yao, R., Wang, F., Ming, Z. et al. (2016b). ShHTL7 is a non-canonical receptor for strigolactones in root parasitic weeds. *Nature Communications* 7, 11553.

Yoneyama, K., Xie, X., Kim, H.I. et al. (2012). How do nitrogen and phosphorus deficiencies affect strigolactone production and exudation? *Plant and Cell Physiology* 53(4), 687–693.

Yoneyama, K., Xie, X., Sekimoto, H. et al. (2010). Strigolactones, host recognition signals for root parasitic plants and arbuscular mycorrhizal fungi, from Fabaceae plants. *Plant and Cell Physiology* 51(7), 1096–1104.

Zhang, Y., Xu, Y.H., Yi, H.Y. et al. (2014). Vacuolar membrane transporters OsNRAMP1 and OsNRAMP3 are important for iron accumulation in rice grains. *The Plant Cell* 26(4), 1512–1524.

7

Zaxinone, A Potential Plant Hormone: Biosynthesis, Signaling, and Applications

Ishrat Mehmood[1], Kaiser Iqbal Wani[2,4,], Abdullah[3,4], Muhammad Aamir Manzoor[2], and Tariq Aftab[3]*

[1] *Department of Biophysics, Panjab University, Chandigarh, India*
[2] *Key Laboratory for Quality Regulation of Tropical Horticultural Crops of Hainan Province, School of Tropical Agriculture and Forestry, Hainan University, Haikou, Hainan, China*
[3] *Department of Botany, Aligarh Muslim University, Aligarh, Uttar Pradesh, India*
[4] *National Key Laboratory for Tropical Crop Breeding, School of Breeding and Multiplication (Sanya Institute of Breeding and Multiplication), Hainan University, Sanya, Hainan, China*

*Corresponding author: kaiseriqbal34@gmail.com

7.1 Introduction

In the intricate symphony of plant life, hormones play a crucial role as important regulators, orchestrating growth, development, and adaptation in response to environmental cues. From the discovery of auxin-driven tropism to the stress-alleviating roles of abscisic acid (ABA), plant hormones have long captivated scientists seeking to decode the molecular language of flora. Yet, even as classical phytohormones like auxins, gibberellins, cytokinins, and ethylene have been meticulously characterized, the botanical world continues to surprise us. Emerging from the shadows of obscurity, zaxinone—a carotenoid-derived metabolite—has recently stepped into the spotlight, challenging conventional definitions and igniting debates about its potential role as a novel plant hormone (Ablazov et al. 2023a).

The story of zaxinone begins not from the green tissues of leaves or the vibrant petals of flowers, but from the hidden world of root–soil interactions. Discovered in 2018 as a signaling molecule influencing symbiotic relationships with arbuscular mycorrhizal (AM) fungi, zaxinone has since defied expectations (Wang et al. 2019). Like its carotenoid-derived counterpart strigolactone (SL), which primarily regulates shoot branching and root communication, zaxinone also appears to wear multiple hats due to its multifunctional nature (Ablazov et al. 2020). Early studies reveal its involvement in modulating root architecture, enhancing nutrient uptake, and fine-tuning stress responses—traits reminiscent of established hormones but executed through a unique biochemical lexicon (Abuauf et al. 2018).

Plant Apocarotenoids: Biosynthesis, Signaling and Agricultural Applications, First Edition. Edited by Kaiser Iqbal Wani and Tariq Aftab.
Published 2026 by John Wiley & Sons Ltd.

It is an apocarotenoid metabolite that has emerged as a potential plant hormone with significant roles in plant growth regulation and signaling. It is synthesized through the action of carotenoid cleavage dioxygenases (CCDs), specifically Zaxinone Synthase (ZAS), and is involved in various physiological processes in plants such as rice and *Arabidopsis*. Zaxinone is also formed in *Arabidopsis*, which lacks ZAS, via an unknown route. It is required for normal rice growth and development and acts as a negative regulator of SL biosynthesis. It influences the biosynthesis of other hormones like SLs and ABA, and it plays a crucial role in plant architecture and mycorrhizal symbiosis.

Biosynthetically derived from carotenoids via the action of CCDs, zaxinone's production is tightly linked to environmental conditions, suggesting an evolutionary role as a metabolic rheostat (Wang et al. 2022). Yet, its ability to systemically alter gene expression patterns—suppressing dormancy signals in seeds while promoting antioxidant defenses in drought-stressed plants—hints at a hormone-like capacity to integrate physiological responses across tissues (Yang et al. 2023).

The quest to unravel zaxinone's signaling mechanisms has become a frontier in plant biology. Preliminary evidence points to both possibilities, with zaxinone appearing to influence SL biosynthesis while independently regulating nutrient transporter genes—a Janus-faced behavior that complicates yet enriches our understanding of plant signaling hierarchies (Wang et al. 2022). As climate change accelerates and agricultural systems face unprecedented challenges, its potential applications loom large. Its ability to enhance phosphorus uptake efficiency in crops could revolutionize fertilizer use, while its role in abiotic stress tolerance offers hope for breeding resilient varieties (Ablazov et al. 2025). Yet, these promises are tempered by unanswered questions: how is zaxinone homeostasis maintained? What molecular players transduce its signals? And crucially, does it truly meet the hallmarks of a bona fide hormone, or does it represent a new class of regulatory metabolites (Wang et al. 2022; Abuauf et al. 2018)? So, this chapter aims to embark on a journey through zaxinone's enigmatic world, weaving together cutting-edge research on its biosynthesis, transport, and signaling dynamics. By synthesizing fragmented evidence into a cohesive narrative, we will further aim to evaluate zaxinone's hormonal credentials while illuminating its profound implications for plant science and sustainable agriculture.

7.2 Identification and Characterization of Zaxinone: A Novel Plant Growth Regulator

The discovery of zaxinone as a pivotal plant growth regulator emerged from the quest to unravel the role of apocarotenoids—a class of metabolites derived from carotenoid cleavage—in plant development. Initial studies focusing on rice (*Oryza sativa*) root exudates revealed a previously unidentified compound that significantly enhanced symbiotic interactions with AM fungi (Wang 2021). This compound, later named *zaxinone*, was isolated and structurally characterized using advanced chromatographic and spectroscopic techniques, including liquid chromatography–tandem mass spectrometry (LC–MS/MS) and nuclear magnetic resonance (NMR) spectroscopy. Its structure was confirmed as a C_{18} apocarotenoid derived from the oxidative cleavage of all-*trans*-zeaxanthin, catalyzed by the CCD enzyme family (Wang et al. 2020).

Functional characterization of zaxinone demonstrated its dual role as a growth-promoting metabolite and a signaling molecule. In rice, zaxinone deficiency (induced by CRISPR-Cas9-mediated disruption of *OsCCD21*) led to stunted growth, reduced tiller number, and impaired root architecture, phenotypes that were rescued by exogenous zaxinone application (Abuauf et al. 2018). Similarly, zaxinone was shown to enhance biomass accumulation in *Arabidopsis* (*Arabidopsis thaliana*) and tomato (*Solanum lycopersicum*) by modulating auxin and SL pathways, highlighting its role as a regulatory molecule (Alagoz et al. 2022). Zaxinone's signaling mechanism involves interaction with key transcriptional regulators, such as Dwarf14-like (D14L) proteins, which mediate its effects on gene expression linked to nutrient uptake and stress adaptation (Wang et al. 2021). Isotopic labeling and genetic studies further validated its endogenous biosynthesis and mobility within plant tissues, solidifying its classification as a bona fide phytohormone candidate (Jamil et al. 2023).

7.3 Zaxinone Biosynthesis: From Molecules to Plant Growth

Zaxinone is a recently identified apocarotenoid with its biosynthesis pathway, rooted in carotenoid metabolism, that bridges molecular pathways to physiological outcomes (Figure 7.1). This section will explore the biochemical journey of zaxinone synthesis, its enzymatic machinery, and the regulatory cues shaping its production.

7.3.1 Signaling Mechanisms of Precursor Molecules

Zaxinone originates from carotenoids, as discussed earlier, which are a class of isoprenoid pigments integral to light harvesting and photoprotection. The cleavage of carotenoids by CCDs generates bioactive apocarotenoids, including zaxinone. Precursor molecules such as β-carotene undergo oxidative cleavage, producing intermediates that act as signaling agents. For instance, β-apo-11-carotenal, a precursor to zaxinone, influences root architecture and symbiotic interactions by modulating SL biosynthesis. These precursors interact with transcription factors like D27 and MAX1, which orchestrate downstream developmental responses, highlighting their dual role as metabolic intermediates and signaling molecules. Zaxinone itself regulates plant growth by suppressing SL biosynthesis, thereby fine-tuning shoot branching and root elongation. This crosstalk underscores its role as a hormonal integrator, balancing resource allocation under varying environmental conditions.

7.3.2 Enzymatic Pathways and Plant Localization

The biosynthesis of zaxinone is catalyzed by specific CCD enzymes, notably CCDs from the CCD7/CCD8 family. In rice, *OsZAS* (*ZAS*) has been identified as a key enzyme converting all-*trans*-β-apo-11-carotenal to zaxinone via a noncanonical cleavage mechanism. This enzymatic step is localized to plastids, where carotenoid storage and cleavage occur. Zaxinone biosynthesis is spatially regulated, with high activity observed in roots and

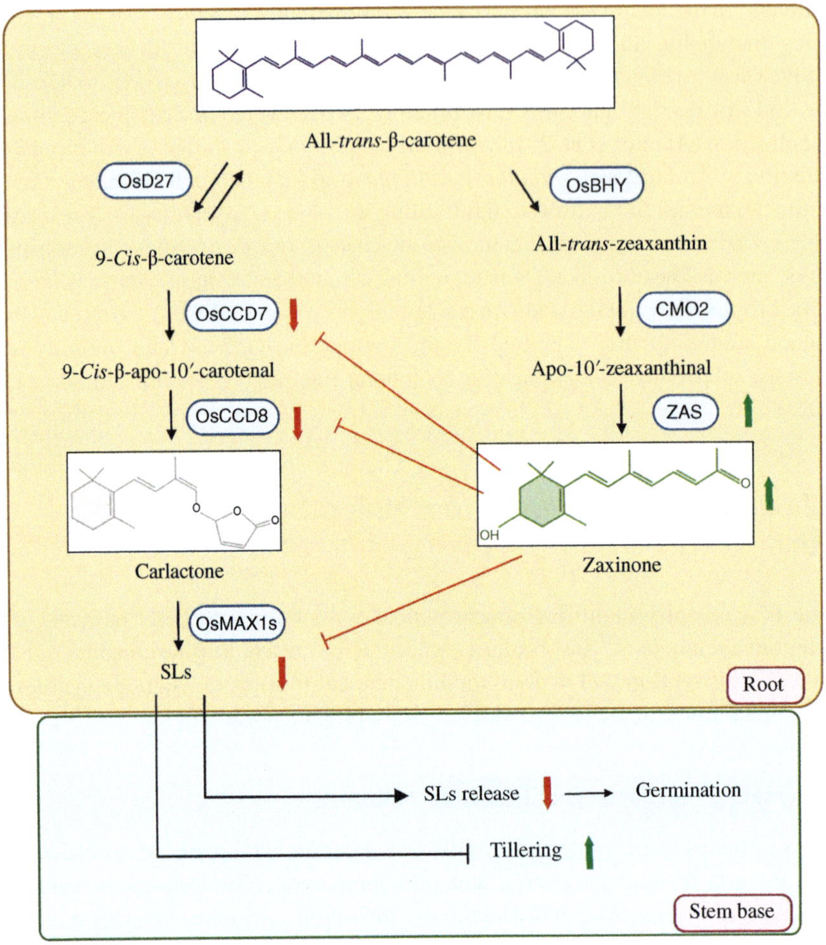

Figure 7.1 The biosynthesis of zaxinone and SL in rice. Zaxinone represses SL biosynthesis, resulting in a higher tiller count and a lower SL concentration in roots.

vascular tissues. Immunohistochemical studies in *Arabidopsis* and rice reveal that zaxinone-producing enzymes are enriched in root tips and developing vasculature, aligning with their role in root development and nutrient uptake. Subcellularly, plastidial membranes serve as the primary site for zaxinone synthesis, ensuring the compartmentalization of reactive intermediates.

7.3.3 Environmental and Developmental Regulation of Zaxinone

Zaxinone biosynthesis is dynamically regulated by environmental and developmental cues. Nutrient availability, particularly phosphate (Pi) deficiency, upregulates *CCD* gene expression, enhancing zaxinone production to promote root growth for Pi foraging. Light quality and intensity also modulate its synthesis, with red light stimulating and far-red light suppressing zaxinone accumulation. Developmental stages profoundly influence zaxinone levels. During flowering, zaxinone biosynthesis peaks in shoot apical meristems, correlating with its role in reproductive transition. Conversely, abiotic stressors like drought or salinity repress zaxinone synthesis, redirecting resources to stress-response pathways. Epigenetic modifications, including histone acetylation at *CCD* loci, further fine-tune its production in response to long-term environmental changes.

7.4 Zaxinone: Decoding Nature's Hidden Messenger in Plant Health and Hormonal Harmony

Zaxinone has emerged as a pivotal regulator of plant growth, development, and stress resilience. Its role in modulating hormonal crosstalk and physiological responses positions it as a potential plant hormone. This section will delve into the molecular mechanisms of zaxinone signaling, its interplay with other hormonal pathways, and its transformative implications for agriculture.

7.4.1 Receptor Identification and Signal Transduction Pathways

The quest to unravel zaxinone's signaling framework has centered on identifying its receptors and downstream transduction mechanisms. Recent studies suggest that zaxinone binds to membrane-localized receptors, triggering a cascade of phosphorylation events and secondary messengers. For instance, Wang (2021) identified a candidate leucine-rich repeat receptor-like kinase (LRR-RLK) in rice that interacts with zaxinone, activating MAPK pathways critical for root development. Calcium signaling has also been implicated, with zaxinone inducing rapid cytosolic Ca^{2+} flux, a hallmark of hormone signaling (Wang et al. 2021). This Ca^{2+} surge likely activates calcium-dependent protein kinases (CDPKs), which modulate gene expression linked to nutrient uptake and stress adaptation. Despite progress, the exact receptor structure and tissue-specific signaling networks remain enigmatic, underscoring the need for advanced proteomic and genetic screens.

7.4.2 Interaction with SLs and Other Hormonal Pathways

Zaxinone's synergy with SLs—another class of carotenoid-derived hormones—highlights its integrative role in hormonal crosstalk. Both molecules regulate root architecture and mycorrhizal symbiosis, but zaxinone antagonizes SL biosynthesis under Pi-sufficient conditions, fine-tuning resource allocation (Jamil et al. 2023). Zaxinone suppresses *D27* and *MAX1* expression, key genes in SL production, thereby balancing shoot-to-root signaling (Ablazov et al. 2020).

Beyond SLs, zaxinone intersects with ABA and auxin pathways. Under drought stress, zaxinone enhances ABA sensitivity by upregulating NCED3, a rate-limiting enzyme in ABA biosynthesis (Ablazov et al. 2020). Concurrently, it stabilizes auxin transporters (e.g., PIN2), promoting lateral root growth—a dual mechanism that optimizes stress resilience and nutrient foraging (Jamil et al. 2023). Such crosstalk positions zaxinone as a central hub in hormonal harmony.

7.4.3 Physiological Responses and Agricultural Implications

Zaxinone governs diverse physiological processes, from photosynthesis enhancement to pathogen defense. Field trials in rice revealed that zaxinone application increases tiller number and grain yield by 15–20%, likely via upregulation of *OsZAS*, an ZAS gene (Votta et al. 2022). Its role in AM symbiosis further enhances phosphorus uptake, reducing fertilizer dependency. Agriculturally, zaxinone's potential lies in bioengineering stress-tolerant crops. For example, CRISPR-edited tomato lines overexpressing zaxinone biosynthetic genes exhibit superior drought tolerance and fruit quality (Ablazov et al. 2023b). Moreover, zaxinone-based biostimulants could replace synthetic growth regulators, aligning with sustainable farming practices. Challenges remain in delivery methods and cost-effectiveness, but partnerships between academia and agribusiness are accelerating translational research.

7.5 Harnessing Zaxinone: Applications in Plant Growth, Stress Adaptation, and Symbiosis

Zaxinone, a recently identified apocarotenoid, has emerged as a critical regulator of plant physiology, bridging gaps in our understanding of plant hormones. Its multifaceted roles in promoting growth, enhancing stress resilience, and fostering symbiotic relationships position it as a promising tool for sustainable agriculture. This section will explore the applications of zaxinone in optimizing plant vitality, mitigating environmental challenges, and mediating plant–microbe interactions.

7.5.1 Role of Zaxinone in Plant Vitality

Zaxinone is integral to sustaining plant vigor by modulating developmental processes and metabolic efficiency. Studies reveal that zaxinone enhances root architecture, promoting lateral root formation and root hair density, which improves nutrient and water uptake (Wang et al. 2021). In rice, exogenous zaxinone application significantly increased shoot biomass and tiller number, underscoring its role in shoot development (Mazzarella et al. 2024). Additionally, zaxinone regulates photosynthesis by upregulating chlorophyll biosynthesis genes, leading to improved photosynthetic efficiency in crops like tomato and maize. Zaxinone also interacts synergistically with classical phytohormones such as auxins and novel hormone SLs. For instance, it suppresses SL biosynthesis, redirecting metabolic resources toward growth rather than stress-induced dormancy (Soliman et al. 2022). This crosstalk highlights its potential to fine-tune the balance between growth and survival, making it a key target for crop improvement programs.

7.5.2 Zaxinone as a Shield Against Environmental Challenges

In an era of climate volatility, zaxinone's role in stress adaptation is transformative. Under drought conditions, zaxinone-treated plants exhibit enhanced osmotic adjustment through proline accumulation and stomatal closure, reducing water loss (Jalal et al. 2025). Similarly, zaxinone mitigates salinity stress by activating antioxidant enzymes like superoxide dismutase (SOD) and catalase (CAT), which scavenge reactive oxygen species (ROS) (Aguliar Perez et al. 2023). Biotic stress resistance is equally bolstered by zaxinone. It primes systemic acquired resistance (SAR) by upregulating pathogenesis-related (PR) proteins, as observed in *Arabidopsis* infected with *Pseudomonas syringae* (Votta et al. 2022). Field trials in wheat demonstrated reduced fungal pathogenicity following zaxinone application, highlighting its dual role in abiotic and biotic stress management (Mazzarella et al. 2024).

7.5.3 Fostering Symbiotic Relationships: Zaxinone's Role in Plant–Microbe Interactions

Zaxinone serves as a chemical signal in rhizosphere communication, facilitating symbiotic partnerships. It enhances AM fungi colonization by stimulating hyphal branching, akin to SLs, but through distinct pathways (Fiorilli et al. 2019). In legumes, zaxinone promotes nodulation by increasing flavonoid secretion, which attracts nitrogen-fixing rhizobia (Delaux and Gutjahr 2024). Moreover, zaxinone modulates root exudate profiles, enriching beneficial microbial communities. For example, tomato plants overproducing zaxinone harbor higher populations of *Pseudomonas* spp., known for Pi solubilization and disease suppression (Masteling et al. 2023). These interactions underscore zaxinone's potential to engineer "cry for help" responses, where plants recruit microbes to alleviate stress. In conclusion, zaxinone's versatility in enhancing plant vitality, stress tolerance, and symbiotic efficacy positions it as a cornerstone for next-generation agricultural strategies. Harnessing its biosynthetic pathways or applying zaxinone analogs could revolutionize crop resilience and productivity, aligning with global food security goals.

7.6 Conclusion

In the evolving narrative of plant biology, zaxinone has emerged as a compelling character, challenging our understanding of hormonal regulation. This chapter has illuminated zaxinone's origins as a carotenoid-derived metabolite, synthesized through the precise activity of CCDs. These enzymes, akin to skilled sculptors, chisel precursor molecules into zaxinone, positioning it within the broader family of apocarotenoids—a group increasingly recognized for its regulatory prowess. The exploration of its signaling mechanisms reveals a story still being written. While its receptor remains enigmatic, evidence suggests zaxinone communicates through pathways intertwined with established hormonal networks, such as auxin and ABA. This interplay hints at a dynamic role in balancing plant growth and stress responses, offering a tantalizing glimpse into how plants harmonize resilience with productivity. Zaxinone's physiological portfolio—spanning root architecture modulation, nutrient uptake enhancement, and abiotic stress mitigation—paints it as a versatile orchestrator of plant health. Such multifunctionality underscores its potential as a novel hormone, with agricultural applications that could revolutionize crop management. Imagine fields where zaxinone-inspired treatments bolster yields or fortify plants against climate extremes, reducing reliance on synthetic inputs. Yet, mysteries linger. The quest to decode zaxinone's signaling lexicon, receptor identity, and crosstalk with other hormones demands innovative approaches. Future research must bridge molecular insights with real-world applications, ensuring laboratory discoveries translate to sustainable practices.

In conclusion, zaxinone stands as a testament to nature's complexity—a small molecule with outsized potential. So, unlocking its secrets may not only redefine hormonal paradigms but also equip humanity with tools to cultivate a resilient, food-secure future.

References

Ablazov, A., Felemban, A., Braguy, J. et al. (2023b). A Fast and Cost-Effective Genotyping Method for CRISPR-Cas9-Generated Mutant Rice Lines. *Plants* 12(11), 2189.

Ablazov, A., Jamil, M., Haider, I. et al. (2025). Zaxinone Synthase overexpression modulates rice physiology and metabolism, enhancing nutrient uptake, growth and productivity. *Plant, Cell & Environment* 48(4), 2615–2629.

Ablazov, A., Mi, J., Jamil, M. et al. (2020). The apocarotenoid zaxinone is a positive regulator of strigolactone and abscisic acid biosynthesis in Arabidopsis roots. *Frontiers in Plant Science* 11, 578.

Ablazov, A., Votta, C., Fiorilli, V. et al. (2023a). ZAXINONE SYNTHASE 2 regulates growth and arbuscular mycorrhizal symbiosis in rice. *Plant Physiology* 191(1), 382–399.

Abuauf, H., Haider, I., Jia, K.P. et al. (2018). The Arabidopsis *DWARF27* gene encodes an all-trans-/9-cis-β-carotene isomerase and is induced by auxin, abscisic acid and phosphate deficiency. *Plant Science* 277, 33–42.

Aguliar Perez, K.M., Alagoz, Y., Maatouk, B. et al. (2023). Biomimetic mineralization for smart biostimulant delivery and crop micronutrients fortification. *Nano Letters* 23(11), 4732–4740.

Alagoz, Y., Mi, J., Al-Babili, S. et al. (2022). Screening for apocarotenoid plant growth regulators in Arabidopsis. In: *Methods in Enzymology* 674: 481–495. Academic Press.

Delaux, P.M. and Gutjahr, C. (2024). Evolution of small molecule-mediated regulation of arbuscular mycorrhiza symbiosis. *Philosophical Transactions B* 379(1914), 20230369.

Fiorilli, V., Wang, J.Y., Bonfante, P. et al. (2019). Apocarotenoids: old and new mediators of the arbuscular mycorrhizal symbiosis. *Frontiers in Plant Science* 10, 1186.

Jalal, A., Abdelaziz, M.E., Salem, M. et al. (2025). Mimics of the growth regulator zaxinone increase saffron yield and improve its nutritional value. *Journal of Plant Growth Regulation* 1–16.

Jamil, M., Lin, P.Y., Berqdar, L. et al. (2023). New series of zaxinone mimics (MiZax) for fundamental and applied research. *Biomolecules* 13(8), 1206.

Masteling, R., Tsega, U., Tessema, T. et al. (2023). Controlling Striga hermonthica by precursor-directed activation of microbial volatile production: from in vitro to greenhouse bioassays. *Leiden Uni ersit* 113.

Mazzarella, T., Chialva, M., de Souza, L.P. et al. (2024). Effect of exogenous treatment with zaxinone and its mimics on rice root microbiota across different growth stages. *Scientific Reports* 14(1), 31374.

Soliman, S., Wang, Y., Han, Z. et al. (2022). Strigolactones in plants and their interaction with the ecological microbiome in response to abiotic stress. *Plants* 11(24), 3499.

Votta, C., Fiorilli, V., Haider, I. et al. (2022). Zaxinone synthase controls arbuscular mycorrhizal colonization level in rice. *The Plant Journal* 111(6), 1688–1700.

Wang, J.Y. (2021). Zaxinone, a natural apocarotenoid, regulates growth and strigolactone biosynthesis in rice. KAUST Research Repository. https://doi.org/10.25781/KAUST-966U5.

Wang, J.Y., Alseekh, S., Xiao, T. et al. (2021). Multi-omics approaches explain the growth-promoting effect of the apocarotenoid growth regulator zaxinone in rice. *Communications Biology* 4(1), 1222.

Wang, J.Y., Haider, I., Jamil, M. et al. (2019). The apocarotenoid metabolite zaxinone regulates growth and strigolactone biosynthesis in rice. *Nature Communications* 10(1), 810.

Wang, J.Y., Jamil, M., Hossain, M.G. et al. (2022). Evaluation of the biostimulant activity of zaxinone mimics (MiZax) in crop plants. *Frontiers in Plant Science* 13, 874858.

Wang, J.Y., Jamil, M., Lin, P.Y. et al. (2020). Efficient mimics for elucidating zaxinone biology and promoting agricultural applications. *Molecular Plant* 13(11), 1654–1661.

8

Zaxinone Mimics: Potential Agricultural Applications for Global Food Security

Shamshad Ahmad Khan[1], *Nauf Al Araimi*[1], *and Priyanka Verma*[2]*

[1] *Department of Applied Biotechnology, University of Technology and Applied Sciences, Sur, Oman*
[2] *(Ex-Pool Scientist) Division of Biochemical Sciences, CSIR-National Chemical Laboratory (NCL), Pune, Maharashtra, India*

*Corresponding author: drpv.invitro@gmail.com

8.1 Introduction

As demand for food is increasing day by day, an approach of sustainable agricultural practices is needed to achieve higher yields from the plants. It requires the stability of the plant systems as well. The arbuscular mycorrhizal fungi (AMFs) have been seen as potent tools in this field as they can work as biostimulants to increase the overall plant growth with an eco-friendlier approach. At the rhizosphere level, AMFs provide minerals to the plant system, even at a level beneath the depletion zones, and help in inducing secondary metabolite production, thus acting as biofertilizers. The plant hormone level in plants is also maintained by the AMFs, as they work as a bioregulator, which leads to controlled overall plant development. AMFs have also been seen as potential candidates to work as bioprotectors, as they provide tolerance to the plant against environmental and soil-related stresses. For an effective use of AMFs precise selection of a fungal strain is considered important, which depends upon the plant host on which this strain has to be used. The effective use of AMFs may help in the lesser use of biocides and chemicals on the soil when properly in a way taking into account the beneficial farming practices and inoculation of AMF strains (Rouphael et al. 2015).

For effective stimulation of plant growth, improving nutrient uptake, and reducing fertilizer requirements, different kinds of biostimulators have been applied to various crops. These biostimulators improve water usage efficiency by plants and also play a crucial role in enhancing the tolerance of plants to stress conditions. As mentioned, biostimulants are of different types, ranging from plant or algal type to acids such as fulvic and humic acids, hydrolysates of proteins, as well as other combinations. The effective use of biostimulators

Plant Apocarotenoids: Biosynthesis, Signaling and Agricultural Applications, First Edition. Edited by Kaiser Iqbal Wani and Tariq Aftab.
Published 2026 by John Wiley & Sons Ltd.

can improve the crops in several ways, which include improving the overall yield of the crops, improving the commercial quality of the products by various changes such as reducing the undesirable nitrates in leafy plants, and enhancing the sugar, protein, and other compounds. Biostimulators are also very effective in providing an edge to the organic and sustainable farming practices. Their use increases the resilience of the plant and helps in increasing the yield from organic farming practices. Apart from it, nutrient use efficiency can also be enhanced by the use of biostimulators. The use of biostimulators improves the overall soil profile and biological activity in the soil. Zaxinone, a recently reported apocarotenoid (Wang et al. 2019) from the rice plant, has emerged as a strong biostimulant as it improves the root growth, plant height, and flowering, increases branch numbers, fruit size, and yield in various plant species (Wang et al. 2020, 2021, 2022, 2023a, b).

In the sub-Saharan region, *Striga* species is considered the foremost threat to legumes and cereals. It majorly affects cowpea, pearl millet, sorghum, and maize, with *Striga asiatica* and *Striga hermonthica* as the prominent species, causing damage to crops in these regions. More than 50 million ha of land belonging to agricultural activities have been reported to be affected by *Striga* infestation, causing damage to cereal crops. Zaxinone and its mimics are known to lower strigolactones, resulting in a decrease in *Striga* seed germination (Wang et al. 2019; Jamil et al. 2023).

8.2 Apocarotenoids and Zaxinone Identification

All the photosynthetic organisms synthesize carotenoids, which are isoprenoid-type pigments containing a conjugated double bond system. The presence of carotenoid polyene helps in oxidative cleavage, which results in the formation of apocarotenoids. As such, carotenoids are present in most living species, but the plant carotenoids are considered important as they include phytohormones such as strigolactones and abscisic acid (ABA). These carotenoids also include signaling molecules such as β-cyclocitral. The compound β-cyclocitral is derived from β-carotene oxidation and is volatile. It has a mediator role in the response of plant cells to singlet oxygen stress (Felemban et al. 2019). Other apocarotenoids, such as ABA, have an important role in plants in responding to the stressed abiotic environment. Apart from it, ABAs are also involved in various processes included in the development of plants. The other apocarotenoids, strigolactones, have a role in providing signaling molecules during the plant and rhizosphere interactions.

As such, carotenoids are a huge group consisting of more than 600 compounds of isoprenoid-type pigments. Generally, carotenoids are known for their presence in all kinds of organisms, but they are mostly present in photosynthetic organisms. In some other cases, carotenoids were also found in certain bacteria and fungi that grow non-photosynthetically. The role of carotenoids in the plant itself is as a pigment for photoprotection, and they also assist in the harvesting of light as they are considered a necessary part of the photosynthetic machinery. Apart from it, plant carotenoids also provide yellowish, orangish, or reddish colors to several of the flowers and fruits. For animals, carotenoids are important as a precursor for vitamin A and for the synthesis of antioxidants, but as such, animals cannot synthesize these carotenoids. Therefore, animals depend on nutritional diets to complete these requirements.

For the development and growth of plants, chemical signals and hormones are very important for adapting to the changes in the environment. Secondary metabolite pathways are the major sources of these chemical signals, and most of the signals are related to carotenoid biosynthesis, which is a source for the synthesis of strigolactones and ABA (Al-Babili and Bouwmeester 2015). The root and shoot development in plants and nutrient availability are regulated by strigolactones (Wani et al. 2023). They also act as rhizospheric signals, which are secreted by roots under the deprived conditions of phosphates, which help in arbuscular mycorrhizal (AM) symbiotic association with the roots. In rice, a carotenoid cleavage dioxygenase (CCD) subfamily has been identified, which has the ability to form signaling molecules and hormones (Ahrazem et al. 2016). One CCD has been recognized as Zaxinone Synthase (ZAS), and it produces zaxinone, which is reported as a new metabolite of apocarotenoid type. Study on rice shows that mutants that had a loss of function exhibited restricted growth and higher levels of strigolactones. It was further revealed that the use of zaxinone may result in decreased strigolactone content and promotion of growth in the wild seedlings, and can be seen as an important regulator for the development of rice and its biotic interactions. Zaxinones can also enhance plant growth and contain the potential to combat *Striga* (Wang et al. 2019).

8.3 Zaxinone Mimics

The organic synthesis of zaxinone is complicated, and the low presence of this important growth regulator led to the synthesis of its mimics, which are known as the mimics of zaxinone (MiZax). MiZax are synthesized based on the activity–structure relationship-based study with the help of an array of apocarotenoid series. Initially, five MiZax were produced, namely MiZax1–MiZax5 (Wang et al. 2019). They differ in structures, methylation, phenyl ring, and carbon chain length (Table 8.1).

MiZax1 has one shorter C atom (five C atoms) in comparison to zaxinone, which lies between the ketone group and the phenyl ring. This could be the reason behind the non-significant impact on the strigolactones level when treated with MiZax1. The other four mimics were synthesized with six C atoms between the ketone and phenyl ring to overcome the impact of shorter chain length encountered during the synthesis of MiZax1. In MiZax4 and MiZax5, the zaxinone isoprenoid chain was substituted by phenyl rings, while in MiZax2 and MiZax3, it was partially replaced by a phenoxy ring. To increase the hydrophobicity, the hydroxy group was methylated in MiZax3 and MiZax5 (Wang et al. 2019). Out of the five generated mimics, MiZax3 and MiZax5 were found to be the best in terms of biostimulant activity (by increasing the growth of roots, shoots, and fruit yield) as well as in controlling the *Striga* infestation (by lowering strigolactones and reducing *Striga* seed germination). MiZax3 was found to be more stable than MiZax5.

Later, derivatives of MiZax3 and MiZax5 were also synthesized to further explore their potential toward higher activities (Jamil et al. 2023). MiZax3 derivatives are MZ3-1–MZ3-4 and MZ3-6–MZ3-8. Similarly, the MiZax5 derivatives are MZ5-2–MZ5-8. Out of all 14 derivatives, MZ3-2, MZ3-7, and MZ5-6 showed higher activity in terms of root number, length, and root biomass, but none of them outperformed the parent molecule, especially MiZax3. Therefore, MiZax3 emerged as a potential molecule both as a biostimulant and

Table 8.1 Chemical structures and functions of various MiZax.

Zaxinone/Mimic	Structure	Molecular formula	Chemical name	Functions	Derivatives
Zaxinone	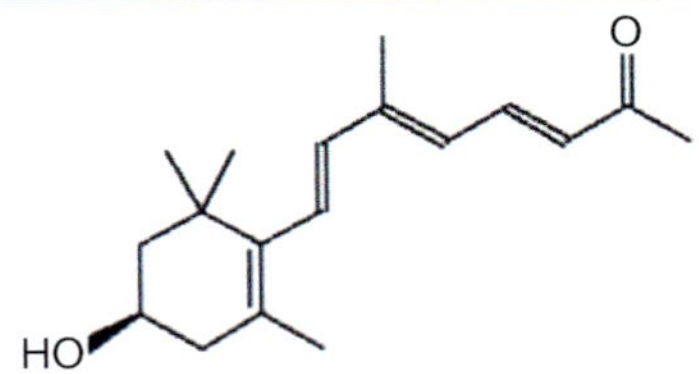	$C_{18}H_{26}O_2$	(3E,5E,7E)-8-(4-Hydroxy-2,6,6-trimethylcyclohex-1-en-1-yl)-6-methylocta-3,5,7-trien-2-one	Biostimulant; lowers the strigolactone synthesis; plant growth regulator	—
MiZax1	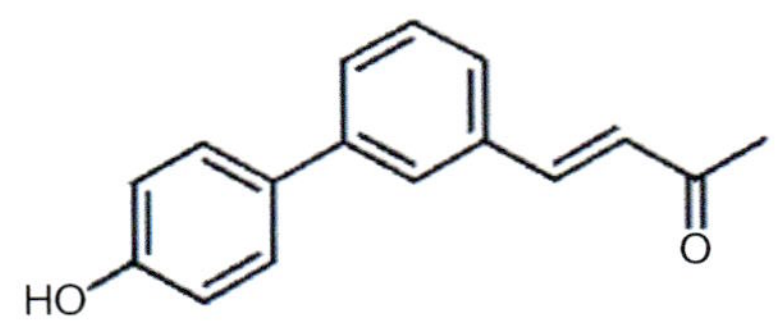	$C_{16}H_{15}O_2$	(E)-4-(4′-Hydroxy-[1,1′-biphenyl]-3-yl)but-3-en-2-one	Does not significantly reduce strigolactones	—
MiZax2	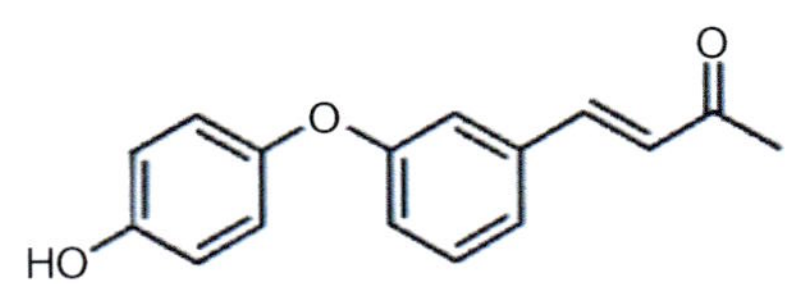	$C_{16}H_{15}O3$	(E)-4-(3-(4-Hydroxyphenoxy)phenyl)but-3-en-2-one	MiZax2 shows a significant decrease in strigolactones and the rate of *Striga* germination	—
MiZax3	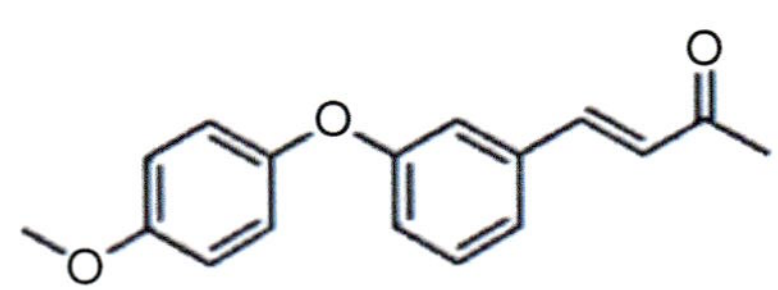	$C_{17}H_{17}O_3$	(E)-4-(3-(4-Methoxyphenoxy)phenyl)but-3-en-2-one	Reduction in *Striga* plant emergence is observed with MiZax3 (71%) in comparison to zaxinone (42%); MiZax3 is found to be more active than zaxinone in increasing the root length of wild-type plants	MZ3-1–MZ3-8

MiZax4	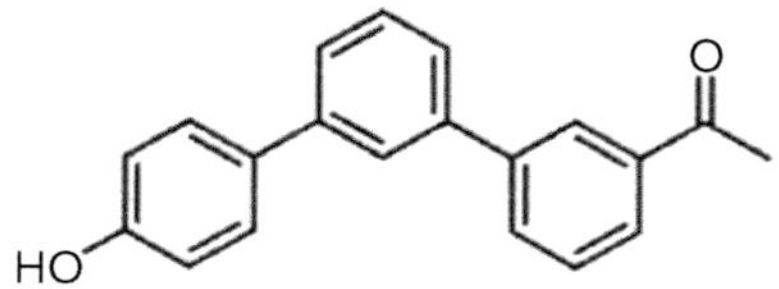	$C_{20}H_{17}O_2$	1-(4″-Hydroxy-[1,1′:3′,1″-terphenyl]-3-yl)ethan-1-one	MiZax4 shows a reduction in strigolactones	—
MiZax5	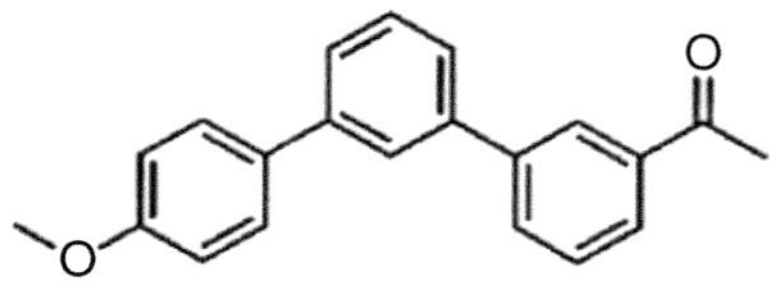	$C_{21}H_{19}O_2$	1-(4″-Methoxy-[1,1′:3′,1″-terphenyl]-3-yl)ethan-1-one	*Striga* plant emergence reduction is observed to be lower than MiZax3 but higher than zaxinone; root growth promoter	MZ5-2–MZ5-8

for lowering the *Striga* infestation. MiZax are seen as biostimulants which may potentially help in the growth of important crops, which in turn may contribute to food security at the global level.

8.4 Role of MiZax in Rice

Detailed works on zaxinone in rice and other plants have been done by various researchers (Table 8.2). The precursors for ABA, strigolactones, and zaxinone are carotenoids. Among these, strigolactones and ABA are considered very important regulators for the development and growth of the plant itself. These metabolites also have a role in responding to various biotic/abiotic elicitors and adapting to different environmental conditions. For the normal growth of rice, zaxinone, which is found commonly in mycorrhizal associations, is produced from the plant with the help of ZAS enzyme. This zaxinone is known to negatively regulate the biosynthesis of strigolactones (Wang et al. 2019).

CCDs help in catalyzing the cleavage of carotenoids, which releases signaling molecules and provides precursors for the plant hormone to act on. An in-depth study found that the genome of rice contains three homologs of *OsZAS*, namely *OsZAS1b*, *OsZAS1c*, and *OsZAS2*, but their exact functions are still not known. A detailed study on the function of *OsZAS2* has revealed that it can form apocarotenoid zaxinone under *in vitro* conditions. Also, the location of *OsZAS2* was found to be mainly in plastids, and a phosphate starvation phenomenon helps in the expression of this enzyme. It is interesting to note that during the mycorrhization process, the expression level of this enzyme increased especially in arbuscule-containing cells. The mutants of *OsZAS2*, which were created through CRISPR, showed less presence of zaxinone in roots and responded with having less biomass in root and shoot. These mutants also exhibited the presence of fewer tillers, and the level of strigolactones was higher. Zaxinone, when applied exogenously, decreased the strigolactone levels and partially helped in halting the retardation of plant growth in these mutants. A comparative study showed that *OsZAS2* has a role in the formation of zaxinone, which helps in rice growth, mycorrhization, and maintaining strigolactone content (Votta et al. 2022).

In a similar kind of study, a mutant rice with a loss of function for *ZAS*, having less zaxinone, was studied. It was observed that rice exhibited restricted growth patterns and higher levels of strigolactones, which helped in determining the architecture of the plant, mediation in mycorrhization, and, more importantly, promoted germination of seeds that are root-parasitic. When the *zas* mutants were treated with zaxinone, it resulted in downregulation of strigolactone biosynthesis and better growth of roots in the wild rice seedlings. Further, a multi-omics analysis depicted the effect of zaxinone to be positive in upregulating the sugar metabolism and induction of photosynthesis is a way that leads to phenotypical changes in the roots of rice. The level of plant growth hormone cytokinin was also enhanced upon treatment with zaxinone, as shown by the transcriptomic analysis (Wang et al. 2021; Ablazov et al. 2023).

To assess whether zaxinone has a role in counteracting strigolactone biosynthesis, a study was conducted in which a co-application of zaxinone was done (Wang et al. 2022). A rice mutant that was deficient in strigolactones was supplemented with the synthetic

Table 8.2 Studies conducted on different plant systems in reference to zaxinone.

S. no.	Plant systems	Objectives	Findings	Reference
1	Rice and *Arabidopsis*	Identification of the role of strigolactone biosynthetic enzymes and also their role as signaling molecules and hormones	Four different enzymes were identified as strigolactone biosynthetic enzymes, followed by a *cis/trans*-carotene isomerase identification	Al-Babili and Bouwmeester (2015)
2	Plants, algae, fungi, and bacteria	Identification of CCDs from various sources and their functions in these organisms	CCD1 and CCD2 subfamilies (plants), CCD7 and CCD8 homologs (algae)	Ahrazem et al. (2016)
3	Plants	Identification of apocarotenoids in plants and the role of carotenoid cleavage oxygenase enzymes	Variation in the apocarotenoids due to precursors, the availability of specific cleavage sites The main precursors of apocarotenoids were zeaxanthin, lycopene, and β-carotene	Beltran and Stange (2016)
4	Rice	Role of strigolactones in root and shoot development in rice	Identified ZAS, which produced zaxinone, a new metabolite of apocarotenoid type. It worked as a plant growth regulator and biostimulant	Wang et al. (2019)
5	Rice	Synthesis of MiZax for cheap and easy use	Two new mimics were synthesized (MiZax3 and MiZax5) and showed positive activity in rescuing the roots of mutant rice; they also showed growth promotion and lowered the strigolactone content. Also depicted was the suppression of strigolactone biosynthesis at the transcript level. It had shown no negative effect on mycorrhization	Wang et al. (2020)
6	*Arabidopsis*	Investigation of zaxinone activity of strigolactones and ABA biosynthesis in *Arabidopsis*	Application of zaxinone resulted in enhanced transcript levels of ABA and strigolactone biosynthetic genes, and also resulted in increased strigolactone release. Also, increased ABA content led to low hypocotyl growth	Ablazov et al. (2020)

(*Continued*)

Table 8.2 (Continued)

S. no.	Plant systems	Objectives	Findings	Reference
7	Rice	A mutant rice with a loss of function for *ZAS*, having less zaxinone, was studied	Rice exhibited restricted growth patterns and more strigolactone content, which helped in determining the architecture of the plant, mediation in mycorrhization, and assistance in germinating seeds that are root-parasitic	Wang et al. (2021)
8	Tomato, capsicum, squash, and date palm	Effect of MiZax on overall growth and yield of tomato, capsicum, squash, and date palm	Induced better plant height, improved flowering, increased branch numbers, and an increase in fruit size, with overall fruit yield increment with the use of less amount of MiZax compared to commercially available biostimulant (humic acid)	Wang et al. (2022)
9	Rice	How OsZAS and exogenous zaxinone treatment regulated mycorrhization	OsZAS impacted the level of AM colonization, working as a factor of a regulatory network that encompasses strigolactones	Votta et al. (2022)
10	Rice	To assess the role of zaxinone in counteracting strigolactone biosynthesis	Indicated the role of zaxinone in the regulation of shoot and root growth, which helped in the growth of rice	Wang et al. (2023a)
11	Rice	Effect of a new series of MiZax on plant growth in certain crops	By the structural modification of MiZax3 and MiZax5, 14 new biostimulants were designed and tested for their activity. But none of them outperformed the parent mimics, especially MiZax3	Jamil et al. (2023)
12	Rice	Study of three homologs of *OsZAS* (*OsZAS1b*, *OsZAS1c*, and *OsZAS2*) and their functions	The function of *OsZAS2* had revealed that it can form apocarotenoid zaxinone under *in vitro* conditions. Also, the location of *OsZAS2* was found to be in plastids The phosphate starvation phenomenon helped in the expression of this enzyme	Ablazov et al. (2023)
13	Potato and strawberry	Assessing the effect of zaxinone and its mimics on cereals and vegetable crops	MiZax (MiZax3 and MiZax5) were found to enhance the overall yield and agronomic traits of the crops in greenhouse and field conditions	Wang et al. (2023b)

analog of strigolactones and zaxinone, which demonstrated that the mutant-type rice variety was rescued. This study positively indicates the role of zaxinone in the regulation of shoot and root growth, which helps in the growth of rice. The strigolactones and ABA are considered very important in the growth and overall development of plants, helping plants in adapting to certain changes in the environment and also in responding to different kinds of biotic and abiotic stresses. It has also been mentioned previously that zaxinone is produced in rice in the presence of the enzyme ZAS, which is a downregulator of strigolactones and required by rice plants in their normal growth. Among MiZax, MiZax3 is considered more effective in downregulating the *Striga* infestation compared to MiZax5 (Wang et al. 2020). As such, both mimics were also tested with some structural modifications and their effects were analyzed on the growth of the plant and *Striga* infestation. There were no observable changes that could be shown by these structural modifications in both mimics, but certain activities were found to increase in them (Wang et al. 2023a).

8.5 MiZax in Other Crops

Further investigation regarding zaxinone has revealed that the biological activity of this metabolite has also been found in *Arabidopsis thaliana* through an unknown pathway despite the absence of the ZAS enzyme. A detailed study was focused on investigating the zaxinone bioactivity in *Arabidopsis* to assess its effect on the biosynthesis of ABA and strigolactones. The application of zaxinone to the seedlings of *Arabidopsis* resulted in increased transcript levels of both strigolactones and ABA, which resulted in higher release of both these signaling molecules. The study has further elaborated that the application of zaxinone led to a reduction of hypocotyl growth, which may be because of increased levels of ABA. The results further indicated that in *Arabidopsis* plants, zaxinone helps in triggering the biosynthesis of strigolactones and ABA, both of which are hormones derived from carotenoids, whereas in non-mycotrophic *Arabidopsis* plants, zaxinone was not able to enhance growth (Ablazov et al. 2020).

Apart from rice, the effects of MiZax on the overall growth and yield of other crops were also tested. Among the crops were dicot plants such as tomato (*Solanum lycopersicum*), capsicum (*Capsicum annuum*), squash (*Cucurbita pepo*), and monocot date palm (*Phoenix dactylifera*) that were chosen for this study. The MiZax showed positive effects on these crops by inducing better plant height, improved flowering, increased branch numbers, and an increase in fruit size with overall fruit yield increment. More importantly, it was found that the use of less amount of MiZax was effective for these growth parameter improvements as compared to commercially available humic acid, which is a known biostimulant (Wang et al. 2022).

An assessment of the effects of zaxinone and its mimics on cereals and vegetable crops was also conducted to get more information for possible use of MiZax on commercial crops. For this purpose, two important crops, namely potato and strawberry, were selected at the greenhouse level. Two of the active MiZax (MiZax3 and MiZax5) were tested on these greenhouse-established plants, and the results showed an enhanced overall yield and agronomic traits in both the crops (Wang et al. 2023b).

8.6 MiZax and *Striga* Infestation

Striga is a root-parasite type of plant belonging to the family Orobanchaceae. It needs signaling chemicals secreted by the roots of the host plant for its seed germination. The seeds of *Striga* may be in a dormant state for many years, and they only germinate in the presence of strigolactone, a plant hormone secreted by the roots, hence enabling the survival of these weed plants. Specific weeds such as *S. hermonthica* are more crop-threatening as they affect crops in sub-Saharan Africa and affect more than 100 million people (Pennisi 2010). This species of weed works as a biotic threat and is known to have affected around 50 million ha of area in the Savannah and Sahel regions of 25 different countries in the African region, leading to a loss of about $7 billion (Parker 2012). Strigolactones are secreted as a plant growth hormone in plants for their better survival and competition with other surrounding plants (Wani et al. 2024). These hormones help plants in growth processes like increasing root hair growth, stem thickness increment, enhancing the growth of internodal segments, and retarding shoot branching of plants. More importantly, strigolactones are involved in increasing mycorrhization association in plants by helping the hyphal growth and branching of the fungus, which in turn helps in enhancing the symbiotic relationship between plant and fungus (Akiyama et al. 2005). *Striga hermonthica* affects different crops including sorghum, rice, maize, and pearl millet, while the species of *Orobanche* weeds specifically affects crops such as oats, wheat, sugarcane, barley, potato, sweet potato, clover, tobacco, melons, and hemp. Apart from these cereal crops, legumes are also affected by *Striga* infestations, which include lentils, beans, peanuts, and peas. Zaxinone and its mimics help in decreasing the seed germination of *Striga*, which helps in reducing infestation by this crop-threatening weed. Zaxinone specifically lowers the strigolactone biosynthesis. While its mimics MiZax2–MiZax5 are known to lower the strigolactone level in the plants, the maximum reduction in *Striga* plant emergence is observed with MiZax3 (71%) followed by MiZax5 (55%) in comparison to zaxinone (42%) (Wang et al. 2019). When the effects of mimics MiZax2–MiZax5 on *Striga* seed-germinating activity were compared, it was concluded that methylation has a significant encouraging effect on the activity of MiZax. The higher activity registered by MiZax3 and MiZax5 has been attributed to increased hydrophobicity triggered by the methyl group, which in turn improves uptake and transport of the mimics (Wang et al. 2019).

8.7 Prospects of Zaxinone and Its Mimics in Crop Improvement

Organic manuring, use of biostimulators, and supplementation have always been important components in terms of soil productivity and overall growth of the plants. Zaxinone and its derivatives or mimics are seen as novel compounds not only for improving the plant growth by acting at the rhizosphere level, regulation of strigolactone biosynthesis, but also as an agent to reduce the *Striga* infestations.

Field-level studies of these mimics are seen positively, as works conducted on certain crops have shown improvements in both dicot and monocot plants. Additionally, MiZax have shown an edge over other commercial biostimulators used, and this has provided a handy possibility in the area of crop improvement through these strategies (Wang et al. 2022). Recently, Jamil et al. (2023) stated that different biostimulants were generated from two of the most important MiZax (MiZax3 and MiZax5) to improve their activity. However, these compounds were not able to act superior to their parent mimics, hence till now MiZax3 and MiZax5 are considered the most promising ones for use in crop productivity. More importantly, a detailed study is also required to study the effect of the use of zaxinone and its mimics on the soil microbes and the fate of these molecules after their use. The dose optimization of MiZax for their safe and efficient use is also one major research area that needs to be focused on if these mimics are to be used as potent biostimulators for crop improvement (Jamil et al. 2023). MiZax, such as MiZax3 and MiZax5, are potential plant growth regulators that have shown their affirmative effect on certain tree species, cereals, as well as the fruit-producing crops. A more inclusive understanding of the molecular level will provide better in-depth knowledge on using these mimics in wider species on a larger scale.

Acknowledgments

The authors greatly acknowledge the financial assistance provided by the Ministry of Higher Education, Oman, and UTAS-Sur, Oman, during the writing of this chapter.

References

Ablazov, A., Mi, J., Jamil, M. et al. (2020). The apocarotenoid zaxinone is a positive regulator of strigolactone and abscisic acid biosynthesis in Arabidopsis roots. *Frontiers in Plant Science* 11, 578. https://doi.org/10.3389/fpls.2020.00578.

Ablazov, A., Votta, C., Fiorilli, V. et al. (2023). *ZAXINONE SYNTHASE 2* regulates growth and arbuscular mycorrhizal symbiosis in rice. *Plant Physiology* 191, 382–399. https://doi.org/10.1093/plphys/kiac472.

Ahrazem, O., Gomez-Gomez, L., Rodrigo, M.J. et al. (2016). Carotenoid cleavage oxygenases from microbes and photosynthetic organisms: features and functions. *International Journal of Molecular Sciences* 17, 1781. https://doi.org/10.3390/ijms17111781.

Akiyama, K., Matsuzaki, K.-I. and Hayashi, H. (2005). Plant sesquiterpenes induce hyphal branching in arbuscular mycorrhizal fungi. *Nature* 435(7043), 824–827. https://doi.org/10.1038/nature03608.

Al-Babili, S. and Bouwmeester, H.J. (2015). Strigolactones, a novel carotenoid-derived plant hormone. *Annual Review of Plant Biology* 66, 161–186.

Beltran, J.C.M. and Stange, C. (2016). Apocarotenoids: a new carotenoid-derived pathway. In: *Subcellular Biochemistry: Vol. 79: Carotenoids in Nature* (ed. C. Stange), 239–272. Springer. https://doi.org/10.1007/978-3-319-39126-7_9.

Felemban, A., Braguy, J., Zurbriggen, M.D. et al. (2019). Apocarotenoids involved in plant development and stress response. *Frontiers in Plant Science* 10, 1168. https://doi.org/10.3389/fpls.2019.01168.

Jamil, M., Lin, P.-Y., Berqdar, L. et al. (2023). New series of zaxinone mimics (MiZax) for fundamental and applied research. *Biomolecules* 13(8), 1206. https://doi.org/10.3390/biom13081206.

Parker, C. (2012). Parasitic weeds: a world challenge. *Weed Science* 60, 269–276.

Pennisi, E. (2010). Armed and dangerous. *Science* 327, 804–805.

Rouphael, Y., Franken, P., Schneider, C. et al. (2015). Arbuscular mycorrhizal fungi act as biostimulants in horticultural crops. *Scientia Horticulturae* 196, 91–108. https://doi.org/10.1016/j.scienta.2015.09.002.

Votta, C., Fiorilli, V., Haider, I. et al. (2022). Zaxinone synthase controls arbuscular mycorrhizal colonization level in rice. *The Plant Journal* 111(6), 1688–1700. https://doi.org/10.1111/tpj.15917.

Wang, J.Y., Alseekh, S., Xiao, T. et al. (2021). Multi-omics approaches explain the growth-promoting effect of the apocarotenoid growth regulator zaxinone in rice. *Communications Biology* 4(1), 1222. https://doi.org/10.1038/s42003-021-02740-8.

Wang, J.Y., Braguy, J. and Al-Babili, S. (2023a). Does zaxinone counteract strigolactones in shaping rice architecture? *Plant Signaling & Behavior* 18, 2184127.

Wang, J.Y., Haider, I. and Jamil, M. (2019). The apocarotenoid metabolite zaxinone regulates growth and strigolactone biosynthesis in rice. *Nature Communications* 10, 810. https://doi.org/10.1038/s41467-019-08461-1.

Wang, J.Y., Jamil, M., AlOtaibi, T.S. et al. (2023b). Zaxinone mimics (MiZax) efficiently promote growth and production of potato and strawberry plants under desert climate conditions. *Scientific Reports* 13, 17438. https://doi.org/10.1038/s41598-023-42478-3.

Wang, J.Y., Jamil, M., Hossain, M.G. et al. (2022). Evaluation of the biostimulant activity of zaxinone mimics (MiZax) in crop plants. *Frontiers in Plant Science* 13, 874858. https://doi.org/10.3389/fpls.2022.874858.

Wang, J.Y., Jamil, M., Lin, P.-Y. et al. (2020). Efficient mimics for elucidating zaxinone biology and promoting agricultural applications. *Molecular Plant* 13(11), 1654–1661. https://doi.org/10.1016/j.molp.2020.08.009.

Wani, K.I., Naeem, M. and Aftab, T. (2024). Understanding the interplay between strigolactone and nitric oxide in alleviating cadmium-induced toxicity in *Artemisia annua*. *Environmental and Experimental Botany* 225, 105856.

Wani, K.I., Zehra, A., Choudhary, S. et al. (2023). Exogenous strigolactone (GR24) positively regulates growth, photosynthesis, and improves glandular trichome attributes for enhanced artemisinin production in *Artemisia annua*. *Journal of Plant Growth Regulation* 42(8), 4606–4615.

9

β-Cyclocitral: Signaling During Stress Acclimatization in Plants

*Vinay Kumar Dhiman[1], Sudarshna[2], Vivek Kumar Dhiman[3], and Devendra Singh[4]**

[1] *Department of Food Science and Technology, Khalsa College, Amritsar, Punjab, India*
[2] *Department of Bioscience and Bioengineering, Indian Institute of Technology Jodhpur, Jodhpur, Rajasthan, India*
[3] *Centre for Herbal Pharmacology and Environmental Sustainability, Chettinad Hospital and Research Institute, Chettinad Academy of Research and Education, Kelambakkam, Tamil Nadu, India*
[4] *Faculty of Biotechnology, Institute of Biosciences and Technology, Shri Ramswaroop Memorial University, Barabanki, Uttar Pradesh, India*

*Corresponding author: devendrasingh.ibst@srmu.ac.in

9.1 Introduction

Carotenoids are isoprenoid compounds naturally found in almost all the colored bio-pigments present in primitive microorganisms such as archaea, now-evolved bacteria, algae, fungi, higher plants, and animals (Maoka 2020; Martínez-Cámara et al. 2021). They are formed of hydrocarbons, thus being soluble in fats (lipophilic due to the presence of long unsaturated aliphatic chains), making them hydrophobic (Burri 2013). Carotenoids are chemically tetraterpenes (eight isoprene units) made up of a carbon skeleton with around 40–50 C atoms, having two main types, α- and β-carotene, in plants along with lycopene, lutein, and zeaxanthin and related pigments (Dumay and Morançais 2016; Zia-Ul-Haq 2021). They contain a polyene chain of 9–11 double bonds and terminal rings with unsubstituted β-ionone rings (Britton 2022). They are potentially related to retinol, having vitamin A activity (provitamin), and can also act as antioxidants. β-Carotene, a polyphenol carotenoid, is the most often occurring carotene in nature and gives yellow and orange pigmentation (Hou et al. 2016; Harrison 2022). It is a precursor for forming its derivative carbonyl products known as apocarotenoids through enzymatic cleavage or the reactive oxygen species (ROS) reaction by direct oxidation (Moreno et al. 2021; Li and Kim 2023). Both carotenoids and their derivatives, apocarotenoids, are important natural pigments that are diversely distributed among living organisms. The abundance of apocarotenoids is due to the considerable amount of carotenoid needed as a

Plant Apocarotenoids: Biosynthesis, Signaling and Agricultural Applications, First Edition. Edited by Kaiser Iqbal Wani and Tariq Aftab.
Published 2026 by John Wiley & Sons Ltd.

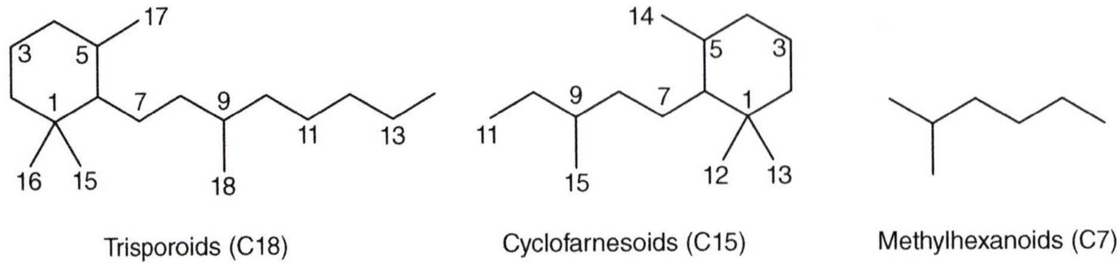

Figure 9.1 Apocarotenoids classification based on the C skeleton.

precursor, which is oxidized at different sites with varied functional groups (Felemban et al. 2019). Apocarotenoid is essential in plant development; it acts as a phytohormone and signals in response to stress. There are numerous naturally occurring apocarotenoids with varied chemical structures: trisporoids, cyclofarnesoids (earlier called apotrisporoids), and methylhexanoids (Sun et al. 2022; Hartati and Paramita 2023). These compounds are formed due to the loss of C atoms in the side chain of carotenes due to oxidation caused by enzymes or free radicals (Singh et al. 2024a, c; Tyagi et al. 2024; Yanykin et al. 2024). Oxidation results in double bond cleavage of β-carotene, resulting in the formation of these three apocarotenoid families (Harrison and Bugg 2014; Bogacz-Radomska and Harasym 2018; Dhiman et al. 2024; Sharma et al. 2024a). These families include aliphatic CH chains having unsaturated chains with carboxyl, hydroxyl, hydroxymethyl, and aldehyde groups, singlet oxygen $(^1O_2)$ bonded at the end of side chains, and structural rings at different conformations that, upon exposure to sunlight, change the color of pigment (Alcaíno et al. 2016; Rathva et al. 2024; Singh et al. 2024c; Sharma et al. 2024b). The modification of structures with different functional groups in these compounds causes variations in yellow, orange, red, blue, purple, or mixed to pale forms of coloration in plants and animals (Maoka 2009). Important plant apocarotenoids for stress response in plants include abscisic acid (ABA), strigolactones, and β-cyclocitral, which also regulate growth and seed dormancy (Felemban et al. 2019; Dhiman et al. 2022; Singh et al. 2024b; Verma et al. 2024). β-Cyclocitral is a carotenoid-derived novel signaling molecule that helps in the regulation of different metabolites during stress and has a role in shaping the roots' structure (Ke et al. 2022). β-Cyclocitral is a volatile apocarotenoid that is prominent in providing plants with resistance to environmental stresses. β-Cyclocitral is abundantly present in plant leaves, roots, flowers, and fruit. It is a volatile molecule that acts as an odorous repellent for food gazers (Rosas-Saavedra and Stange 2016; Simkin 2021). It enhances cell division in emerging lateral roots, regulates plant growth, and enhances branching. Plants produce different metabolites and use them in developing immunity responses to environmental damage caused by insects and pests, and the formation of molecules like β-cyclocitral is an example of this response (Figure 9.1) (Singh et al. 2022b, c).

9.2 β-Cyclocitral: Formation and Derivatives

Many biologically important apocarotenoids are produced during enzymatic reaction pathways catalyzed by carotenoid cleavage dioxygenases (CCDs) and 9-*cis*-epoxy-dioxygenase (NCED) (Zheng et al. 2021). CCD-1 and CCD-4 are involved in the formation of β-cyclocitral, a volatile apocarotenoid (Auldridge et al. 2006; Wang et al. 2020). The studies observed for

CCD-1 showed β-carotene cleavage at the 7,8 (7′8′) position to form β-cyclocitral (C10) in the cytosol of grape wine cells acting as aroma and flavor compound (Rubio-Moraga et al. 2014). In citrus planta, it was observed that a form of CCD-4 was able to cleave β-carotene into β-apo-8′-carotenal (C30) and β-cyclocitral (C10) and α-carotene into ε-apo-8′-carotenal (C30) and β-cyclocitral (C10) due to dioxygenase mechanism in plastid (Simkin 2021). In plants, the chlorophyll antennae consist of β-carotene, which oxidizes, especially due to $^1O_2$ production in the PSII reaction center. $^1O_2$, being an electrophilic molecule, has high-affinity molecule unsaturated bonds in β-carotene, thereby causing oxidative reactions (Ramel et al. 2012; Dhami et al. 2022). As carotenes quenches the reactive O_2 in chloroplast, oxidation of carotene leads to the formation of these short-chained compounds (Ivanov et al. 2011). β-Cyclocitral is accumulated in chloroplast along with other volatile apocarotenoid derivatives (β-ionone, α-ionone, and dihydroactinidiolide) during the oxidation of β-carotene via $^1O_2$ formed during photosynthetic reaction (Figures 9.2 and 9.3) (Fiedor et al. 2005; Winterhalter and Rouseff 2005; Mitra and Gershenzon 2022).

β-Cyclocitral has been naturally detected in various plant species, and its presence suggests that it has played a significant role in the survival of these plants and is conserved on an evolutionary basis (Fischer et al. 2013). The accumulated β-cyclocitral gets rapidly oxidized to β-cyclocitric acid as found in *Arabidopsis* leaves. β-Cyclocitric acid improves plants' ability to withstand drought stress by stimulating β-cyclocitral-dependent signaling

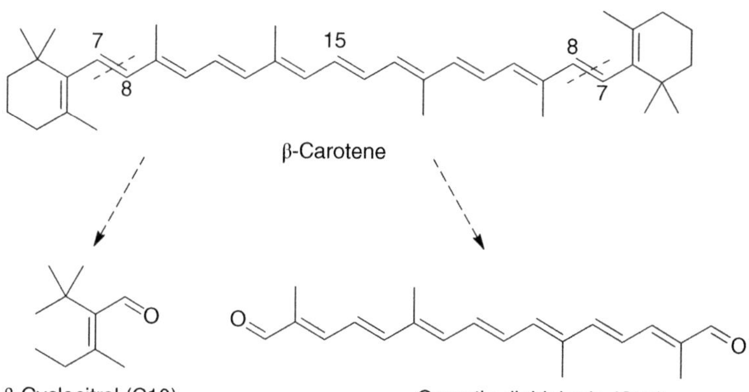

Figure 9.2 Scheme for β-carotene cleavage at 7,8 (7′8′) by CCD (CCD-1) in cytosol.

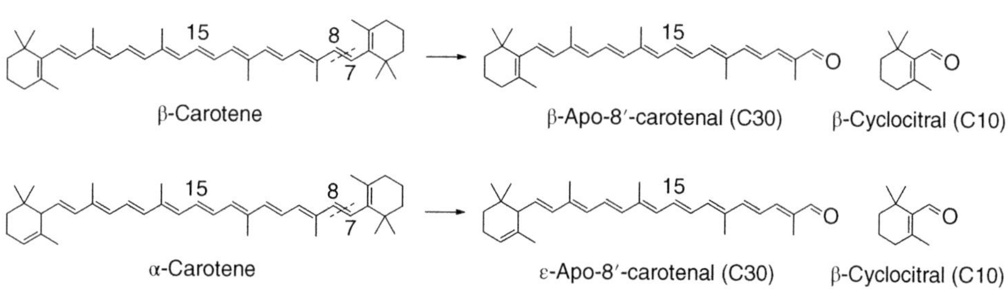

Figure 9.3 Scheme for α-carotene and β-carotene cleavage at 7,8 (7′8′) by CCD (CCD-4) in plastid.

pathway. The plants experiencing long-term drought stress exhibit higher concentrations of β-cyclocitric acid in comparison to the amounts of β-cyclocitral present (Tomita et al. 2016). β-Cyclocitric acid was observed to be high in tomato leaves combined with temperature and light changes, causing a drought-protective effect and significantly benefiting plant productivity (D'Alessandro et al. 2019). As β-cyclocitral is naturally converted to β-cyclocitric acid in the presence of water, a catalytic enzyme may be involved in producing the acid by different monooxygenases (van Berkel et al. 2006). Apocarotenoid glycosylation impacts plant signaling. Glycosylation of β-cyclocitral occurs in plants by glycosyl transferases; glycosylated β-cyclocitral helps in signal mediation and is found in lower fraction compared to β-cyclocitral (Mi et al. 2019; Moreno et al. 2021). Exposure to UV light converts β-cyclocitral into 2,6-dimethyl-cyclohexanol and 2,2,6-trimethyl-cyclohexanone, which may be found during algal bloom. β-Cyclocitral derivative 2,2,6-trimethyl-cyclohexanone also functions in stress tolerance (Kim et al. 2019). Different chemically stable bioactive lactones have been made via chemical synthesis that are β-cyclocitral derivatives; dihydroactinidiolide has been found to be formed during oxidative stress by photons.

9.3 β-Cyclocitral and Salicylic Acid Interplay

Phytohormones are essential mediators that play a role in adapting to light and directing communication within the plants (Singh et al. 2022a). Phytohormones such as salicylic acid (SA) and jasmonic acid (JA) play their roles in response to light-induced stress (Singh et al. 2022d; Sharma et al. 2023a). The expression of defense genes during the light reaction is involved in triggering pathways that also alleviate SA (Ramel et al. 2013; Bali et al. 2023). β-Cyclocitral is also a vital compound in upregulating SA signaling. SA helps β-cyclocitral in the reduction of ROS production during the early stages of light reaction; in turn, β-cyclocitral increases the accumulation of SA during the process. The isochorismate pathway is involved in increasing SA concentration via *isochorismate synthase 1* (*ICS1*). *ICS1* expression is upregulated by β-cyclocitral, which upregulates SA biosynthesis. The *ICS1*-mediated isochorismate pathway is mainly responsible for SA production in response to improved light acclimation in the presence of β-cyclocitral (Gawroński et al. 2013; Lv et al. 2015). In turn, enhanced disease susceptibility 1 (EDS1) regulates *ICS1* gene expression, which is further involved in SA accumulation in the presence of β-cyclocitral. The localization of the *no-express or pathogenesis-related gene 1* (*NPR-1*) from the cytoplasm to the nucleus is important to activate the defense-related gene, β-cyclocitral, which induces NPR-1 translocation during light reaction. SA, in the presence of β-cyclocitral, also translocates NPR-1 to the nucleus. β-Cyclocitral is also responsible for modulating genes that initiate cell survival during acclimation (Lv et al. 2015). NPR-1 triggers a cascade of alterations in gene expression for the transcription of *glutathione-S-transferase 5* (*GST5*) and *GST13*; their transcriptional expression is responsible for the detoxification mechanism. This pathway shows a potential molecular regulatory mechanism for increased adaptation to light stress through β-cyclocitral (Bobik and Burch-Smith 2015; Lv et al. 2015; Sharma et al. 2023b). The level of ROS in chloroplasts reduces with the aid of phytohormones such as SA, which provide resistance to different stresses and have complex signaling in association with stress during acclimation, such that β-cyclocitral regulating SA signaling during such stress (Jiang and Dehesh 2021).

9.4 Could It Be a New Agricultural Tool?

Among many phytohormones produced, β-cyclocitral plays several important roles, particularly in plant defense, growth regulation, and stress tolerance. It is a multifaceted compound that contributes significantly to plant health and resilience. It is found to be produced during herbivory and pathogen defense, has a role in root architecture and shoot development, and stimulates signaling in the presence of ROS, thereby inducing accumulation of necessary phytohormones (Havaux 2020). β-Cyclocitral plays a crucial role in aroma. β-Cyclocitral is involved in producing appealing and nuanced scents along with other volatile compounds to create synergistic effects, which may also be a foul odor to protect plants to enhance the overall aroma of the fruit or plant (Shi et al. 2020; Lin et al. 2024). Balance in β-cyclocitral and other volatiles determines the final aroma profile. The insect-repellent properties reduce the likelihood of insect attacks on the plant surfaces or deter feeders due to the unpleasant odor. They also regulate oviposition in insects from laying eggs and disrupt the navigational abilities of insects (Yuan et al. 2022). In agriculture, the formulation of natural insect repellents can reduce the use of synthetic chemical pesticides. Incorporation of β-cyclocitral into Integrated Pest Management strategies can provide an environmentally friendly response against pests. Selective breeding techniques can be employed to naturally employ higher levels of β-cyclocitral in plant varieties.

9.5 β-Cyclocitral Signaling in Plants Under Abiotic Stress

9.5.1 Impact on Expression of Stress-responsive Genes

In *Arabidopsis* plants, a high level of β-cyclocitral initiates changes in the expression of more than 1,000 genes that are used to form signaling pathways. Most activated genes encode proteins involved in environmental interaction, stress response, and intracellular transport for signaling. Many of the suppressed genes are associated with development, growth, and cellular structure formation (Xiao et al. 2013; Chauhan et al. 2023; Singh et al. 2023; Yadav et al. 2023). β-Cyclocitral thus acts as a stress signal, altering gene expression and shifting plant cells from active growth to cellular defense in response to stress (Table 9.1).

The synthesis and catabolism of carotenoid-derived apocarotenoids are crucial in the mechanisms coping with ROS. Singlet oxygen ($^1O_2$) triggers the mechanism involved in the expression of genes during photoacclimatization (Ramel et al. 2013). High-light stress mediator pathways are largely unknown; recently, the MBS (Methylene Blue Sensitivity)

Table 9.1 β-Cyclocitral-facilitated changes in *Arabidopsis* under abiotic stress.

Stress	Plant	Effect in plants
Oxidative stress	*Arabidopsis thaliana*	Gene expression is influenced, induction of Zn/Cu SOD
Light stress		Increased lipid peroxidation
Water stress		The genetic responses exhibit overlaps, highlighting the influence of $^1O_2$

protein has been found to be involved in signaling to induce tolerance. MBS1 is involved in $^1O_2$ crosstalk with β-cyclocitral, which acts as a mediator in signaling $^1O_2$-induced light stress to the nucleus for photoacclimation (Shumbe et al. 2017; Rac et al. 2021). High-light stress also stimulates *UDP-glucosyltransferase* (*UGT*) genes in many plant species for glycosylation (Gharabli et al. 2023).

Lipid peroxidation is also linked to oxidative stress in plants and produces bioactive and toxic compounds such as reactive carbonyl species (RCS), phytoprostanes, and phytofurans (Altomare et al. 2021). In *Arabidopsis* plants under photooxidative stress, the effects of these secondary lipid oxidation effects were observed using luminescence techniques (Rac et al. 2021). *ALKENAL REDUCTASE* (*AER*) gene promoter was used as the control for the bioluminescent marker *LUCIFERASE* (*LUC*) gene to observe the detoxification activity in autoluminescence imaging for the transcription regulator SCARECROW-LIKE 14 (SCL14). 4-Hydroxynonenal (HNE) was identified as particularly harmful and capable of triggering *AER* gene expression in wild-type parent, *SCL14*-overexpressing transgenic line, and *SCL14*-deficient mutants. HNE was equally damaging across different genetic lines when applied exogenously (Birtic et al. 2011; Havaux and Ksas 2022). *Arabidopsis* plants pretreated with apocarotenoid β-cyclocitral in an *SCL14*-overexpressing transgenic line were able to nullify the effect of HNE. Boosting the SCL14 detoxification pathway with β-cyclocitral pretreatment mitigated HNE's damage, indicating a safener-type action targeting toxic lipid peroxidation products such as HNE (Rac et al. 2021).

9.6 JA-induced β-Cyclocitral and Protection Against Bacterial Blight in Rice

JA, another phytohormone, responds to pathogenic rice defense. JA is responsible for inducing resistance against rice bacterial blight caused by the bacterial pathogen *Xanthomonas oryzae* pv. *oryzae* (Xoo). β-Cyclocitral apocarotenoid is responsive to JA, providing induced resistance against bacterial blight. It also negatively affects ABA biosynthesis, inhibiting ABA and providing dual action-boosting defense, which may contribute to rice's ability to fend off bacterial blight (Taniguchi et al. 2023). The accumulation of β-cyclocitral induced by JA is controlled by OsJAZ8; it acts as a suppressor of JA signaling in rice. The β-cyclocitral helps in the development of resistance to *X. oryzae* pv. *oryzae* (Xoo) and increasing the expression of defense-related genes in rice. JA and β-cyclocitral together also help in the activation of ABA-responsive genes and ABA-biosynthesis genes, which results in a reduction of ABA levels in rice (Yamada et al. 2012). β-Cyclocitral does not affect the ABA-dependent interactions between OsPYL/RCAR5 and OsPP2C49 in yeast cells. The rice CCD-4b (OsCCD4b) also responds to JA, is located in the chloroplast, and generates β-cyclocitral (Taniguchi et al. 2023).

9.7 Role in Regulation of Root Architecture

β-Cyclocitral stimulates the growth of roots and the development of the root system, thereby increasing the division of cells in the roots. Under saline conditions, the presence of β-cyclocitral in *Arabidopsis* plants accelerates root growth and improves the plant's ability to withstand stress. Primary roots and lateral roots are increased by exogenously applied β-cyclocitral due to induced cell division of root meristems (Dickinson et al. 2019; Ke et al. 2022). It acts as a root growth promoter and has been applied in tomato and rice research. It affects the architectural pattern of roots, the number of roots, and the direction of deeper roots. β-Cyclocitral greatly enhances the development of both roots and shoots (Deshpande et al. 2021). Naturally produced molecules such as β-cyclocitral can enhance crop vitality, particularly in challenging environmental conditions. It has a lot of potential when applied exogenously in plants. Even lower concentrations of β-cyclocitral stimulate changes in signaling of plants (D'Alessandro et al. 2019).

9.8 β-Cyclocitral and Nonvascular Plants: Allelochemical Response

β-Cyclocitral, as an allelochemical, has the potential to significantly impact the growth and development of nonvascular plants through various mechanisms. β-Cyclocitral acts as an allelochemical and influences the growth, behavior, and distribution of other plants (Shan et al. 2023). Nonvascular plants lack xylem and phloem, which are found in higher plants and include mosses, liverworts, and hornworts. They rely on diffusion to transport water and nutrients and are typically found in moist environments. β-Cyclocitral inhibits the growth of nonvascular plants by affecting cell division and elongation, leading to reduced biomass and rather slower colonization rates (Havaux 2020). Exposure to β-cyclocitral might cause morphological changes in nonvascular plants, such as thallus structure or pigmentation alterations, impacting their photosynthetic capacity and reproductive abilities. Nonvascular plants respond to β-cyclocitral by activating stress mechanisms that may lead to protective metabolite synthesis or metabolic changes. Additionally, β-cyclocitral can hinder spore germination, reducing establishment rates (Faizan et al. 2022). Methyl 2,6,6-trimethyl-1-cyclohexene-1-carboxylate is a volatile organic compound derivative of β-cyclocitral that is formed in mosses for competitive survival in species (Vicherová et al. 2020; Ninkovic et al. 2021). *Hamatocaulis vernicosus* and *Sphagnum fexuosum* are known moss species that increase the release of such molecules (Vicherová et al. 2020).

9.9 β-Cyclocitral and Herbivory

Insects feed on plants and damage them, thereby weakening their growth progress and alarming the plant's defense mechanisms. The upregulation of defense metabolites is essential during herbivory. In *Arabidopsis*, the herbivore-triggered signal β-cyclocitral hinders the activity of the initially formed enzyme in the MEP (2-*C*-methyl-D-erythritol-4-phosphate) pathway, which is responsible for the synthesis of photosynthetic pigments (Dhiman et al. 2021; Sangma et al. 2022; Sharma et al. 2022; Al-Khayri et al. 2023). The decrease in MEP pathway flow is followed by an increase in the plant's defense abilities against a chewing herbivore. In *Arabidopsis*, plant-insect attacks suppress the MEP pathway involved in photosynthesis, metabolite synthesis, and are crucial for plant growth. The herbivory effect of *Spodoptera littoralis* causes a decrease in the flow of substances involved in the pathway, which reduces the concentration of intermediate 1-deoxy-D-xylulose

5-phosphate (DXP) in the pathway, thereby affecting the process (Mitra et al. 2021; Yuan et al. 2022). Herbivory also increases the ROS content, which in turn increases β-cyclocitral from apocarotenoids. β-Cyclocitral downregulates the MEP pathway as it leads to the formation of terpenoid chlorophylls, carotenoids, plastoquinones, phylloquinone, etc., that play a role in photosystem (Vranová et al. 2012). The process also affects DXP pool reduction due to a diminished supply of substrates to the pathway, due to a decrease in the photosynthetic rate indicating that β-cyclocitral controls flux in such crucial pathway under normal growth conditions (González-Cabanelas et al. 2015; Mitra et al. 2021).

9.10 β-Cyclocitral Role in Metabolite Formation

β-Cyclocitral upregulates many metabolites having a role in growth and defense mechanisms in plants. Growth requirements are essential to upregulate β-cyclocitral specifically for those metabolites, as seen with the exogenous application of β-cyclocitral in *Arabidopsis* plants. To combat drought and enhance resistance to insect herbivores, β-cyclocitral shows various effects on plant physiology via the priming of stress-responsive genes and defense pathways (Deshpande et al. 2021). Reduction in C flux during growth and defense tradeoff is observed in plants (Biswal 1995). β-Cyclocitral-treated plants have been found to be significantly influenced by amino acid metabolism, along with the accumulation of phytophenols. The levels of tryptophan and its precursor shikimate were found to be upregulated in the presence of β-cyclocitral, which is also a precursor for phenylalanine. It also increases the level of other linked exudates in plant systems that influence growth and photosynthetic machinery. Glutamate, glycine, aspartate, etc. play an important role in regulating chlorophyll content, root hair growth, energy metabolism, and signaling molecule interplay and are part of many crucial biochemical pathways (Singh and Verma 2023; Singh et al. 2024d). Leucine is involved in serine/threonine protein kinases and other proteins that require leucine to facilitate cell signaling. Aspartate is the precursor of four essential amino acids by forming a common intermediate (Deshpande and Mitra 2023). β-Cyclocitral induces the accumulation of these amino acids; many of the amino acids also play an important role in stress, along with building plant structure. The formation of other C sources, such as sugars and organic acids, is also enhanced. Exogenous application of β-cyclocitral in citrus can enhance the provitamin A carotenoids, phytoene, and phytofluene (Zheng et al. 2020).

9.11 Conclusion

β-Cyclocitral is one of the derivative apocarotenoids that is under study due to its potential for stimulating and signal-altering activities during stress. Many areas of plant growth are linked to phytohormones, and β-cyclocitral is among the crucial molecules that provide different defensive roles to plant physiology, such as during herbivory. It also induces stress signaling during photosynthesis reactions in chloroplasts and plastid. It regulates different crucial pathways that are linked to the triggering of phytohormones like jasmonate and SA. Gene upregulation, as well as downregulation of different phytohormones, was observed during exogenous application of β-cyclocitral. The exogenously applied β-cyclocitral was known to enhance stress tolerance, upregulate defensive mechanisms, and balance the C flux of different terpenoid formations. Overall, many applications of β-cyclocitral can be explored in vascular as well as nonvascular plants. The odor of different plant species is controlled by the level of β-cyclocitral in them as well as in their fruits. Moreover, the evolutionary basis of different organisms in retaining β-cyclocitral as an important molecule in nature is evident. Its role in communication in nonvascular plants, its origin of volatile communication, and its repellent nature in some plants are all different ways of communication that are linked to the survival of organisms.

References

Alcaíno, J., Baeza, M. and Cifuentes, V. (2016). Carotenoid distribution in nature. In: *Subcellular Biochemistry: Vol. 79: Carotenoids in Nature: Biosynthesis, Regulation, and Function* (ed. C. Stange), 3–33. Springer.

Al-Khayri, J.M., Rashmi, R., Toppo, V. et al. (2023). Plant secondary metabolites: the weapons for biotic stress management. *Metabolites* 13(6), 716.

Altomare, A., Baron, G., Gianazza, E. et al. (2021). Lipid peroxidation derived reactive carbonyl species in free and conjugated forms as an index of lipid peroxidation: limits and perspectives. *Redox Biology* 42, 101899.

Auldridge, M.E., Block, A., Vogel, J.T. et al. (2006). Characterization of three members of the Arabidopsis carotenoid cleavage dioxygenase family demonstrates the divergent roles of this multifunctional enzyme family. *The Plant Journal* 45(6), 982–993.

Bali, S., Gautam, A., Dhiman, A. et al. (2023). Salicylate and jasmonate intertwine in ROS-triggered chloroplast-to-nucleus retrograde signaling. *Physiologia Plantarum* 175(5), e14041.

Birtic, S., Ksas, B., Genty, B. et al. (2011). Using spontaneous photon emission to image lipid oxidation patterns in plant tissues. *The Plant Journal* 67(6), 1103–1115.

Biswal, B. (1995). Carotenoid catabolism during leaf senescence and its control by light. *Journal of Photochemistry and Photobiology B: Biology* 30(1), 3–13.

Bobik, K. and Burch-Smith, T.M. (2015). Chloroplast signaling within, between, and beyond cells. *Frontiers in Plant Science* 6, 781.

Bogacz-Radomska, L. and Harasym, J. (2018). β-Carotene—properties and production methods. *Food Quality and Safety* 2(2), 69–74.

Britton, G. (2022). Getting to know carotenoids. In: *Methods in Enzymology* (ed. E.T. Wurtzel), 1–56. Academic Press.

Burri, B. (2013). Carotenoids: chemistry, sources and physiology. In: *Reference Module in Biomedical Science: Vol. 1: Encyclopedia of Human Nutrition* (ed. B. Caballero), 283–291. Academic Press.

Chauhan, D., Singh, D., Pandey, H. et al. (2023). Impact of transcription factors in plant abiotic stress: a recent advancement for crop improvement. In: *Plant Transcription Factors: Contribution in Development, Metabolism and Environmental Stress* (eds. V. Srivastava, S. Mishra, S. Mehrotra et al.), 271–286. Academic Press.

D'Alessandro, S., Mizokami, Y., Légeret, B. et al. (2019). The apocarotenoid β-cyclocitric acid elicits drought tolerance in plants. *iScience* 19, 461–473.

Deshpande, S., Manoharan, R. and Mitra, S. (2021). Exogenous β-cyclocitral treatment primes tomato plants against drought by inducing tolerance traits, independent of abscisic acid. *Plant Biology* 23, 170–180.

Deshpande, S. and Mitra, S. (2023). β-Cyclocitral-mediated metabolic changes optimize growth and defense responses in *Solanum lycopersicum* L. *Metabolites* 13(3), 329.

Dhami, N., Pogson, B.J., Tissue, D.T. et al. (2022). A foliar pigment-based bioassay for interrogating chloroplast signalling revealed that carotenoid isomerisation regulates chlorophyll abundance. *Plant Methods* 18(1), 18.

Dhiman, V.K., Chauhan, V., Kanwar, S.S. et al. (2021). Purification and characterization of actinidin from *Actinidia deliciosa* and its utilization in inactivation of α-amylase. *Bulletin of the National Research Centre* 45, 213. https://doi.org/10.1186/s42269-021-00673-0.

Dhiman, V.K., Rana, N., Dhiman, V.K. et al. (2022). Effect of rhizobial isolates and nitrogen fertilizers on nursery performance, nodulation behavior and nitrogenase activity of *Dalbergia sissoo* Roxb. seedlings. *Plant Stress* 4, 100080.

Dhiman, V.K., Singh, D., Dhiman, V.K. et al. (2024). Multi-omics: an advanced bioinformatics approach for crop improvement in agriculture. In: *Bioinformatics for Plant Research and Crop Breeding* (ed. J.-T. Chen), 75–98. Wiley.

Dickinson, A.J., Lehner, K., Mi, J. et al. (2019). β-Cyclocitral is a conserved root growth regulator. *Proceedings of the National Academy of Sciences of the United States of America* 116(21), 10563–10567.

Dumay, J. and Morançais, M. (2016). Proteins and pigments. In: *Seaweed in Health and Disease Prevention* (eds. J. Fleurence and I. Levine), 275–318. Academic Press.

Faizan, M., Tonny, S.H., Afzal, S. et al. (2022). β-Cyclocitral: emerging bioactive compound in plants. *Molecules* 27(20), 6845.

Felemban, A., Braguy, J., Zurbriggen, M.D. et al. (2019). Apocarotenoids involved in plant development and stress response. *Frontiers in Plant Science* 10, 1168.

Fiedor, J., Fiedor, L., Haeßner, R. et al. (2005). Cyclic endoperoxides of β-carotene, potential pro-oxidants, as products of chemical quenching of singlet oxygen. *Biochimica et Biophysica Acta (BBA)—Bioenergetics* 1709(1), 1–4.

Fischer, B.B., Hideg, É. and Krieger-Liszkay, A. (2013). Production, detection, and signaling of singlet oxygen in photosynthetic organisms. *Antioxidants & Redox Signaling* 18(16), 2145–2162.

Gawroński, P., Górecka, M., Bederska, M. et al. (2013). Isochorismate synthase 1 is required for thylakoid organization, optimal plastoquinone redox status, and state transitions in *Arabidopsis thaliana*. *Journal of Experimental Botany* 64(12), 3669–3679.

Gharabli, H., Della Gala, V. and Welner, D.H. (2023). The function of UDP-glycosyltransferases in plants and their possible use in crop protection. *Biotechnology Advances* 67, 108182.

González-Cabanelas, D., Wright, L.P., Paetz, C. et al. (2015). The diversion of 2-*C*-methyl-D-erythritol-2, 4-cyclodiphosphate from the 2-*C*-methyl-D-erythritol 4-phosphate pathway to hemiterpene glycosides mediates stress responses in *Arabidopsis thaliana*. *The Plant Journal* 82(1), 122–137.

Harrison, E.H. (2022). Carotenoids, β-apocarotenoids, and retinoids: the long and the short of it. *Nutrients* 14(7), 1411.

Harrison, P.J. and Bugg, T.D. (2014). Enzymology of the carotenoid cleavage dioxygenases: reaction mechanisms, inhibition and biochemical roles. *Archives of Biochemistry and Biophysics* 544, 105–111.

Hartati, I. and Paramita, V. (2023). Apocarotenoids: sources, classification and their potential application as food additives. *Research in Chemical Engineering (RiCE)* 2(2), 44–50.

Havaux, M. (2020). β-Cyclocitral and derivatives: emerging molecular signals serving multiple biological functions. *Plant Physiology and Biochemistry* 155, 35–41.

Havaux, M. and Ksas, B. (2022). Imaging of lipid peroxidation-associated chemiluminescence in plants: spectral features, regulation and origin of the signal in leaves and roots. *Antioxidants* 11(7), 1333.

Hou, X., Rivers, J., León, P. et al. (2016). Synthesis and function of apocarotenoid signals in plants. *Trends in Plant Science* 21(9), 792–803.

Ivanov, B., Kozuleva, M. and Mubarakshina, M. (2011). Oxygen metabolism in chloroplast. In: *Cell Metabolism: Cell Homeostasis and Stress Response* (ed. P. Bubulya), 39–72. InTech.

Jiang, J. and Dehesh, K. (2021). Plastidial retrograde modulation of light and hormonal signaling: an odyssey. *New Phytologist* 230(3), 931–937.

Ke, D., Guo, J., Li, K. et al. (2022). Carotenoid-derived bioactive metabolites shape plant root architecture to adapt to the rhizospheric environments. *Frontiers in Plant Science* 13, 986414.

Kim, T., Kim, T.K. and Zoh, K.D. (2019). Degradation kinetics and pathways of β-cyclocitral and β-ionone during UV photolysis and UV/chlorination reactions. *Journal of Environmental Management* 239, 8–16.

Li, M. and Kim, C. (2023). Singlet oxygen in plants: from genesis to signaling. In: *Advances in Botanical Research* (eds. R. Mittler and F. Van Breusegem), 1–42. Academic Press.

Lin, Y., Cheng, C., Dai, Y. et al. (2024). The origins of odor (β-cyclocitral) under different water nutrient conditions: algae or submerged plants? *Science of The Total Environment* 931, 173024.

Lv, F., Zhou, J., Zeng, L. et al. (2015). β-Cyclocitral upregulates salicylic acid signalling to enhance excess light acclimation in *Arabidopsis*. *Journal of Experimental Botany* 66(15), 4719–4732.

Maoka, T. (2009). Recent progress in structural studies of carotenoids in animals and plants. *Archives of Biochemistry and Biophysics* 483(2), 191–195.

Maoka, T. (2020). Carotenoids as natural functional pigments. *Journal of Natural Medicines* 74(1), 1–6.

Martínez-Cámara, S., Ibañez, A., Rubio, S. et al. (2021). Main carotenoids produced by microorganisms. *Encyclopedia* 1(4), 1223–1245.

Mi, J., Jia, K.P., Balakrishna, A. et al. (2019). An LC-MS profiling method reveals a route for apocarotene glycosylation and shows its induction by high light stress in Arabidopsis. *Analyst* 144(4), 1197–1204.

Mitra, S., Estrada-Tejedor, R., Volke, D.C. et al. (2021). Negative regulation of plastidial isoprenoid pathway by herbivore-induced β-cyclocitral in *Arabidopsis thaliana*. *Proceedings of the National Academy of Sciences of the United States of America* 118(10), e2008747118.

Mitra, S. and Gershenzon, J. (2022). Effects of herbivory on carotenoid biosynthesis and breakdown. In: *Methods in Enzymology* (ed. E.T. Wurtzel), 497–517. Academic Press.

Moreno, J.C., Mi, J., Alagoz, Y. et al. (2021). Plant apocarotenoids: from retrograde signaling to interspecific communication. *The Plant Journal* 105(2), 351–375.

Ninkovic, V., Markovic, D. and Rensing, M. (2021). Plant volatiles as cues and signals in plant communication. *Plant, Cell & Environment* 44(4), 1030–1043.

Rac, M., Shumbe, L., Oger, C. et al. (2021). Luminescence imaging of leaf damage induced by lipid peroxidation products and its modulation by β-cyclocitral. *Physiologia Plantarum* 171(2), 246–259.

Ramel, F., Birtic, S., Cuiné, S. et al. (2012). Chemical quenching of singlet oxygen by carotenoids in plants. *Plant Physiology* 158(3), 1267–1278.

Ramel, F., Ksas, B., Akkari, E. et al. (2013). Light-induced acclimation of the *Arabidopsis chlorina1* mutant to singlet oxygen. *The Plant Cell* 25(4), 1445–1462.

Rathva, H., Pandey, A.K., Suthar, K. et al. (2024). Genetic relatedness analysis in sapota using SSR markers. *Ecological Genetics and Genomics* 31, 100234.

Rosas-Saavedra, C. and Stange, C. (2016). Biosynthesis of carotenoids in plants: enzymes and color. In: *Subcellular Biochemistry:* Vol. 79: *Carotenoids in Nature: Biosynthesis, Regulation and Function* (ed. C. Stange), 35–69. Springer.

Rubio-Moraga, A., Rambla, J.L., Fernández-de-Carmen, A. et al. (2014). New target carotenoids for CCD4 enzymes are revealed with the characterization of a novel stress-induced carotenoid cleavage dioxygenase gene from *Crocus sativus. Plant Molecular Biology* 86, 555–569.

Sangma, R.R., Manpoong, C., Sharma, A. et al. (2022). Performance of black rice (*Oryza sativa*) varieties grown in Namsai district of Arunachal Pradesh, India. *Research on Crops* 23(1), 11–14.

Shan, Z., Zhou, S., Shah, A. et al. (2023). Plant allelopathy in response to biotic and abiotic factors. *Agronomy* 13(9), 2358.

Sharma, A., Hazarika, M., Heisnam, P. et al. (2024a). Controlled environment ecosystem: a cutting-edge technology in speed breeding. *ACS Omega* 9(27), 29114–29138.

Sharma, A., Hazarika, M., Heisnam, P. et al. (2024b). Spontaneous formation, gene regulation of *Trichoderma* and slow decomposition in cocopeat. *Next Sustainability* 4, 100051.

Sharma, A., Pandey, H., Manpoong, C. et al. (2023a). Biometric analysis, biosynthetic pathway and multipurpose uses of buckwheat local varieties in Namsai district of eastern Himalayas in India. *Trends in Food Science & Technology* 136, 251–267.

Sharma, A., Pandey, H., Nampoothiri Devadas, V.A.S. et al. (2023b). Regulations and factors affecting of light emitting diodes artificial light in crop production under soilless culture. *ACS Agricultural Science & Technology* 3(8), 616–630.

Sharma, S.K., Singh, D., Pandey, H. et al. (2022). An overview of roles of enzymatic and nonenzymatic antioxidants in plant. In: *Antioxidant Defense in Plants* (eds. T. Aftab and K.R. Hakeem), 1–13. Springer. https://doi.org/10.1007/978-981-16-7981-0_1.

Shi, J., Cao, C., Xu, J. et al. (2020). Research advances on biosynthesis, regulation, and biological activities of apocarotenoid aroma in horticultural plants. *Journal of Chemistry* 2020(1), 2526956.

Shumbe, L., D'Alessandro, S., Shao, N. et al. (2017). METHYLENE BLUE SENSITIVITY 1 (MBS1) is required for acclimation of Arabidopsis to singlet oxygen and acts downstream of β-cyclocitral. *Plant, Cell & Environment* 40(2), 216–226.

Simkin, A.J. (2021). Carotenoids and apocarotenoids in planta: their role in plant development, contribution to the flavour and aroma of fruits and flowers, and their nutraceutical benefits. *Plants* 10(11), 2321.

Singh, D., Dhiman, V.K., Pandey, H. et al. (2022a). Crosstalk between salicylic acid and auxins, cytokinins and gibberellins under biotic stress. In: *Signaling and Communication in Plants: Auxins, Cytokinins and Gibberellins Signaling in Plants* (ed. T. Aftab), 249–262. Springer International Publishing.

Singh, D., Pandey, H., Dhiman, V.K. et al. (2022b). Secondary metabolite engineering for plant immunity against various pathogens. In: *Metabolic Engineering in Plants* (eds. T. Aftab and K.R. Hakeem), 123–143. Springer Nature Singapore.

Singh, D., Pandey, H., Shrivastava, N.K. et al. (2022c). Green synthesized gold and silver nanoparticles for antimicrobial applications. In: *Encyclopedia of Green Materials* (eds. C. Baskar, S. Ramakrishna and A. Daniela La Rosa), 1–12. Springer Nature Singapore.

Singh, D., Pandey, H., Thakur, K. et al. (2022d). Aquaporins and their functions in water transportation in different plant species. *Environmental Sustainability* 5, 443–456. https://doi.org/10.1007/s42398-022-00251-8.

Singh, D., Sharma, A., Verma, S.K. et al. (2024a). Impact of nanoparticles on plant physiology, nutrition, and toxicity: a short review. *Next Nanotechnology* 6, 100081.

Singh, D., Singh, A. and Kumar, S. (2023). Probiotics: friend or foe to the human immune system. *Bulletin of the National Research Centre* 47(1), 126.

Singh, D. and Verma, S.K. (2023). Impacts of particulate matter pollution on plants. In: *Environmental Science and Engineering: New Frontiers in Plant-Environment Interactions* (ed. T. Aftab), 483–500. Springer. https://doi.org/10.1007/978-3-031-43729-8_17.

Singh, D., Verma, S.K. and Kumar, S. (2024b). Genetic manipulation for secondary metabolite production. In: *Secondary Metabolites and Biotherapeutics: Developments in Applied Microbiology and Biotechnology* (eds. A. Kumar and S. Kumar), 161–188. Academic Press.

Singh, D., Verma, S.K. and Kumar, S. (2024c). Trends in secondary metabolites production from plant sources. In: *Secondary Metabolites and Biotherapeutics: Developments in Applied Microbiology and Biotechnology* (eds. A. Kumar and S. Kumar), 103–126. Academic Press.

Singh, D., Verma, S.K. and Shyam, P. (2024d). Green materials and purification of plant secondary metabolite as medicinal raw materials. In: *Secondary Metabolites and Biotherapeutics: Developments in Applied Microbiology and Biotechnology* (eds. A. Kumar and S. Kumar), 9–38. Academic Press.

Singh, D., Verma, S.K., Wani, K.I. et al. (2024e). An overview of metal and metalloid toxicity and signaling in plants. In: *Signaling and Communication in Plants: Metals and Metalloids in Plant Signaling* (ed. T. Aftab), 1–17. Springer. https://doi.org/10.1007/978-3-031-59024-5_1.

Sun, T., Rao, S., Zhou, X. et al. (2022). Plant carotenoids: recent advances and future perspectives. *Molecular Horticulture* 2(1), 3.

Taniguchi, S., Takeda, A., Kiryu, M. et al. (2023). Jasmonic acid-induced β-cyclocitral confers resistance to bacterial blight and negatively affects abscisic acid biosynthesis in rice. *International Journal of Molecular Sciences* 24(2), 1704.

Tomita, K., Hasegawa, M., Arii, S. et al. (2016). Characteristic oxidation behavior of β-cyclocitral from the cyanobacterium *Microcystis. Environmental Science and Pollution Research* 23, 11998–12006.

Tyagi, S., Dhiman, V.K., Dhiman, V.K. et al. (2024). Plant defense strategies and biomarkers against heavy metal-induced stress: a comprehensive review. *ACS Agricultural Science & Technology* 4(2), 129–143.

van Berkel, W.J., Kamerbeek, N.M. and Fraaije, M. (2006). Flavoprotein monooxygenases, a diverse class of oxidative biocatalysts. *Journal of Biotechnology* 124(4), 670–689.

Verma, S.K., Singh, D. and Shyam, P. (2024). Elicitation of secondary metabolites from plants. In: *Secondary Metabolites and Biotherapeutics: Developments in Applied Microbiology and Biotechnology* (eds. A. Kumar and S. Kumar), 127–159. Academic Press.

Vicherová, E., Glinwood, R., Hájek, T. et al. (2020). Bryophytes can recognize their neighbours through volatile organic compounds. *Scientific Reports*10(1), 7405.

Vranová, E., Coman, D. and Gruissem, W. (2012). Structure and dynamics of the isoprenoid pathway network. *Molecular Plant* 5(2), 318–333.

Wang, J., Zhang, N., Zhao, M. et al. (2020). Carotenoid cleavage dioxygenase 4 catalyzes the formation of carotenoid-derived volatile β-ionone during tea (*Camellia sinensis*) withering. *Journal of Agricultural and Food Chemistry* 68(6), 1684–1690.

Winterhalter, P. and Rouseff, R. (2005). Carotenoid-derived aroma compounds: an introduction. In: *ACM Symposium Series: Carotenoid-Derived Aroma Compounds* (eds. P. Winterhalter and R.L. Rouseff), 1–17. American Chemical Society.

Xiao, Y., Wang, J. and Dehesh, K. (2013). Review of stress specific organelles-to-nucleus metabolic signal molecules in plants. *Plant Science* 212, 102–107.

Yadav, G., Prajapati, P., Singh, D. et al. (2023). Emerging trends in plant metabolomics and hormonomics to study abiotic stress tolerance associated with rhizospheric probiotics. In: *Plant Hormones in Crop Improvement* (eds. M. Iqbal, R. Khan, A. Singh et al.), 283–306. Academic Press.

Yamada, S., Kano, A., Tamaoki, D. et al. (2012). Involvement of OsJAZ8 in jasmonate-induced resistance to bacterial blight in rice. *Plant and Cell Physiology* 53(12), 2060–2072.

Yanykin, D., Paskhin, M., Ashikhmin, A.A. et al. (2024). Carotenoid-dependent singlet oxygen photogeneration in light-harvesting complex 2 of *Ectothiorhodospira haloalkaliphila* leads to the formation of organic hydroperoxides and damage to both pigments and protein matrix. *PeerJ* 12, e16615.

Yuan, G.G., Zhao, L.C., Du, Y.W. et al. (2022). Repellence or attraction: secondary metabolites in pepper mediate attraction and defense against *Spodoptera litura*. *Pest Management Science* 78(11), 4859–4870.

Zheng, X., Yang, Y. and Al-Babili, S. (2021). Exploring the diversity and regulation of apocarotenoid metabolic pathways in plants. *Frontiers in Plant Science* 12, 787049.

Zheng, X., Zhu, K., Ye, J. et al. (2020). The effect of β-cyclocitral treatment on the carotenoid content of transgenic Marsh grapefruit (*Citrus paradisi* Macf.) suspension-cultured cells. *Phytochemistry* 180, 112509.

Zia-Ul-Haq, M. (2021). Historical and introductory aspects of carotenoids. In: *Carotenoids: Structure and Function in the Human Body* (eds. M. Zia-Ul-Haq, S. Dewanjee and M. Riaz), 1–42. Springer.

10

β-Ionone: A Key Apocarotenoid in Plants and Wine

Alice Vilela[1]* *and Berta Gonçalves*[2]

[1] *Chemistry Research Centre–Vila Real (CQ-VR), Department of Agronomy, School of Agrarian and Veterinary Sciences (ECAV), University of Trás-os-Montes e Alto Douro, Quinta de Prados, Vila Real, Portugal*
[2] *Centre for the Research and Technology of Agro-Environmental and Biological Sciences (CITAB), Institute for Innovation, Capacity Building and Sustainability of Agri-Food Production (Inov4Agro), Department of Biology and Environment, School of Life and Environmental Sciences (ECVA), University of Trás-os-Montes e Alto Douro, Quinta de Prados, Vila Real, Portugal*

*Corresponding author: avimoura@utad.pt

10.1 Introduction

Norisoprenoids are a class of compounds derived from the oxidative cleavage of carotenoids (El Hadi et al. 2013; Li et al. 2024). They are known for significantly impacting the aroma and flavor profiles of various fruits, vegetables, and beverages, such as wine and juice.

Norisoprenoids have been identified in various sources, including orange juice and wine. In Valencia orange juice, four norisoprenoids—α-ionone, β-ionone, β-cyclocitral, and β-damascenone—were identified, contributing significantly to the floral aroma category (Mahattanatawee et al. 2005). In wine, norisoprenoids such as β-damascenone, β-ionone, and TDN (1,1,6-trimethyl-1,2-dihydronaphthalene) have been noted for their low olfactory perception thresholds, making them highly impactful on wine aroma (Mendes-Pinto 2009).

The biosynthesis of norisoprenoids involves the cleavage of carotenoids, a process facilitated by enzymes such as carotenoid cleavage dioxygenases (CCDs) (Paparella et al. 2021). In carrots, the enzyme DcCCD1 was found to cleave cyclic carotenes to generate α-ionone and β-ionone, indicating its role in flavor biosynthesis (Yahyaa et al. 2013). In grapes, the genetic control of norisoprenoid content has been linked to specific single-nucleotide polymorphisms (SNPs) and genes such as *VvDXS1* and *VvGGPPS*, which are part of the isoprenoid metabolic pathway (Sun et al. 2023).

Environmental factors and hormonal treatments can influence norisoprenoid levels. For instance, abscisic acid (ABA) and synthetic auxin (NAA) treatments in grape berries were found to increase norisoprenoid concentrations, suggesting that these hormones play

Plant Apocarotenoids: Biosynthesis, Signaling and Agricultural Applications, First Edition. Edited by Kaiser Iqbal Wani and Tariq Aftab.
Published 2026 by John Wiley & Sons Ltd.

a role in the regulation of norisoprenoid biosynthesis during ripening (He et al. 2021). Additionally, nitrogen fertilization in vineyards has been shown to affect norisoprenoid levels in Riesling wines, with varying impacts on different norisoprenoid compounds depending on the vintage and environmental conditions (Linsenmeier and Löhnertz 2016).

Transcription factors also play a crucial role in regulating norisoprenoid biosynthesis. The transcription factor VvWRKY70 was identified as a repressor of norisoprenoid and flavonol production in grape berries, with its expression being downregulated by light and high-temperature treatments (Wei et al. 2023). This highlights the complex regulatory networks that control the production of these vital aroma compounds.

10.2 Role in Plants

β-Ionone is a volatile apocarotenoid derived from the oxidative cleavage of carotenoids, playing essential roles in plant physiology, development, and ecological interactions (Paparella et al. 2021; Felemban et al. 2024). This compound, characterized by its floral and fruity aroma, is widely distributed in various plant species, particularly in flowers and fruits, where it contributes to aroma and flavor profiles (Simkin 2021). However, beyond its role in plant secondary metabolism, β-ionone also exerts significant biological effects that influence plant growth, stress responses, and plant–insect interactions (Isah 2019; Liu et al. 2022; Zhou et al. 2022).

10.2.1 Contribution to Floral and Fruit Aroma

One of the primary functions of β-ionone in plants is its contribution to floral and fruit scent. It is a key volatile compound in many flowers, including violets (*Viola* spp.) (Anca et al. 2009), petunias (*Petunia* spp.) (Simkin et al. 2004), tulips (Mostafa et al. 2022), roses (*Rosa* spp.), brown boronia (*Boronia megastigma*), henna (*Lawsonia inermis*), sweet acacia (*Acacia farnesiana*), and yellow oleander (*Thevia peruviana*), where it enhances pollinator attraction (Paparella et al. 2021). The presence of β-ionone in floral scent bouquets has been shown to influence the behavior of pollinators such as bees and butterflies, thereby facilitating cross-pollination and ensuring reproductive success (Rachersberger et al. 2019; Llorens et al. 2023).

In fruits, β-ionone contributes to the characteristic aroma of grapes (*Vitis vinifera*) (Lin et al. 2019), raspberries (*Rubus idaeus*) (Aprea et al. 2015), mangoes (*Mangifera indica*), apricots (*Prunus armeniaca*), and melons (*Cucumis melo*) (Paparella et al. 2021). This volatile compound enhances fruit appeal to seed-dispersing animals and plays a role in post-harvest ripening processes (Distefano et al. 2022; Liu et al. 2022).

10.2.2 Role in Plant Defense

The apocarotenoid family comprises precursors of key phytohormones, including ABA and strigolactones. These molecules function as signaling molecules regulating plant growth, development, and defense against pathogens and herbivores (Moreno et al. 2021; Wani et al. 2021). Beyond its contribution to aroma, β-ionone functions as a plant defense signal (Paparella et al. 2021).

Studies have demonstrated that β-ionone can act as an allelochemical, influencing herbivore behavior and providing protection against insect pests. For instance, β-ionone is involved in the induced defense of canola (*Brassica napus*) and is emitted in response to herbivore-inflicted damage (Caceres et al. 2016). Certain herbivores and pathogens perceive β-ionone as a deterrent, reducing feeding and infestation rates. Moreover, β-ionone attracts parasitoids, such as *Microplitis pallidipes* Szépligeti (Hymenoptera: Braconidae), suggesting its potential to enhance parasitoid-based biocontrol of insect herbivores (Zhang et al. 2022). Additionally, β-ionone can induce systemic resistance mechanisms in plants, triggering defense-related gene expression and enhancing resilience against biotic stressors. In *Escherichia coli* cells, β-ionone induces oxidative stress (PkatG and Pdps promoters) through a specific response mediated by the OxyR/OxyS regulon (Plyuta et al. 2024). β-Ionone has also been reported to inhibit the growth of various pathogens, including *Peronospora tabacina*, *Candida albicans*, *Aspergillus niger*, and *Colletotrichum musae* (Felemban 2022).

The dual role of β-ionone as both an insect attractant and repellent exemplifies its versatility in plant–insect interactions. This volatile apocarotenoid plays a crucial role in maintaining ecological balance by influencing pollinator attraction, herbivore deterrence, and indirect plant defense. Understanding the mechanisms underlying these effects can inform agricultural practices, such as developing natural pest control strategies and pollination enhancement techniques.

10.2.3 Involvement in Stress Responses

Plants exposed to environmental stressors such as drought, salinity, high temperature, and oxidative stress often undergo metabolic adjustments that produce volatile apocarotenoids, including β-ionone. Recent research has shown that β-ionone plays a role in abiotic stress responses and contributes significantly to plant defense against biotic stressors (Cao et al. 2011; Felemban et al. 2024). Felemban and co-workers demonstrated that the application of β-ionone to *Arabidopsis thaliana* plants at micromolar concentrations caused a global reprogramming of gene expression, affecting thousands of transcripts involved in stress tolerance, growth, hormone metabolism, pathogen defense against *Botrytis cinerea*, and photosynthesis. Indeed, this compound has been implicated in modulating stress-responsive pathways, possibly by acting as a signaling molecule that interacts with phytohormones such as ABA and jasmonic acid (JA) (Ramel et al. 2012; Felemban et al. 2024). These interactions can enhance plant tolerance to biotic and abiotic stresses by regulating some key processes, such as stomatal closure, osmolyte accumulation, and antioxidant enzyme activity (Asselbergh et al. 2008; Cao et al. 2011; Chandrasekaran et al. 2021; Aslam et al. 2022). For instance, according to Chandrasekaran et al. (2021), arbuscular mycorrhizal fungi (AMFs) colonization, which results in increased β-ionone production, has been associated with enhanced activities of antioxidant enzymes such as superoxide dismutase (SOD), catalase (CAT), peroxidase (POD), and ascorbate peroxidase (APX) under salt stress conditions.

Zhang et al. (2025) showed that treating broccoli with β-ionone significantly maintained its antioxidant capacity during storage. The treatment upregulated genes involved in glucosinolate biosynthesis and the ascorbate–glutathione cycle, leading to higher levels of these antioxidants. Moreover, Ramel et al. (2012) have shown that β-ionone can increase plant tolerance to photooxidative stress. *Arabidopsis* plants exposed to β-ionone exhibited increased protection against chlorophyll bleaching and lipid peroxidation under high light stress conditions.

10.2.4 Potential Role in Plant Growth Regulation

Recent evidence suggests that β-ionone may play a role in plant development by influencing cell signaling pathways (Aloum et al. 2020; Moreno et al. 2021; Liu et al. 2022). It has been hypothesized that β-ionone derivatives can interact with plant hormones to regulate processes such as germination, flowering, and senescence (Ahrazem et al. 2016; Nambara and Van Wees 2021; Gómez-Sagasti et al. 2023). However, the precise molecular mechanisms underlying these effects remain to be fully elucidated. Indeed, the interaction between β-ionone and plant hormones appears complex and multifaceted. For instance, β-ionone has been shown to influence the jasmonate signaling pathway, which plays a crucial role in plant development and defense responses. The formation of the COI1–JAZ complex, a key component of jasmonate signaling, is affected by jasmonoyl-isoleucine (JA-Ile), a metabolite that shares structural similarities with β-ionone (Katsir et al. 2008; Pratiwi et al. 2017; Felemban et al. 2024). This suggests that β-ionone may modulate jasmonate-dependent developmental processes (Felemban et al. 2024; Qiu et al. 2024).

As research in this field progresses, it becomes increasingly clear that β-ionone and other apocarotenoids play crucial roles in plant biology beyond their traditional associations with pigmentation and aroma. The complex interplay between these compounds and plant hormones represents an exciting frontier in plant science, with potential implications for agriculture, biotechnology, and our understanding of plant evolution and adaptation.

10.3 β-Ionone Signaling and Mycorrhization

Mycorrhizal associations play a critical role in plant nutrient acquisition and stress tolerance (Azcón-Aguilar and Barea 2015; Begum et al. 2019), and emerging evidence suggests that volatile apocarotenoids such as β-ionone contribute to the regulation of these symbiotic interactions (Ramos et al. 2011; Fiorilli et al. 2019). β-Ionone may act as a signaling molecule that influences the establishment and function of mycorrhizal relationships by modulating plant and fungal communication pathways (Fiorilli et al. 2019).

Fiorilli et al. (2019) have shown that β-ionone emission can alter root exudation patterns, potentially influencing AMFs recruitment. By modifying the chemical composition of the rhizosphere, β-ionone may enhance the colonization efficiency of beneficial fungi such as Glomeromycota, improving plant access to essential nutrients like phosphorus and nitrogen. The interplay between β-ionone, phytohormones, and fungal symbionts remains a promising study area for understanding plant resilience under changing environmental conditions.

In summary, β-ionone is a multifunctional apocarotenoid with significant roles in plant ecology and physiology (Table 10.1). This volatile compound exemplifies the intricate connections between plant metabolism and environmental interactions, from attracting pollinators and seed dispersers to mediating defense responses, stress adaptation, and mycorrhizal interactions. Further research into its biosynthesis, regulation, and signaling mechanisms will deepen our understanding of its contribution to plant fitness and survival in dynamic ecosystems.

Table 10.1 Roles of β-ionone in several plants.

Role	Function	Example plants	References
Floral and fruit aroma	Attracts pollinators and seed dispersers	Violets, grapes, mangoes	Anca et al. (2009), Lin et al. (2019), Paparella et al. (2021)
Insect interaction	Attracts pollinators, repels herbivorous insects	Petunias, orchids, roses	Simkin et al. (2004), Paparella et al. (2021), Perkins et al. (2023)
Plant defense	Induces resistance against pests and pathogens	Tomatoes, cotton, basil	Sun et al. (2014), Teliban et al. (2021), Weinblum et al. (2021)
Stress response	Enhances tolerance to abiotic stress factors	Rice, wheat, maize	Markovic et al. (2019), Ficke et al. (2022), Votta et al. (2024)
Growth regulation	Modulates germination, flowering, and senescence	*Arabidopsis*, sunflowers	Paparella et al. (2021), Felemban et al. (2024)
Mycorrhization	Enhances fungal symbiosis and nutrient uptake	Rice, legumes, maize, soybeans	Sharma et al. (2017), Fiorilli et al. (2019), Votta et al. (2024)

10.4 Key Enzymes Involved in β-Ionone Synthesis

The synthesis of β-ionone involves several key enzymes, primarily from the CCD family (CCD1 [Carotenoid Cleavage Dioxygenase 1] and CCD4 [Carotenoid Cleavage Dioxygenase 4]), which cleave carotenoids like β-carotene to produce β-ionone (Figure 10.1). Additionally, lutein cleavage dioxygenase (EhLCD) and cytochrome P450 enzymes (CYP2B6 and CYP1A2) play roles in the synthesis and modification of β-ionone and its derivatives. These enzymes are found in various plants and microorganisms, each contributing uniquely to the biosynthesis pathway (Table 10.2).

10.5 Heterologous Production of β-Ionone in Microbial Cell Factories

Microbial cell factories are engineered microorganisms that produce a wide range of products, including fuels, chemicals, pharmaceuticals, and food ingredients, from renewable resources. This microbial approach integrates synthetic and systems biology tools to improve substrate utilization, growth performance, stress tolerance, and product export capacity. Enhancing physiological functions aims to overcome the bottlenecks in low-efficiency cell factories (Liu et al. 2021). Numerous works have been published about the use of microbial cell factories. Using PubMed and searching using the sentence "microbial cell factories," 4,829 results (from 1995 to 2025) were found, and a bibliometric map was generated using VOSviewer software version 1.6.19, arranging authors according to the number of publications and their works' timeline (Figure 10.2).

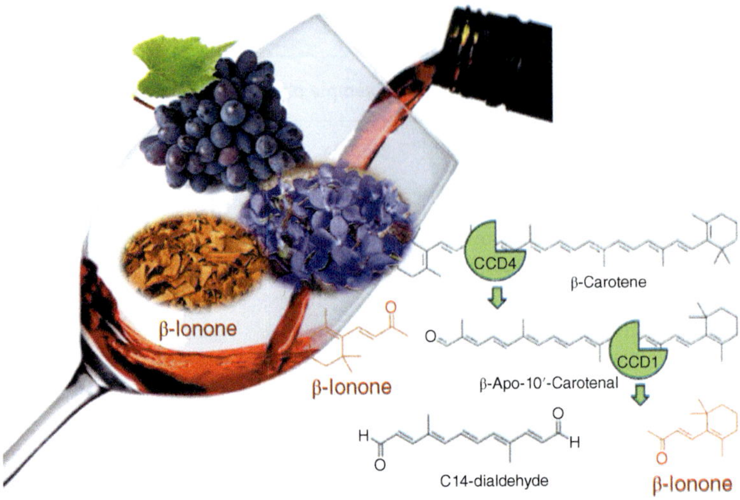

Figure 10.1 Schematic representation of the synthesis of β-ionone and the action of CCD1 and CCD4 enzymes.

Microbial cell factories are being developed to synthesize plant natural products (PNPs) traditionally extracted from plants. Protein and metabolic engineering can enhance biological networks and boost the production of target compounds. Protein engineering, in particular, helps generate desirable enzymes, facilitating the biosynthesis of value-added products (Xu et al. 2020). Despite challenges like unknown biosynthetic pathways and high fermentation costs, progress is being made in pathway mining and metabolic engineering for industrial-scale production (Liu et al. 2017).

The concept of microbial cell factories has evolved since the 1980s, with advancements in recombinant DNA technologies. Initially, simple strains like *E. coli* and *Saccharomyces cerevisiae* were used, but now more complex and diverse microbial species are being explored for industrial applications (Lee et al. 2012). Efforts are being made to standardize methods and platform organisms for bioproduction. New-to-industry microorganisms, such as *Rhodobacter sphaeroides*, are being investigated for their potential in the bioeconomy (Orsi et al. 2020; Nikel and Mattanovich 2021).

In conclusion, microbial cell factories are pivotal in advancing sustainable industrial bioprocesses. While challenges remain, ongoing research in metabolic and protein engineering and developing robust microbial strains holds promise for expanding their applications and improving their efficiency. This is important for accelerating the production of important enzymes, for instance.

The heterologous production of valuable compounds, like enzymes, in microbial cells has gained attention. In PubMed, searching for “enzymes production by microbial cell factories” retrieved 1,802 articles from 2002 to 2025. A bibliometric map was generated using VOSviewer software version 1.6.19, arranging authors according to the number of publications and their works’ timeline (Figure 10.3).

The production of β-ionone in microbial systems has become increasingly appealing because of the challenges associated with traditional chemical synthesis and plant

Table 10.2 Key enzymes involved in β-ionone synthesis.

Enzyme name	Enzyme action	Reference
CCD1	CsCCD1a and CsCCD1b, found in *Crocus sativus*, cleave β-carotene at the 9,10 (9′,10′) positions to yield β-ionone CsCCD1a is constitutively expressed, while CsCCD1b is specific to stigma tissue	Rubio et al. (2008)
	In *Petunia hybrida*, PhCCD1 is responsible for the β-ionone formation, with its expression regulated by light and circadian rhythms	Simkin et al. (2004)
	The enzyme NtCCD1-3, from *Nicotiana tabacum*, cleaves β-carotene and other carotenoids at the 9,10 (9′,10′) positions, significantly enhancing β-ionone production in yeast	Gong et al. (2022)
CCD4	CsCCD4a and CsCCD4b are enzymes from *Crocus sativus*. They are localized in plastids and highly expressed during stigma development, contributing to β-ionone synthesis	Rubio et al. (2008)
	The enzyme CsCCD4, in *Camellia sinensis*, catalyzes the formation of β-ionone during the tea withering process	Wang et al. (2020b)
	Transcript level of *PpCCD4* showed a positive relationship with the accumulation of β-ionone during the ripening process of peach	Liu et al. (2022)
EhLCD	The enzyme EhLCD, from *Enterobacter hormaechei* YT-3, converts lutein to 3-hydroxy-β-ionone, indicating its role in β-ionone derivative synthesis	Long et al. (2021)
Cytochrome P450	Human liver enzymes CYP2B6 and CYP1A2 catalyze the hydroxylation of β-ionone, converting it to 4-hydroxy-β-ionone	Marumoto et al. (2016)

extraction methods. This process entails converting β-carotene into β-ionone using engineered microbial hosts, like yeast and bacteria. However, a PubMed search for "β-ionone production microbial cell factories" yielded only 10 results, indicating that this topic requires more attention.

10.5.1 Host Strain Selection and Genetic Modifications

Yeast strains, particularly *S. cerevisiae* and *Yarrowia lipolytica*, are commonly used for β-ionone production due to their established genetic tools and ability to accumulate high precursors like β-carotene (Figure 10.4a, b). According to Groenewald et al. (2013), mating two carotenoid-overproducing *Y. lipolytica* strains creates a progeny with rich genetic diversity. White colonies have lost the trait to produce colored carotenoids (Figure 10.4a).

Regarding *S. cerevisiae*, in the work of López et al. (2015), an *S. cerevisiae* platform was constructed to synthesize β-ionone by combining two genetic engineering approaches to

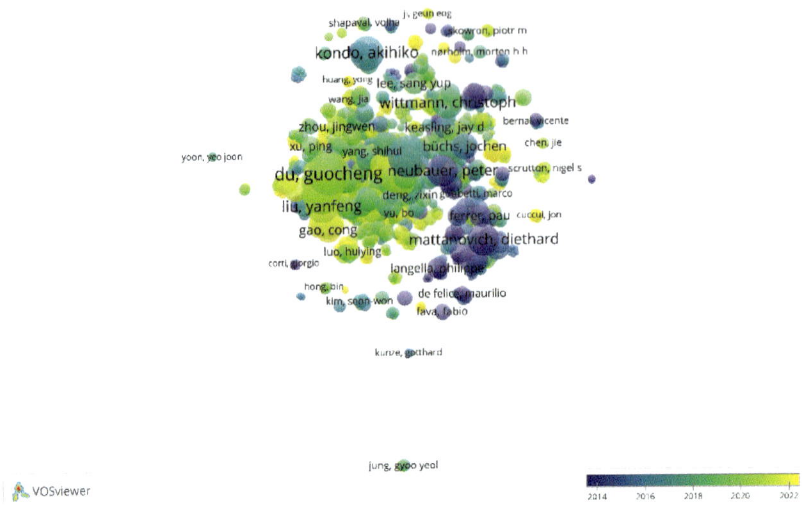

Figure 10.2 A bibliometric map showing authors according to the number of publications and their works' timeline regarding the topic "microbial cell factories." The colors on the map reflect different periods, with darker colors pointing to earlier dates and lighter colors indicating more recent times.

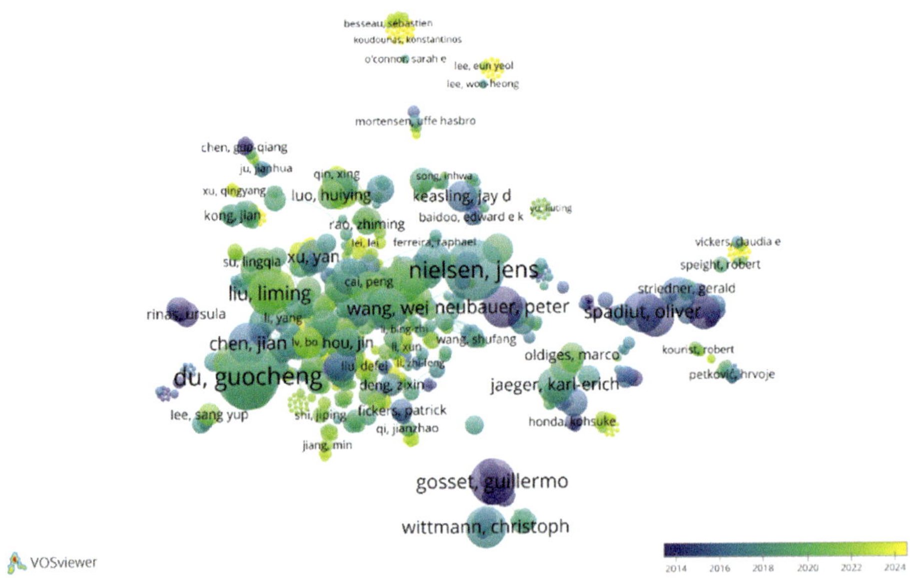

Figure 10.3 A bibliometric map showing authors according to the number of publications and their works' timeline regarding the topic "enzymes production by microbial cell factories." The colors on the map reflect different periods, with darker colors pointing to earlier dates and lighter colors indicating more recent times.

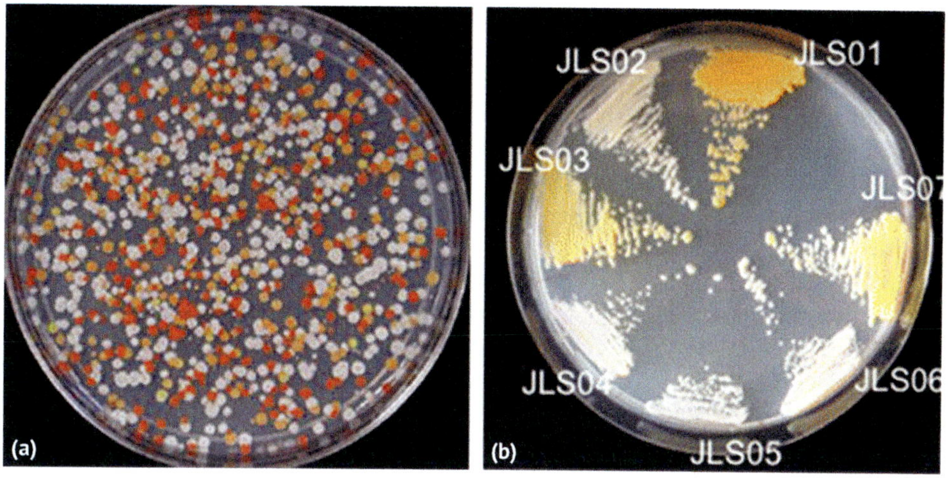

Figure 10.4 Carotenoid production in (a) *Y. lipolytica* and (b) *S. cerevisiae*. *Source:* (a) Groenewald et al. (2013)/Taylor & Francis Group; (b) López et al. (2015)-B/Springer Nature/CC0 1.0/CC BY 4.0. Colonies with different colors produce different carotenoids and/or quantities of carotenoids.

increase protein expression: (i) USER cloning-compatible integrative vectors and high-copy-number episomal expression systems; and (ii) by overexpressing a truncated version of the *HMG1* gene (*tHMG1*) and the endogenous GGPP synthase gene *BTS1*, together with the *crtYB* and the *crtI* genes from *Xanthophyllomyces dendrorhous* and the *CCD1* (*carotene cleavage dioxygenase 1*) gene from *Petunia hybrida* (*PhCCD1*) in an FPP-overproducing strain (SCGIS22). The authors generated a β-ionone-producing microbial cell factory, reaching a maximal concentration of (0.63 ± 0.02) mg/g biomass in shake flask cultures. In Figure 10.4b, the color of the different strains is constructed. The color changed from strain JLS01, which only expressed the *BST1* gene from *S. cerevisiae* and the carotenogenic genes *crtYB* and *crtI* from *X. dendrorhous*, to JLS02, that additionally expressed the *P. hybrida* gene *PhCCD1*. Color of the different strains expressed episomal versions of the *crtYB*/*crtI*/*PhCCD1* genes. All the strains expressed the integrated *BTS1*, *crtYB*, *crtI*, and *PhCCD1* genes and were subsequently transformed with episomal vectors containing *crtYB* (JLS3), *crtI* (JLS04), *PhCCD1* (JLS05), *crtYB–crtI* (JLS06), and *crtYB–PhCCD1* (JLS07) (López et al. 2015).

Genetic modifications, such as integrating carotenogenic genes from other organisms and optimizing gene expression, have significantly improved β-ionone yields (Czajka et al. 2018; López et al. 2020).

Escherichia coli has also been engineered for β-ionone production. By incorporating enzyme engineering strategies and optimizing metabolic pathways, *E. coli*-based systems have achieved high levels of β-ionone and other apocarotenoids (Chen et al. 2019).

The integration of carotenogenic genes (*crtE*, *crtYB*, and *crtI*) from *X. dendrorhous* into *S. cerevisiae* using CRISPR/Cas9 technology has been shown to increase β-carotene production significantly. Further expression of the *CCD1* gene from *P. hybrida* led to β-ionone production, achieving up to 33 mg/L in shake flasks (López et al. 2020). Fine-tuning gene expression and alleviating pathway bottlenecks, such as those catalyzed by hydroxymethylglutaryl-CoA reductase (HMGR1), resulted in the highest reported carotenoid concentration in *S. cerevisiae* (López et al. 2020).

Introduction of carotenogenic genes from *Mucor circinelloides* and *CCD1* from *P. hybrida* into *Y. lipolytica*, combined with enhancements in cytosolic acetyl-CoA supply and mevalonate pathway flux, led to a β-ionone titer of 358 mg/L in shake-flask fermentation and approximately 1 g/L in fed-batch fermentation (Lu et al. 2020). Overexpression of both heterologous and native genes in the mevalonate pathway, along with the introduction of *CCD1* from *Osmanthus fragrans*, resulted in β-ionone amounts of 68 mg/L in shaking flasks and 380 mg/L in a 2-L fermenter (Czajka et al. 2018).

Regarding protein engineering and enzyme optimization, mutations in the membrane-binding domains of *CCD1*, such as the K38A mutant of *NtCCD1-3* from *Nicotiana tabacum*, significantly enhanced β-ionone production in *S. cerevisiae*. Overexpression of NADH kinase Pos5 also improved yields (Gong et al. 2022). Single-point mutations and adding membrane destination peptides to PhCCD1 increased β-ionone yields up to fourfold compared to the native enzyme (Werner et al. 2019).

Future directions for the *de novo* synthesis of β-ionone are currently being studied. The nonspecific cleavage activity of CCD1 enzymes remains a bottleneck in β-ionone production. Engineering efforts should focus on improving the specificity and activity of these enzymes (Czajka et al. 2018; López et al. 2020). Moreover, both β-carotene- and β-ionone-producing strains showed decreased performance after successive subcultures, indicating a need for improved strain stability for industrial applications (Czajka et al. 2018).

10.6 Industrial Applications of β-Ionone

β-Ionone is widely used in the fragrance, flavor, and cosmetic industries and has potential uses in pharmaceuticals and agriculture. Due to its pleasant floral scent, often described as violet-like, β-ionone is a key ingredient in the fragrance industry used in perfumes and cosmetics (Costa et al. 2017; Czajka et al. 2018). It is also valued in the flavor industry for its contribution to the aroma and quality of products like tea (Wang et al. 2020a).

β-Ionone and its derivatives have been identified as promising eco-friendly agents for pest management, particularly in controlling aphid populations. They exhibit repellent activity and can be used as botanical-based products for sustainable agriculture (Pan et al. 2024).

Recent studies have highlighted β-ionone's potential in the biomedical field, particularly for its anticancer, antioxidant, and antibacterial properties (Paparella et al. 2021; Xu et al. 2024). These properties make it a compound of interest for developing new therapeutic agents. β-Ionone has demonstrated significant antiproliferative and apoptosis-inducing properties *in vitro* and *in vivo*. It affects various malignant cells by targeting distinct signaling pathways, potentially overcoming resistance to conventional anticancer agents (Janakiram et al. 2008; Ansari and Emami 2016). The compound mediates its effects through several mechanisms, including activating olfactory receptors, regulating cell cycle proteins, and modulating proapoptotic and antiapoptotic proteins (Aloum et al. 2020). It also enhances reactive oxygen species (ROS) generation, leading to apoptosis in cancer cells (Yang et al. 2020).

In specific cancer studies, β-ionone has demonstrated the ability to inhibit cell growth and induce apoptosis in colon cancer cells. It significantly suppresses aberrant crypt foci formation in rat models (Janakiram et al. 2008). Additionally, β-ionone inhibits epithelial–mesenchymal transition (EMT) in prostate cancer cells by negatively regulating the Wnt/β-catenin pathway, which reduces migration and invasion (Fang et al. 2022).

In mammary cancer, β-ionone has been shown to suppress mammary carcinogenesis in rat models, decreasing tumor incidence and the induction of apoptosis (Liu et al. 2008). This molecule has also been found to decrease telomerase activity and induce apoptosis in leukemia cells, highlighting its potential as a treatment option for specific types of leukemia (Faezizadeh et al. 2015).

10.7 β-Ionone in Wine

β-Ionone is a significant aroma compound in wine, contributing to its sensory profile with violet and woody notes. Understanding the presence and impact of β-ionone in wine is crucial for authenticity control and enhancing wine's organoleptic properties.

It is an aromatic compound found in grapes and wines, varying in concentration across different wine types and regions. For instance, Burgundy Pinot noir wines have shown the highest levels of β-ionone among the tested samples (Kotseridis et al. 1999; Langen et al. 2016).

β-Ionone significantly contributes to the aroma profile of wines, particularly influencing floral and berry notes. Its impact varies depending on the wine matrix, affecting descriptors such as floral, red berry, or dark berry aromas. Its presence allows individuals to distinguish between wines with varying concentrations of this compound.

The odor threshold of β-ionone in wine is relatively low, making it a potent aroma compound even at minimal concentrations. This threshold is crucial for determining its sensory significance in different wine styles (Kotseridis et al. 1999). It can improve the organoleptic properties of wine, making it more desirable to consumers (Timmins et al. 2020, 2022, 2023). However, sensitivity to β-ionone varies among individuals due to genetic differences, specifically the rs6591536 variant in the *OR5A1* gene. This genetic variation affects individuals' perception of the aroma, influencing their food and beverage preferences. Those with heightened sensitivity describe β-ionone as "fragrant" and "floral," which can affect their emotional associations and product choices (Jaeger et al. 2013).

β-Ionone concentrations in wines vary significantly. Authentic wines typically have β-ionone levels below the detection limit of 1 μg/L, while commercial wines with intense flavors can have concentrations up to 3.6 μg/L (Langen et al. 2016). In a specific study, a 1996 St-Emilion Merlot wine contained 84.1 ng/L of β-ionone (Kotseridis et al. 1998). Burgundy Pinot noir wines exhibited the highest levels of β-ionone among various red wines analyzed (Kotseridis et al. 1999). Elevated levels of β-ionone and other compounds like α-ionone and β-damascenone can suggest potential adulteration, especially in wines with suspiciously intense flavors (Langen et al. 2016).

Nevertheless, incorporating β-ionone in wine beverages can significantly enhance their sensory profile, making them more appealing to consumers. Its floral aroma is particularly valued in creating novel beverage flavors, providing a unique and pleasant sensory experience (Jaeger et al. 2013; Long et al. 2021). It is a potential flavor to be added as an aromatic reconstituent of dealcoholized wines. Dealcoholization often significantly reduces volatile compounds, which are essential for the wine's aroma and flavor. For instance, osmotic distillation can preserve some esters, but a substantial decrease in volatile compounds is common, especially when alcohol content is reduced significantly (Motta et al. 2017; Corona et al. 2019; Liguori et al. 2019).

10.8 Final Remarks

β-Ionone is an important apocarotenoid in plants, derived from carotenoids. It plays key roles in attracting pollinators, deterring herbivores, and acting as a signaling molecule. Additionally, it contributes to floral scents, has potential pharmaceutical benefits, and is widely used in the fragrance and food industries.

The heterologous production of β-ionone in microbial factories, mainly in yeast species like *S. cerevisiae* and *Y. lipolytica*, has shown promising results through advanced genetic engineering and pathway optimization. However, challenges like enzyme specificity, strain stability, and efficient biotransformation remain. Continued research and development are essential for achieving commercially viable production levels.

β-Ionone exhibits significant potential as an anticancer agent due to its ability to inhibit cancer cell proliferation, induce apoptosis, and enhance the efficacy of other anticancer treatments. Its diverse mechanisms of action and effectiveness across different cancer types make it a promising candidate for further cancer prevention and therapy development.

β-Ionone plays a crucial role in the aroma profile of wines, with its presence and concentration being key factors in enhancing sensory qualities and ensuring authenticity. Advances in biotechnology and analytical methods continue to improve our understanding and manipulation of this critical aroma compound in the wine industry.

References

Ahrazem, O., Gómez-Gómez, L., Rodrigo, M.J. et al. (2016). Carotenoid cleavage oxygenases from microbes and photosynthetic organisms: features and functions. *International Journal of Molecular Sciences* 17(11), 1781. https://doi.org/10.3390/ijms17111781.

Aloum, L., Alefishat, E., Adem, A. et al. (2020). Ionone is more than a violet's fragrance: a review. *Molecules* 25(24), 5822. https://doi.org/10.3390/molecules25245822.

Anca, T., Philippe, V., Ilioara, O. et al. (2009). Composition of essential oils of *Viola tricolor* and *V. arvensis* from Romania. *Chemistry of Natural Compounds* 45, 91–92. https://doi.org/10.1007/s10600-009-9244-y.

Ansari, M. and Emami, S. (2016). β-Ionone and its analogs as promising anticancer agents. *European Journal of Medicinal Chemistry* 123, 141–154. https://doi.org/10.1016/j.ejmech.2016.07.037.

Aprea, E., Biasioli, F. and Gasperi, F. (2015). Volatile compounds of raspberry fruit: from analytical methods to biological role and sensory impact. *Molecules* 20(2), 2445–2474. https://doi.org/10.3390/molecules20022445.

Aslam, M.M., Waseem, M., Jakada, B.H. et al. (2022). Mechanisms of abscisic acid-mediated drought stress responses in plants. *International Journal of Molecular Sciences* 23(3), 1084. https://doi.org/10.3390/ijms23031084.

Asselbergh, B., De Vleesschauwer, D. and Höfte, M. (2008). Global switches and fine-tuning—ABA modulates plant pathogen defense. *Molecular Plant-Microbe Interactions* 21(6), 709–719. https://doi.org/10.1094/MPMI-21-6-0709.

Azcón-Aguilar, C. and Barea, J.M. (2015). Nutrient cycling in the mycorrhizosphere. *Journal of Soil Science and Plant Nutrition* 15(2), 372–396. https://dx.doi.org/10.4067/S0718-95162015005000035.

Begum, N., Qin, C., Ahanger, M.A. et al. (2019). Role of arbuscular mycorrhizal fungi in plant growth regulation: implications in abiotic stress tolerance. *Frontiers in Plant Science* 10, 1068. https://doi.org/10.3389/fpls.2019.01068.

Caceres, L.A., Lakshminarayan, S., Yeung, K. et al. (2016). Repellent and attractive effects of α-, β-, and dihydro-β-ionone to generalist and specialist herbivores. *Journal of Chemical Ecology* 42(2), 107–117. https://doi.org/10.1007/s10886-016-0669-z.

Cao, F.Y., Yoshioka, K. and Desveaux, D. (2011). The roles of ABA in plant–pathogen interactions. *Journal of Plant Research* 124(4), 489–499. https://doi.org/10.1007/s10265-011-0409-y.

Chandrasekaran, M., Boopathi, T. and Manivannan, P. (2021). Comprehensive assessment of ameliorative effects of AMF in alleviating abiotic stress in tomato plants. *Journal of Fungi* 7(4), 303. https://doi.org/10.3390/jof7040303.

Chen, X., Shukal, S. and Zhang, C. (2019). Integrating enzyme and metabolic engineering tools for enhanced α-ionone production. *Journal of Agricultural and Food Chemistry* 67(49), 13451–13459. https://doi.org/10.1021/acs.jafc.9b00860.

Corona, O., Liguori, L., Albanese, D. et al. (2019). Quality and volatile compounds in red wine at different degrees of dealcoholization by membrane process. *European Food Research and Technology* 245, 2601–2611. https://doi.org/10.1007/s00217-019-03376-z.

Costa, M., de Meireles, A. and Gusevskaya, E. (2017). Aerobic palladium-catalyzed oxidations in the upgrading of biorenewables: oxidation of β-ionone and α-ionone. *Asian Journal of Organic Chemistry* 6, 1628–1634. https://doi.org/10.1002/AJOC.201700337.

Czajka, J., Nathenson, J., Benites, V. et al. (2018). Engineering the oleaginous yeast *Yarrowia lipolytica* to produce the aroma compound β-ionone. *Microbial Cell Factories* 17, 136. https://doi.org/10.1186/s12934-018-0984-x.

Distefano, M., Mauro, R.P., Page, D. et al. (2022). Aroma volatiles in tomato fruits: the role of genetic, preharvest, and postharvest factors. *Agronomy* 12(2), 376. https://doi.org/10.3390/agronomy12020376.

El Hadi, M.A., Zhang, F.J., Wu, F.F. et al. (2013). Advances in fruit aroma volatile research. *Molecules* 18(7), 8200–8229. https://doi.org/10.3390/molecules18078200.

Faezizadeh, Z., Gharib, A. and Goudarzi, M. (2015). The effect of β-ionone on telomerase activity in the human leukemia cell line K562. *Journal of Kermanshah University of Medical Sciences* 19, 118–127. https://doi.org/10.22110/JKUMS.V19I3.2160.

Fang, Q., Que, T., Liu, B. et al. (2022). β-Ionone inhibits epithelial-mesenchymal transition (EMT) in prostate cancer cells by negatively regulating the Wnt/β-catenin pathway. *Frontiers in Bioscience* 27(12), 335. https://doi.org/10.31083/j.fbl2712335.

Felemban, A.S. (2022). The apocarotenoid β-ionone is a positive regulator of *Arabidopsis thaliana* response to the pathogen *Botrytis cinereal*. Doctoral dissertation. King Abdullah University of Science and Technology.

Felemban, A., Moreno, J.C., Mi, J. et al. (2024). The apocarotenoid β-ionone regulates the transcriptome of *Arabidopsis thaliana* and increases its resistance against *Botrytis cinerea*. *The Plant Journal* 117(2), 397–415. https://doi.org/10.1111/tpj.16510.

Ficke, A., Asalf, B. and Norli, H.R. (2022). Volatile organic compound profiles from wheat diseases are pathogen-specific and can be exploited for disease classification. *Frontiers in Microbiology* 12, 803352. https://doi.org/10.3389/fmicb.2021.803352.

Fiorilli, V., Wang, J.Y., Bonfante, P. et al. (2019). Apocarotenoids: old and new mediators of the arbuscular mycorrhizal symbiosis. *Frontiers in Plant Science* 10, 1186. https://doi.org/10.3389/fpls.2019.01186.

Gómez-Sagasti, M.T., López-Pozo, M., Artetxe, U. et al. (2023). Carotenoids and their derivatives: a "Swiss Army knife-like" multifunctional tool for fine-tuning plant-environment interactions. *Environmental and Experimental Botany* 207, 105229. https://doi.org/10.1016/j.envexpbot.2023.105229.

Gong, X., Li, F., Liang, Y. et al. (2022). Characteristics of NtCCD1-3 from tobacco, and protein engineering of the CCD1 to enhance β-ionone production in yeast. *Frontiers in Microbiology* 13, 1011297. https://doi.org/10.3389/fmicb.2022.1011297.

Groenewald, M., Boekhout, T., Neuvéglise, C. et al. (2013). *Yarrowia lipolytica*: safety assessment of an oleaginous yeast with a great industrial potential. *Critical Reviews in Microbiology* 40(3), 187–206. https://doi.org/10.3109/1040841X.2013.770386.

He, L., Meng, N., Castellarin, S. et al. (2021). Combined metabolite and transcriptome profiling reveals the norisoprenoid responses in grape berries to abscisic acid and synthetic auxin. *International Journal of Molecular Sciences* 22(3), 1420. https://doi.org/10.3390/ijms22031420.

Isah, T. (2019). Stress and defense responses in plant secondary metabolites production. *Biological Research* 52, 39. https://doi.org/10.1186/s40659-019-0246-3.

Jaeger, S., McRae, J., Bava, C. et al. (2013). A Mendelian trait for olfactory sensitivity affects odor experience and food selection. *Current Biology* 23, 1601–1605. https://doi.org/10.1016/j.cub.2013.07.030.

Janakiram, N., Cooma, I., Mohammed, A. et al. (2008). β-Ionone inhibits colonic aberrant crypt foci formation in rats, suppresses cell growth, and induces retinoid X receptor-α in human colon cancer cells. *Molecular Cancer Therapeutics* 7(1), 181–190. https://doi.org/10.1158/1535-7163.MCT-07-0529.

Katsir, L., Schilmiller, A.L., Staswick, P.E. et al. (2008). COI1 is a critical component of a receptor for jasmonate and the bacterial virulence factor coronatine. *Proceedings of the National Academy of Sciences of the United States of America* 105(19), 7100–7105. https://doi.org/10.1073/pnas.0802332105.

Kotseridis, Y., Baumes, R.L., Bertrand, A. et al. (1999). Quantitative determination of β-ionone in red wines and grapes of Bordeaux using a stable isotope dilution assay. *Journal of Chromatography A* 848 (1–2), 317–325. https://doi.org/10.1016/S0021-9673(99)00422-7.

Kotseridis, Y., Baumes, R. and Skouroumounis, G. (1998). Synthesis of labelled [$^2H_4$]β-damascenone, [$^2H_2$]2-methoxy-3-isobutylpyrazine, [$^2H_3$]α-ionone, and [$^2H_3$]β-ionone, for quantification in grapes, juices and wines. *Journal of Chromatography A* 824, 71–78. https://doi.org/10.1016/S0021-9673(98)00650-5.

Langen, J., Wegmann-Herr, P. and Schmarr, H. (2016). Quantitative determination of α-ionone, β-ionone, and β-damascenone and enantio-differentiation of α-ionone in wine for authenticity control using multidimensional gas chromatography with tandem mass spectrometric detection. *Analytical and Bioanalytical Chemistry* 408, 6483–6496. https://doi.org/10.1007/s00216-016-9767-6.

Lee, S., Mattanovich, D. and Villaverde, A. (2012). Systems metabolic engineering, industrial biotechnology and microbial cell factories. *Microbial Cell Factories* 11, 156. https://doi.org/10.1186/1475-2859-11-156.

Li, X., Ahmad, N., Gao, Y. et al. (2024). Norisoprenoid accumulation under genotype and vintage effects in *Vitis vinifera* L. wine varieties. *Horticulturae* 10, 970. https://doi.org/10.3390/horticulturae10090970.

Liguori, L., Albanese, D., Crescitelli, A. et al. (2019). Impact of dealcoholization on quality properties in white wine at various alcohol content levels. *Journal of Food Science and Technology* 56, 3707–3720. https://doi.org/10.1007/s13197-019-03839-x.

Lin, J., Massonnet, M. and Cantu, D. (2019). The genetic basis of grape and wine aroma. *Horticulture Research* 6, 81. https://doi.org/10.1038/s41438-019-0163-1.

Linsenmeier, A. and Löhnertz, O. (2016). Changes in norisoprenoid levels with long-term nitrogen fertilisation in different vintages of *Vitis vinifera* var. Riesling wines. *South African Journal of Enology and Viticulture* 28, 17–24. https://doi.org/10.21548/28-1-1455.

Liu, H., Cao, X., Azam, M. et al. (2022). Metabolism of carotenoids and β-ionone are mediated by carotenogenic genes and *PpCCD4* under ultraviolet B irradiation and during fruit ripening. *Frontiers in Plant Science* 13, 814677. https://doi.org/10.3389/fpls.2022.814677.

Liu, X., Ding, W. and Jiang, H. (2017). Engineering microbial cell factories for the production of plant natural products: from design principles to industrial-scale production. *Microbial Cell Factories* 16, 125. https://doi.org/10.1186/s12934-017-0732-7.

Liu, H., Qi, Y., Zhou, P. et al. (2021). Microbial physiological engineering increases the efficiency of microbial cell factories. *Critical Reviews in Biotechnology* 41, 339–354. https://doi.org/10.1080/07388551.2020.1856770.

Liu, J., Sun, X., Dong, H., et al. (2008). β-Ionone suppresses mammary carcinogenesis, proliferative activity and induces apoptosis in the mammary gland of the Sprague-Dawley rat. *International Journal of Cancer* 122(12), 2689–2698. https://doi.org/10.1002/ijc.23453.

Llorens, L., Tomàs, J., Ferriol, P. et al. (2023). Floral aroma and pollinator relationships in two sympatric late-summer-flowering Mediterranean *Asparagus* species. *Plants* 12(18), 3219. https://doi.org/10.3390/plants12183219.

Long, Z., Naixin, D., Xue, Y. et al. (2021). Characterization of a novel lutein cleavage dioxygenase, EhLCD, from *Enterobacter hormaechei* YT-3 for the enzymatic synthesis of 3-hydroxy-β-ionone from lutein. *Catalysts* 11(11), 1257. https://doi.org/10.3390/catal11111257.

López, J., Bustos, D., Camilo, C. et al. (2020). Engineering *Saccharomyces cerevisiae* for the overproduction of β-ionone and its precursor β-carotene. *Frontiers in Bioengineering and Biotechnology* 8, 578793. https://doi.org/10.3389/fbioe.2020.578793.

López, J., Essus, K., Kim, I.-k. et al. (2015). Production of β-ionone by combined expression of carotenogenic and plant CCD1 genes in *Saccharomyces cerevisiae. Microbial Cell Factories* 14, 84. https://doi.org/10.1186/s12934-015-0273-x.

Lu, Y., Yang, Q., Lin, Z. et al. (2020). A modular pathway engineering strategy for the high-level production of β-ionone in *Yarrowia lipolytica. Microbial Cell Factories* 19(1), 49. https://doi.org/10.1186/s12934-020-01309-0.

Mahattanatawee, K., Rouseff, R., Valim, M. et al. (2005). Identification and aroma impact of norisoprenoids in orange juice. *Journal of Agricultural and Food Chemistry* 53(2), 393–397. https://doi.org/10.1021/JF049012K.
Markovic, D., Colzi, I., Taiti, C. et al. (2019). Airborne signals synchronize the defenses of neighboring plants in response to touch. *Journal of Experimental Botany* 70(2), 691–700. https://doi.org/10.1093/jxb/ery375.
Marumoto, S., Shimizu, R., Tanabe, G. et al. (2016). *In vitro* regio- and stereoselective oxidation of β-ionone by human liver microsomes. *Planta Medica* 83, 292–299. https://doi.org/10.1055/s-0042-112128.
Mendes-Pinto, M. (2009). Carotenoid breakdown products, the-norisoprenoids-in wine aroma. *Archives of Biochemistry and Biophysics* 483(2), 236–245. https://doi.org/10.1016/J.ABB.2009.01.008.
Moreno, J.C., Mi, J., Alagoz, Y. et al. (2021). Plant apocarotenoids: from retrograde signaling to interspecific communication. *Plant Journal* 105(2), 351–375. https://doi.org/10.1111/tpj.15102.
Mostafa, S., Wang, Y., Zeng, W. et al. (2022). Floral scents and fruit aromas: functions, compositions, biosynthesis, and regulation. *Frontiers in Plant Science* 13, 860157. https://doi.org/10.3389/fpls.2022.860157.
Motta, S., Guaita, M., Petrozziello, M. et al. (2017). Comparison of the physicochemical and volatile composition of wine fractions obtained by two different dealcoholization techniques. *Food Chemistry* 221, 1–10. https://doi.org/10.1016/j.foodchem.2016.10.046.
Nambara, E. and Van Wees, S.C.M. (2021). Plant hormone functions and interactions in biological systems. *The Plant Journal* 105(2), 287–289. https://doi.org/10.1111/tpj.15151.
Nikel, P. and Mattanovich, D. (2021). Microbial cell factories: a biotechnology journey across species. *Essays in Biochemistry* 65(2), 143–145. https://doi.org/10.1042/EBC20210037.
Orsi, E., Beekwilder, J., Eggink, G. et al. (2020). The transition of *Rhodobacter sphaeroides* into a microbial cell factory. *Biotechnology and Bioengineering* 118, 531–541. https://doi.org/10.1002/bit.27593.
Pan, S., Qu, C., Liu, Y. et al. (2024). Sustainable natural resources for aphid management: β-ionone and its derivatives as promising ecofriendly botanical-based products. *Journal of Agricultural and Food Chemistry* 72 (40), 22035–22044. https://doi.org/10.1021/acs.jafc.4c04053.
Paparella, A., Shaltiel-Harpaza, L. and Ibdah, M. (2021). β-Ionone: its occurrence and biological function and metabolic engineering. *Plants* 10(4), 754. https://doi.org/10.3390/plants10040754.
Perkins, J., Hayashi, T., Peakall, R. et al. (2023). The volatile chemistry of orchid pollination. *Natural Product Reports* 40(4), 819–839. https://doi.org/10.1039/D2NP00060A.
Plyuta, V.A., Sidorova, D.E., Koksharova, O.A. et al. (2024). The effect of β-ionone on bacterial cells: the use of specific lux-biosensors. *Research in Microbiology* 175(7), 104214. https://doi.org/10.1016/j.resmic.2024.104214.
Pratiwi, P., Tanaka, G., Takahashi, T. et al. (2017). Identification of jasmonic acid and jasmonoyl-isoleucine and characterization of AOS, AOC, OPR, and JAR1 in the model lycophyte *Selaginella moellendorffii*. *Plant & Cell Physiology* 58(4), 635–646. https://doi.org/10.1093/pcp/pcx031.

Qiu, Z., Li, A., Huang, W. et al. (2024). Metabolic profiling reveal changes in shoots and roots of nitrogen-deficient tea plants (*Camellia sinensis* cv. Jinxuan). *Scientia Horticulturae* 337, 113528. https://doi.org/10.1016/j.scienta.2024.113528.

Rachersberger, M., Cordeiro, G.D., Schäffler, I. et al. (2019). Honeybee pollinators use visual and floral scent cues to find apple (*Malus domestica*) flowers. *Journal of Agricultural and Food Chemistry* 67(48), 13407–13415. https://doi.org/10.1021/acs.jafc.9b05854.

Ramel, F., Birtic, S., Ginies, C. et al. (2012). Carotenoid oxidation products are stress signals that mediate gene responses to singlet oxygen in plants. *Proceedings of the National Academy of Sciences of the United States of America* 109(14), 5535–5540. https://doi.org/10.1073/pnas.1115982109.

Ramos, A.C., Façanha, A.R., Palma, L.M. et al. (2011). An outlook on ion signaling and ionome of mycorrhizal symbiosis. *Brazilian Journal of Plant Physiology* 23(1), 79–89. https://doi.org/10.1590/S1677-04202011000100010.

Rubio, A., Rambla, J., Santaella, M. et al. (2008). Cytosolic and plastoglobule-targeted carotenoid dioxygenases from *Crocus sativus* are both involved in β-ionone release. *Journal of Biological Chemistry* 283, 24816–24825. https://doi.org/10.1074/jbc.M804000200.

Sharma, E., Anand, G. and Kapoor, R. (2017). Terpenoids in plant and arbuscular mycorrhiza-reinforced defence against herbivorous insects. *Annals of Botany* 119(5), 791–801. https://doi.org/10.1093/aob/mcw263.

Simkin, A.J. (2021). Carotenoids and apocarotenoids in planta: their role in plant development, contribution to the flavour and aroma of fruits and flowers, and their nutraceutical benefits. *Plants* 10(11), 2321. https://doi.org/10.3390/plants10112321.

Simkin, A., Underwood, B., Auldridge, M. et al. (2004). Circadian regulation of the PhCCD1 carotenoid cleavage dioxygenase controls emission of β-ionone, a fragrance volatile of petunia flowers. *Plant Physiology* 136, 3504–3514. https://doi.org/10.1104/pp.104.049718.

Sun, Q., He, L., Sun, L. et al. (2023). Identification of SNP loci and candidate genes genetically controlling norisoprenoids in grape berry based on genome-wide association study. *Frontiers in Plant Science* 14, 1142139. https://doi.org/10.3389/fpls.2023.1142139.

Sun, Y., Yu, H., Zhou, J.-J. et al. (2014). Plant volatile analogues strengthen attractiveness to insects. *PLoS ONE* 9(6), e99142. https://doi.org/10.1371/journal.pone.0099142.

Teliban, G.-C., Burducea, M., Zheljazkov, V.D. et al. (2021). The effect of myco-biocontrol-based formulates on yield, physiology, and secondary products of organically grown basil. *Agriculture* 11(2), 180. https://doi.org/10.3390/agriculture11020180.

Timmins, J., Kroukamp, H., Paulsen, I. et al. (2020). The sensory significance of apocarotenoids in wine: importance of carotenoid cleavage dioxygenase 1 (CCD1) in the production of β-ionone. *Molecules* 25(12), 2779. https://doi.org/10.3390/molecules25122779.

Timmins, J., Kroukamp, H., Walker, R. et al. (2022). Comparative evaluation of secreted plant carotenoid cleavage dioxygenase 1 (CCD1) enzymes in *Saccharomyces cerevisiae. Fermentation* 8(8), 395. https://doi.org/10.3390/fermentation8080395.

Timmins, J., van Wyk, N., Kroukamp, H. et al. (2023). Engineering a fermenting yeast able to produce the fragrant β-ionone apocarotenoid for enhanced aroma properties in wine. *FEMS Yeast Research* 23, foad001. https://doi.org/10.1093/femsyr/foad001.

Votta, C., Wang, J.Y., Cavallini, N. et al. (2024). Integration of rice apocarotenoid profile and expression pattern of carotenoid cleavage dioxygenases reveals a positive effect of β-ionone on mycorrhization. *Plant Physiology and Biochemistry* 207, 108366. https://doi.org/10.1016/j.plaphy.2024.108366.

Wang, J., Wu, B., Zhang, N. et al. (2020a). Dehydration-induced carotenoid cleavage dioxygenase 1 reveals a novel route for β-ionone formation during tea (*Camellia sinensis*) withering. *Journal of Agricultural and Food Chemistry* 68(39), 10815–10821. https://doi.org/10.1021/acs.jafc.0c04208.

Wang, J., Zhang, N., Zhao, M. et al. (2020b). Carotenoid cleavage dioxygenase 4 catalyzes the formation of carotenoid derived volatile β-ionone during tea (*Camellia sinensis*) withering. *Journal of Agricultural and Food Chemistry* 68(6), 1684–1690. https://doi.org/10.1021/acs.jafc.9b07578.

Wani, K.I., Zehra, A., Choudhary, S. et al. (2021). Mechanistic insights into strigolactone biosynthesis, signaling, and regulation during plant growth and development. *Journal of Plant Growth Regulation* 40(5), 1836–1852. https://doi.org/10.1007/s00344-020-10234-w.

Wei, Y., Meng, N., Wang, Y. et al. (2023). Transcription factor VvWRKY70 inhibits both norisoprenoid and flavonol biosynthesis in grape. *Plant Physiology* 193(3), 2055–2070. https://doi.org/10.1093/plphys/kiad423.

Weinblum, N., Cna'ani, A., Yaakov, B. et al. (2021). Tomato cultivars resistant or susceptible to spider mites differ in their biosynthesis and metabolic profile of the monoterpenoid pathway. *Frontiers in Plant Science* 12, 630155. https://doi.org/10.3389/fpls.2021.630155.

Werner, N., Ramírez-Sarmiento, C. and Agosin, E. (2019). Protein engineering of carotenoid cleavage dioxygenases to optimize β-ionone biosynthesis in yeast cell factories. *Food Chemistry* 299, 125089. https://doi.org/10.1016/j.foodchem.2019.125089.

Xu, N., Liu, Y., Jiang, H. et al. (2020). Combining protein and metabolic engineering to construct efficient microbial cell factories. *Current Opinion in Biotechnology* 66, 27–35. https://doi.org/10.1016/j.copbio.2020.06.001.

Xu, J., Xia, Y., Shi, Y. et al. (2024). Metabolic engineering of *Candida tropicalis* for the *de novo* synthesis of β-ionone. *ACS Synthetic Biology* 13(8), 2533–2544. https://doi.org/10.1021/acssynbio.4c00286.

Yahyaa, M., Bar, E., Dubey, N. et al. (2013). Formation of norisoprenoid flavor compounds in carrot (*Daucus carota* L.) roots: characterization of a cyclic-specific carotenoid cleavage dioxygenase 1 gene. *Journal of Agricultural and Food Chemistry* 61(50), 12244–12252. https://doi.org/10.1021/jf404085k.

Yang, J., Mu, W., Cao, Y. et al. (2020). Synthesis and biological evaluation of β-ionone oriented proapoptosis agents by enhancing the ROS generation. *Bioorganic Chemistry* 104, 104273. https://doi.org/10.1016/j.bioorg.2020.104273.

Zhang, F., Cao, M., Shen, L. et al. (2025). β-Ionone treatment enhances the antioxidant capacity in postharvest broccoli (*Brassica oleracea* L. var. *Italica*) by maintaining the levels of bioactive substances. *Foods* 14(5), 762. https://doi.org/10.3390/foods14050762.

Zhang, H., Wang, J.-Y., Chen, Y.-J. et al. (2022). β-Ionone is a functional plant volatile that attracts the parasitic wasp, *Microplitis pallidipes. BioControl* 67, 1–13. https://doi.org/10.1007/s10526-021-10117-3.

Zhou, W., Wang, Y., Wang, J. et al. (2022). β-Ionone causes endocrine disruption, hyperpigmentation, and hypoactivity in zebrafish early life stages. *Science of The Total Environment* 834, 155433. https://doi.org/10.1016/j.scitotenv.2022.155433.

Section III

Saffron Apocarotenoids: Special Focus

11

Saffron Apocarotenoids

Braulio Edgar Herrera Cabrera[1], *Rafael Salgado Garciglia*[2], *Luis Germán López Valdez*[3], *Alejandra Hernández Garcia*[2], *Adriana Delgado Alvarado*[1], *Francisco Javier Verduzco Miramón*[4], *and Hebert Jair Barrales-Cureño*[4*]

[1] *Colegio de Postgraduados, Campus Puebla, Cholula, PU, Mexico*
[2] *Universidad Michoacana de San Nicolás de Hidalgo, Morelia, MC, Mexico*
[3] *Universidad Autónoma Chapingo, Texcoco, EM, Mexico*
[4] *Instituto Tecnológico de Estudios Superiores de Zamora, Tecnológico Nacional de México, La Piedad, MC, Mexico*

*Corresponding author: hebert.jair@uiep.edu.mx

11.1 Occurrence and Structure

Saffron is the most expensive spice in the world. It is usually obtained from the stigma of *Crocus sativus* flowers, and it takes between 150,000 and 200,000 flowers to produce 1 kg of saffron. Apocarotenoids are a type of carotenoids that have a skeleton with less than 40 carbon atoms, resulting from the loss of molecules at one or both ends of the original carotenoids. Not all carotenoids consist of eight isoprenoid units, since some, called apocarotenoids, have a skeleton of less than 40 carbon atoms, probably due to the split at one (e.g., β-apo-8′-carotenal, a pigment present in loquat and citrus fruits, among other sources) or both ends of the molecule (e.g., crocetin, a characteristic pigment of saffron) (Carmona et al. 2006).

In the *C. sativus* plant, there are compounds in its stigmas such as the carotenoids crocin and crocetin, the monoterpene aldehydes picrocrocin and safranal, as well as the platelet aggregation activator/inhibitor proteins, lectins that bind mannose, and a cytotoxic proteoglycan (Aissa et al. 2023). The most important active principles that saffron possesses include the types of color, flavor, and odor due to crocin, picrocrocin, and safranal.

11.1.1 Crocin

The essential active coloring principles are crocetin ($C_{20}H_{24}O_4$), a carotenoid characterized by two carboxyl groups, and crocin ($C_{44}H_{64}O_{24}$), an ester composed of two preceding

Plant Apocarotenoids: Biosynthesis, Signaling and Agricultural Applications, First Edition. Edited by Kaiser Iqbal Wani and Tariq Aftab.
Published 2026 by John Wiley & Sons Ltd.

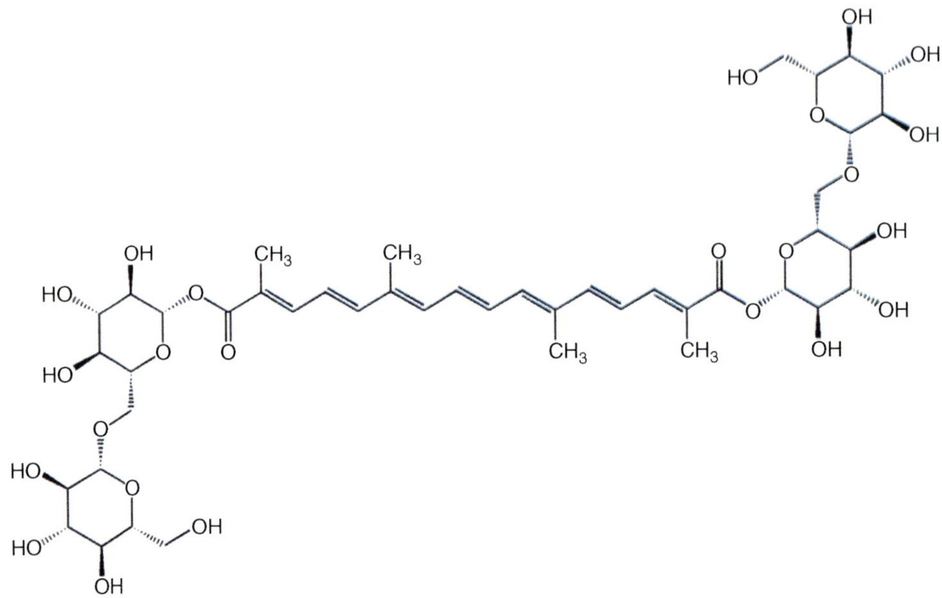

Figure 11.1 Crocin; $C_{44}H_{64}O_{24}$, 976.96 g/mol.

acid molecules (a disaccharide). There are also analogous molecules that are only partially esterified, including sugar glucose. Crocin (Figure 11.1), the main component responsible for the coloring power of saffron, whose molar absorption coefficient is 133,750 L/mol·cm, is soluble in water; therefore, its assimilation, as well as its antioxidant action, is due to free radicals (Rocchi et al. 2024). It is also soluble in alcohol and does not crystallize. In water, crocin tends to hydrolyze spontaneously, forming crocetin, almost insoluble and of a more reddish color. This is one of the reasons why the saffron powder revives its color if it is humid. During storage, the humidity contained in the product leads to a progressive decrease in color and therefore a change of the color to an ever less intense yellow (Fiorito et al. 2022).

The color of saffron comes from crocins: water-soluble pigments derived from carotenoids through a process catalyzed by enzymes known as carotenoid cleavage dioxygenases (CCDs) (Frusciante et al. 2014). Crocins are also found, although in much smaller amounts, in the fruits of *Gardenia jasminoides*, an ornamental plant used in traditional Chinese medicine (Yang et al. 2009). Crocins have a high therapeutic potential, among other things, for their role in protecting neuronal cells against degradation, as well as for their antidepressant, sedative, and antioxidant properties. They also play an important role as natural food colorants. Harvesting and processing saffron stigmas, collected by hand, are labor-intensive. In addition, saffron is only grown in limited areas of the Mediterranean and Asia. Therefore, new biotechnological approaches to produce these compounds in large quantities are in high demand.

11.1.2 Picrocrocin

This monoterpene aldehyde is responsible for the characteristic bitter taste of saffron. The picrocrocin molecule (Figure 11.2) hydrolyzes spontaneously in aqueous solution and also during the preservation of filaments and powder even if the humidity is too high.

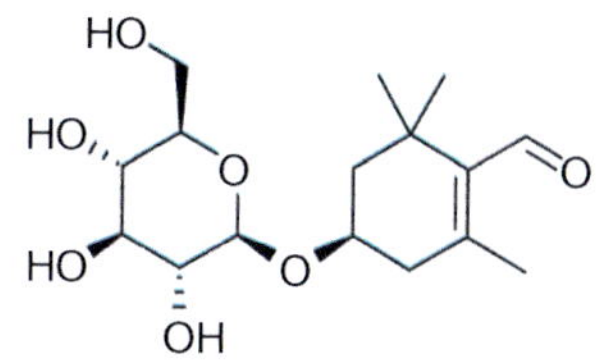

Figure 11.2 Picrocrocin; $C_{16}H_{26}O_7$, 330.37 g/mol.

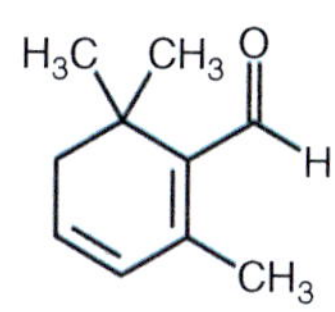

Figure 11.3 Safranal; $C_{10}H_{14}O$, 150.21 g/mol.

Picrocrocin ($C_{16}H_{26}O_7$) is also called 4-(β-D-glucopyranosyloxy)-2,6,6-trimethyl-1-cyclohexene-1-carboxaldehyde and is considered the substance responsible for the bitter taste of saffron. Its ultraviolet–visible (UV–vis) spectrum in aqueous solution shows a maximum at 250 nm, and its molar extinction coefficient is 10,100 L/cm·mol (Buchecker and Eugster 1973). Picrocrocin is considered the precursor of safranal. During the dehydration process, this transformation takes place either by the temperature reached or by the action of glycosidases (Himeno and Sano 1987). During stigma development, crocin and picrocrocin would maintain a constant ratio because HTCC (4-hydroxy-2,6,6-trimethyl-1,3-cyclohexen-1-carbaldehyde) formed by the action of a carotenase would be glycosidized and transformed into picrocrocin. At the time of anthesis, one or more glycosidases would act on picrocrocin, converting it back into HTCC (Husaini et al. 2022). In the case of saffron dried at high temperature, the activity of these enzymes, glycosidases and carotenases, would be short. However, saffron dried at room temperature could maintain the activity for many hours during the several days of the drying process.

11.1.3 Safranal

The characteristic aroma of saffron is due to safranal (Figure 11.3) and other volatile compounds that are formed over time and are in smaller quantities, representing 70% of the total volatiles (Rameshrad et al. 2018). The qualitative and quantitative compositions vary depending on the cultivation land, as well as time. It is a monoterpene aldehyde and a principal component of saffron responsible for its unique odor. Pharmacologically, it shows anti-inflammatory, antioxidant, cardioprotective, nephroprotective, neuroprotective, and gastrointestinal protective properties (Esmaealzadeh et al. 2023).

Saffron stigmas contain water-soluble carotenoids, the so-called crocins (crocetin mono- and diglucosides). Crocins 1–4 stand out, with crocin 1 being the one found in higher proportion and the best studied for its activity. There are also flavonoids (quercetin, kenferol; mainly identified in the flower petals), traces of essential oil, lipids, mucilages, bitter glycosides such as picrocrocoside or picrocrocin, hydrolyzable in glucose, and an intermediate derivative that is transformed into safranal, which is a monoterpene aldehyde, the main component of the essential oil (Ahmed et al. 2016). The color of saffron is due to carotenoids, the bitter taste due to picrocrocin, and the aroma due to safranal.

Crocin and picrocrocin are glycosylated apocarotenoids responsible, respectively, for the unique color and flavor of the saffron spice, known as red gold due to its high price. Several studies have also shown their beneficial health properties (Christodoulou et al. 2015). However, their high costs hinder the wide use of these metabolites in the pharmaceutical sector, considering that saffron apocarotenoids are harvested from the flower stigmas once a year.

11.2 Biosynthetic Pathways of Major Saffron Apocarotenoids

Because of the importance of carotenoids, the enzymes responsible for their biosynthesis, from phytoene to zeaxanthin, have been characterized from many plant species, and the genes encoding them have been isolated (Cunningham and Gantt 1998; Hirschberg 2001). However, only more recently have genes encoding enzymes involved in the oxidative cleavage of carotenoids and producing apocarotenoid derivatives such as abscisic acid and retinol, with important metabolic and hormonal functions, begun to be isolated and studied. Consequently, these enzymes are clear candidates for engineering apocarotenoid metabolism and are essential for achieving the production of certain apocarotenoids in other organisms that do not naturally produce them (Misawa and Shimada 1998). Moreover, since the steps catalyzed by these enzymes can be limiting, manipulation of the expression of the corresponding genes could increase yield as well as change the apocarotenoid profile in the plant under study, as demonstrated by some recent works (Sandmann 2001).

Saffron is mainly known for its use as a culinary additive and as a colorant, and is currently the most expensive spice in the world. In the stigma of *C. sativus* L., the apocarotenoids crocin and crocetin have been identified as the major components of saffron spice, responsible for its color and aroma (Winterhalter and Straubinger 2000). The biogenesis of the compounds responsible for the color and organoleptic properties of saffron is made from the oxidative cleavage of the carotenoid zeaxanthin (3,3′-dihydroxy-β-carotene).

There are a number of models proposed to explain the formation of apocarotenoids in plants, bacteria, and animals. Some authors suggested that apocarotenoid biosynthesis takes place from carotenoids with at least 40 carbon atoms, such as β-carotene and zeaxanthin.

Currently, the two most widely accepted models are those that propose homo- or bicyclic carotenoids and linear carotenoids as the substrates, with the enzymatic cleavage of carotenoids at one or two specific sites of the polyene chain being the key reaction in the biosynthesis of apocarotenoids. It has been suggested that apocarotenoids are formed randomly, as products of photooxidation, by-products of photooxidation, or by co-oxidation triggered by lipoxygenases (Liang et al. 2021). However, it has also been suggested that biologically important apocarotenoids may require more precise mechanisms for their biosynthesis, including specific pathways and enzymes for their formation. There is evidence for such enzymes in the organisms of the five kingdoms, but there is little information on their characterization. Recently, the gene coding for 9-*cis*-epoxy-carotenoid dioxygenase, the enzyme involved in ABA biosynthesis, has been cloned. The identification and characterization of this gene have provided the basis for further identification of carotenoid-degrading enzymes.

11.2.1 Biosynthesis of Apocarotenoids from Homo- or Bicyclic Carotenoids

This model proposes that the formation of apocarotenoids would involve dioxygenase enzymes, which would act on precursors with rings at the ends of the straight chain containing two of the four oxygen molecules in the ABA molecule (5), and that apocarotenoids, such as vitamin A (C18), crocin (C20), β-cyclocitral (C10), mycoradicins, persicaxanthin (C25), and persicachrome (C27), are formed from carotenoids of 40 carbon atoms with a ring at each end, such as β-carotene, 9-*cis*-neoxanthin, 9-*cis*-violaxanthin, and zeaxanthin (Schwartz et al. 2001).

Dioxygenases incorporate O_2 into an organic substrate and are characterized by catalyzing hydroxylation, desaturation, and epoxidation reactions, as well as by their sequence similarity (Schwartz et al. 2001). Dioxygenases act on carotenoids and other polyenes and include plant enzymes that cleave the double bond of neoxanthin, bacterial lignostilbene dioxygenase, and vertebrate RPE65 (Retinal Pigment Epithelium) protein (von Lintig and Vogt 2000). The general characteristic of the reaction is oxidative degradation, by dioxygenase, of a double bond in the polyene chain, in a central position (15,15′ bond), or a lateral double bond (11′,12′, 9′,10′, or 7′,8′ bonds) (Pfander and Schurtenberg 1982) resulting in the formation of two apocarotenoid molecules with equal or different chain lengths (von Lintig and Vogt 2000).

The enzyme that cleaves at the 15,15′ position releases two symmetric retinal, whereas those that cleave at the lateral position produce two molecules of asymmetric length, and depending on the bond produce β-apo-8′-carotenal (bond 7′,8′), β-apo-10′-carotenal (bond 9′,10′), or β-apo-12′-carotenal (bond 11′,12′). After asymmetric cleavage, the additional carbons are removed from the linear product by a mechanism similar to that of β-oxidation, to form a retinoic acid molecule.

In higher plants, an oxidative enzymatic cleavage reaction of a double bond occurs on one end of the chain. Examples of these reactions are the formation of crocetin in saffron, citraurin and other apocarotenoids in citrus fruits and ABA (Qin and Zeevaart 1999). Experiments supporting this model were carried out by Gross and Eckhardt (1981), who reported the presence of two apocarotenoid pigments in Golden apple (*Malus*), citrus fruit (*Citrus sinensis*), and avocado (*Persea americana*), with these being persicaxanthin, a 25-carbon epoxycarotenol, and persicachromeof 27 carbons. On the other hand, Pfander and Schurtenberg (1982) reported that crocetin, a compound of 20 carbon atoms present in saffron, is formed from the degradation of carotenoids of 40 carbon atoms such as β-carotene or xanthophylls, since quantities of β-carotene, phytofluene, tetrahydrolicopene, and phytoene were found in saffron, in addition to the presence of picrocrocin and safranal, which possess the same stereochemistry as zeaxanthin. The enzyme that cuts at the 7,8 bond is a dioxygenase originally named 7,8-carotenase (Pfander and Schurtenberg 1982).

11.2.2 Biosynthesis of Apocarotenoids from Linear Carotenoids

The biosynthesis of apocarotenoids in plants from linear carotenoids has not been described. It has only been reported in bacteria such as *Staphylococcus aureus* and *Streptococcus faecium*, but the mechanism of action and the type of enzymes involved have not been elucidated. For *S. aureus*, the initial substrate is phytoene and the final products are carotenoids with 30 carbon atoms called diapocarotenoids. The compounds formed are phytoene analogs: phytofluene, ζ-carotene, 7,8,11,23-tetrahydrolicopene, and neurosporene; and the so-called diapocarotenoids: diapofitofluene, diapo-ζ-carotene, diapo-7,8,11,12-tetrahydrolicopene, and diaponeurosporene, respectively (Goodwin 1980). In *S. faecium*, the presence of 4,4′-diapofitoene, 4,4′-diapofitofluene, 4,4′-diapo-ζ-carotene, 4,4′-diapo-7,8,11,12-tetrahydrolicopene, 4,4′-diaponeurosporene, 4-hydroxy-4,4′-diaponeurosporene, 4-D-glucopyranosyloxy-4,4′-diaponeurosporene, 4,4′-diaponeurosporen-4-al, and 4,4′-diapolicopen-4-al (Goodwin 1980).

11.3 Impact of Abiotic Stress on Saffron Apocarotenoid Signals

Carotenoids are one of the most diverse and widely distributed groups of natural pigments (Bourgaud et al. 2001). Apocarotenoids are generated by oxidative cleavage of these terpenoid pigments and, together with them, play a fundamental role as protectors against oxidative and abiotic stresses, have a hormonal role, and are, in part, responsible for the diverse colorations and aroma characteristics of the products of some plants such as tomatoes and saffron.

11.4 Crocin and Crocetin Biosyntheses

The stigmas of the *C. sativus* L. plant contain numerous compounds, including carotenoids, such as crocin and crocetin; aldehydes, such as picrocrocin; and aromatic compounds, the main component of which is safranal. Crocin, a chemical compound exclusive to the flowers of *Crocus* and *G. jasminoides* species, is the main carotenoid responsible for the coloring power of the saffron spice. It should be noted that crocin is an ester formed by a disaccharide (gentiobiose, composed of two glucose units joined by β-1,6 glycosidic bond) and a dicarboxylic acid, which is crocetin. The biosynthetic pathways of different saffron apocarotenoids are shown in Figure 11.4.

Regarding picrocrocin, an aldehyde derived from zeaxanthin, it should be noted that it is the chemical compound responsible for the bitter taste of saffron. Picrocrocin is also a compound that favors gastric secretions, just as crocin stimulates the secretion of bile in the liver, which results in better digestion.

The main component of the aromatic fraction of saffron spice is safranal, which represents 70% of this fraction. It is a chemical compound formed during the dehydration process of saffron stigmas. Several works (Lozano et al. 1999) have stated that safranal is synthesized from picrocrocin during the dehydration process by two different ways, which are: chemical synthesis, due to the high temperatures reached, or enzymatic synthesis, carried out by the action of glycosidase enzymes.

Figure 11.5 shows schematically the two proposed synthesis pathways of safranal from picrocrocin. The right-hand side of the image corresponds to the chemical synthesis of

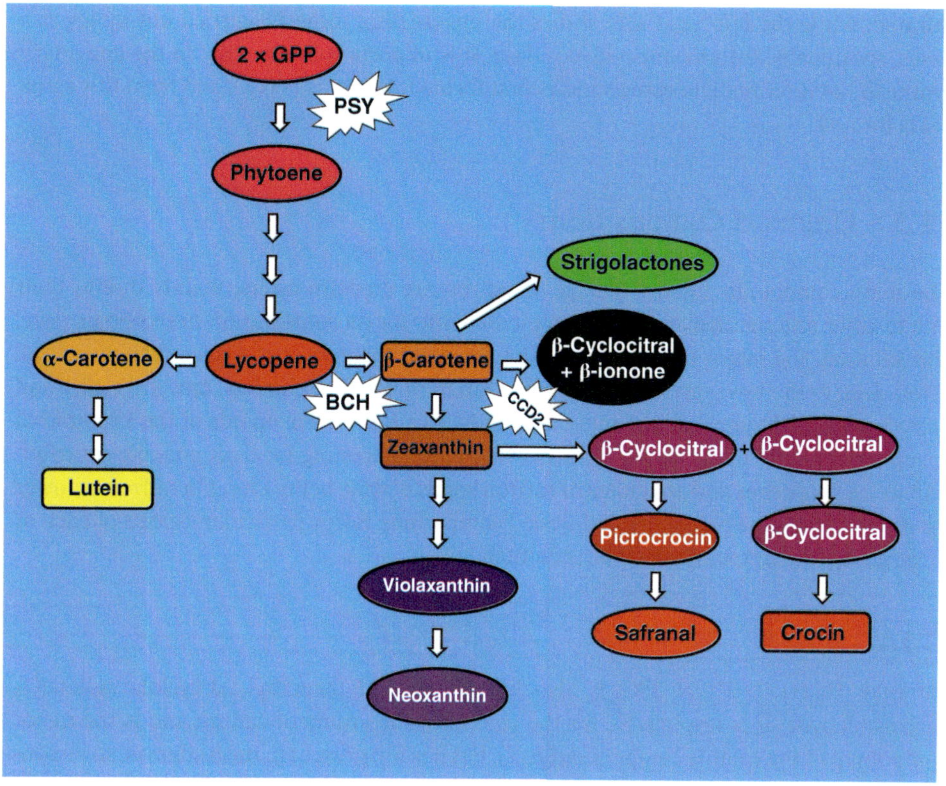

Figure 11.4 Carotenoid and apocarotenoid biosynthetic pathways in *C. sativus*. PSY, phytoene synthase; BCH, β-carotene hydroxylase; and CCD2, carotenoid cleavage dioxygenase 2.

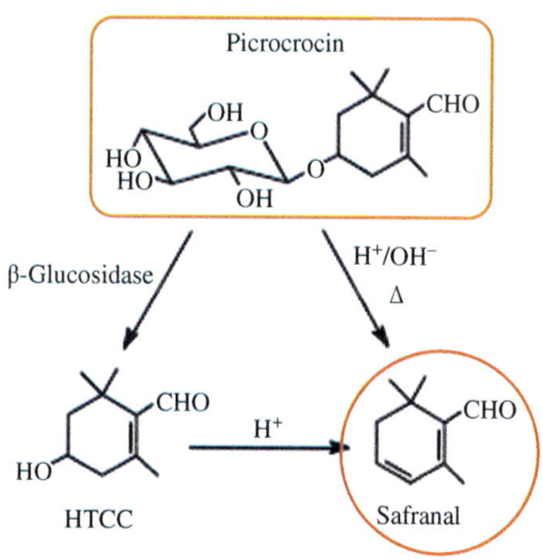

Figure 11.5 Enzymatic and chemical syntheses of safranal from picrocrocin (Himeno and Sano 1987).

safranal, while the left-hand side shows the enzymatic synthesis, carried out by glycosidases, specifically, the enzyme β-glycosidase. It is important to note that in the enzymatic synthesis, an intermediate compound is first formed, which is HTCC, and from this safranal is formed.

11.5 Chemical Compositions

The main compounds for the quality of saffron are crocetin esters, mainly due to their role in color and medicinal properties; picrocrocin, for its contribution to the bitter taste, and safranal, the major compound for the aromatic fraction of the spice (Sánchez et al. 2009). Due to the low water solubility of safranal, in aqueous extracts, crocetin esters and picrocrocin are the most abundant compounds. In addition, flavonoids of the kaempferol group are found in the aqueous extracts, which are interesting for their possible contribution to the properties and attributes of saffron spice, despite being found in saffron in lower proportions than the rest of the mentioned compounds. Some particular aspects of each of these groups of compounds will be presented below.

11.5.1 Safranal

Safranal ($C_{10}H_{14}O$, Mr = 150) is also known as 2,6,6-trimethyl-1,3-cyclohexadiene-1-carboxaldehyde. It is responsible for the characteristic aroma of saffron and is the major component of the volatile compounds (70%), but it is not present in fresh stigmas, so its concentration depends strongly on the roasting and storage conditions. It shows its maximum absorption at 330 nm in aqueous solution, but its solubility in water is limited, 673.07-mg safranal/L water at 20 °C (Garcia-Rodriguez et al. 2014). However, this determination is not the most appropriate since crocin esters of cis-configuration also absorb at this wavelength (Alonso et al. 1996) and interfere with estimating safranal. For this reason, Maggi et al. (2011) established an HPLC methodology to quantify safranal.

11.5.2 Crocetin Esters

These substances are responsible for the color that saffron spice gives to aqueous solutions and belong to the group of carotenoid pigments. A mixture of different glycosylated esters of 8,8′-diapo-ψ,ψ′-carotenedioic acid, called crocetins, is found in aqueous extracts of saffron. The sugars esterifying the ends of crocetin are glucose and gentiobiose. Unlike most carotenoids, these esters are water-soluble and are also known as crocetins. This name comes from the di-(-D-gentiobiosyl) crocetin ester, the major compound in saffron that is called crocin. Its molar absorption coefficients in water at 440 and 257 nm are 89,000 and 63,350 L/cm·mol according to Speranza et al. (1984). Another important aspect of this group of saffron carotenoids is their possible involvement in the formation of aroma, as proposed by Carmona et al. (2006). According to these authors, aroma formation would take place from crocetin esters by the action of a carotenase. The *cis*-crocins, with respect to their trans-namesakes, present an additional absorption band around 334 nm in their UV–vis spectrum and the absorption maximum at 440 nm presents a hypsochromic effect of about 5 nm (Pfister et al. 1996).

11.6 Key Enzymes Involved

Previously, it has been reported the relationship that exists in the expression levels of genes involved in carotenoid biosynthesis routes, and the accumulation of these, as mentioned by Gómez-Gómez et al. (2017) who analyzed the expression of the *carotenoid cleavage dioxygenase 2* gene in saffron (*CaCCD2*), which shows a correlation with the crocetin content in saffron, concluding that the accumulation of this pigment is mainly controlled at the transcriptional level by that gene. It is interesting to mention that similarly in *C. sativus*, it was reported that *CsALDH3* could be involved in the biosynthesis of the apocarotenoid crocin through the oxidation of crocetin dialdehyde (Tan et al. 2019).

KAUST researchers identified a highly efficient CCD enzyme from *G. jasminoides* that produces the crocin precursor, crocetin dialdehyde (Zheng et al. 2022). They have now established a system to investigate the enzymatic activity of CCD in plants and have developed a multigene engineering approach for the sustainable biotechnological production of crocetins in plant tissues. At the enzymatic level, the dioxygenase enzyme production and multigenic engineering strategy could be used to establish a sustainable plant cell factory for crocin production in tissue cultures of different plant species, and in the future, develop crocin-rich functional foods.

The research paves the way for the efficient biotechnological production of crocins and other high-value carotenoid-derived compounds (apocarotenoids) as pharmaceuticals in green tissues as well as in other starch-rich plant organs. It also highlights the contribution of functional diversification among *CCD* genes to the independent evolution of alternative routes of apocarotenoid biosynthesis in different plants. Most of the knowledge on CCD enzymatic activity and substrate specificity comes from experiments with *Escherichia coli* modified to produce different carotenoids (Chae et al. 2010). Functional characterization in plants, for example, using a transgenic approach, is important to deduce the role of CCDs in carotenoid metabolism and unravel their actual contribution to the "carotenoid/apocarotenoid pattern." The platform technology could be used to produce other important carotenoid-derived compounds, including the widely used perfumes and dyes. It could be used to produce safranal and picrocrocin, for example, which give rise to the characteristic flavor and aroma of saffron. They could be used as flavor additives and also have bioactive potential that awaits exploration.

11.7 Micro-RNAs Involved in Apocarotenoid Biosynthesis in Saffron

Currently, candidate micro-RNAs (miRNAs), target genes, transcription factors (TFs), and apocarotenoid biosynthetic pathway genes related to regulation and flowering transition have been determined in *C. sativus*. Bhat et al. (2024) identified significant modulations in the expression of miR156, miR159, miR166, miR172, miR395, miR396, miR399, and miR408 gene families focused on carbohydrate metabolism, stress response, cold, defense, phytohormone signaling, and regulation of floral development.

Zinati et al. (2016) performed a computational analysis to identify miRNAs and their targets using EST (Expressed Sequence Tag) from mature saffron stigmas. A gene co-expression network was constructed to identify genes strongly involved in carotenoid and apocarotenoid production. EST analysis allowed for the identification of two putative miRNAs (miR414 and miR837-5p) together with the corresponding stem precursors. Co-expression network analysis indicated that miR414 and miR837-5p play roles in *C. sativus* metabolic pathways and lead to the identification of candidate genes, six TFs, and a protein kinase that could be involved in carotenoid/apocarotenoid metabolism.

References

Ahmed, N., Anwar, S., Al-Sokari, S.S. et al. (2016). Saffron crocus (*Crocus sativus*) oils. In: *Essential Oils in Food Preservation, Flavor and Safety* (ed. V.R. Preedy), 705–713. Academic Press.

Aissa, R., Ibourki, M., Ait Bouzid, H. et al. (2023). Phytochemistry, quality control and medicinal uses of saffron (*Crocus sativus* L.): an updated review. *Journal of Medicine and Life* 16(6), 822–836.

Alonso, G.L., Salinas, M.R., Esteban-Infantes, F.J. et al. (1996). Determination of safranal from saffron (*Crocus sativus* L.) by thermal desorption–gas chromatography. *Journal of Agricultural and Food Chemistry* 44(1), 185–188.

Bhat, A., Mishra, S., Kaul, S. et al. (2024). Comparative analysis of miRNA expression profiles in flowering and non-flowering tissue of *Crocus sativus* L. *Protoplasma* 261(4), 749–769.

Bourgaud, F., Gravot, A., Milesi, S. et al. (2001). Production of plant secondary metabolites: a historical perspective. *Plant Science* 161(5), 839–851.

Buchecker, R. and Eugster, C.H. (1973). Absolute konfiguration von picrocrocin. *Helvetica Chimica Acta* 56(3), 1121–1124.

Carmona, M., Zalacain, A., Sánchez, A.M. et al. (2006). Crocetin esters, picrocrocin and its related compounds present in *Crocus sativus* stigmas and *Gardenia jasminoides* fruits. Tentative identification of seven new compounds by LC-ESI-MS. *Journal of Agricultural and Food Chemistry* 54(3), 973–979.

Chae, H.S., Kim, K.H., Kim, S.C. et al. (2010). Strain-dependent carotenoid productions in metabolically engineered *Escherichia coli*. *Applied Biochemistry and Biotechnology* 162(8), 2333–2344.

Christodoulou, E., Kadoglou, N.P., Kostomitsopoulos, N. et al. (2015). Saffron: a natural product with potential pharmaceutical applications. *Journal of Pharmacy and Pharmacology* 67(12), 1634–1649.

Cunningham Jr., F.X. and Gantt, E. (1998). Genes and enzymes of carotenoid biosynthesis in plants. *Annual Review of Plant Biology* 49(1), 557–583.

Esmaealzadeh, D., Ghalibaf, A.M., Rad, M.S. et al. (2023). Pharmacological effects of safranal: an updated review. *Iranian Journal of Basic Medical Sciences* 26(10), 1131–1143.

Fiorito, S., Palumbo, L., Epifano, F. et al. (2022). An easy way for the hydrolysis, pre-concentration, and chemical stabilization of crocetin from saffron powder. *Food Chemistry* 377, 132040.

Frusciante, S., Diretto, G., Bruno, M. et al. (2014). Novel carotenoid cleavage dioxygenase catalyzes the first dedicated step in saffron crocin biosynthesis. *Proceedings of the National Academy of Sciences of the United States of America* 111(33), 12246–12251.

García-Rodríguez, M., Serrano-Díaz, J., Tarantilis, P.A. et al. (2014). Determination of saffron quality by high-performance liquid chromatography. *Journal of Agricultural and Food Chemistry* 62(32), 8068–8074.

Gómez-Gómez, L., Parra-Vega, V., Rivas-Sendra, A. et al. (2017). Unraveling massive crocins transport and accumulation through proteome and microscopy tools during the development of saffron stigma. *International Journal of Molecular Sciences* 18(1), 76.

Goodwin, T.W. (1980). *The Biochemistry of the Carotenoids, Volume I Plants* (2nd ed). Chapman and Hall.

Gross, J. and Eckhardt, G. (1981). Structures of persicaxanthin, persicachrome and other apocarotenols of various fruits. *Phytochemistry* 20(9), 2267–2269.

Himeno, H. and Sano, K. (1987). Synthesis of crocin, picrocrocin and safranal by saffron stigma-like structures proliferated *in vitro*. *Agricultural and Biological Chemistry* 51(9), 2395–2400.

Hirschberg, J. (2001). Carotenoid biosynthesis in flowering plants. *Current Opinion in Plant Biology* 4(3), 210–218.

Husaini, A.M., Haq, S.A.U. and Jiménez, A.J.L. (2022). Understanding saffron biology using omics-and bioinformatics tools: stepping towards a better *Crocus* phenome. *Molecular Biology Reports* 49(6), 5325–5340.

Liang, M.H., He, Y.J., Liu, D.M. et al. (2021). Regulation of carotenoid degradation and production of apocarotenoids in natural and engineered organisms. *Critical Reviews in Biotechnology* 41(4), 513–534.

Lozano, P., Castellar, M.R., Simancas, M.J. et al. (1999). A quantitative high-performance liquid chromatographic method to analyse commercial saffron (*Crocus sativus* L.) products. *Journal of Chromatography A* 830(2), 477–483.

Maggi, F., Barboni, L., Caprioli, G. et al. (2011). HPLC quantification of coumarin in bastard balm (*Melittis melissophyllum* L., Lamiaceae). *Fitoterapia* 82(8), 1215–1221.

Misawa, N. and Shimada, H. (1998). Metabolic engineering for the production of carotenoids in non-carotenogenic bacteria and yeasts. *Journal of Biotechnology* 59(3), 169–181.

Pfander, H. and Schurtenberg, H. (1982). Biosynthesis of C 20-carotenoids in *Crocus sativus*. *Phytochemistry* 21(5), 1039–1042.

Pfister, S., Meyer, P., Steck, A. et al. (1996). Isolation and structure elucidation of carotenoid–glycosyl esters in gardenia fruits (*Gardenia jasminoides* Ellis) and saffron (*Crocus sativus* Linne). *Journal of Agricultural and Food Chemistry* 44(9), 2612–2615.

Qin, X. and Zeevaart, J. (1999). The 9-*cis*-epoxycarotenoid cleavage reaction is the key regulatory step of abscisic acid biosynthesis in water-stressed bean. *Proceedings of the National Academy of Sciences of the United States of America* 96(26), 15354–15361.

Rameshrad, M., Razavi, B.M. and Hosseinzadeh, H. (2018). Saffron and its derivatives, crocin, crocetin and safranal: a patent review. *Expert Opinion on Therapeutic Patents* 28(2), 147–165.

Rocchi, R., Di Mattia, C., Gabriele, G. et al. (2024). Influence of solvent polarity on crocin content and surface properties of saffron (*Crocus sativus* L.) extracts. *Molecules* 29(21), 5144.

Sánchez, A.M., Carmona, M., del Campo, C.P. et al. (2009). Solid-phase extraction for picrocrocin determination in the quality control of saffron spice (*Crocus sativus* L.). *Food Chemistry* 116(3), 792–798.

Sandmann, G. (2001). Genetic manipulation of carotenoid biosynthesis: strategies, problems and achievements. *Trends in Plant Science* 6(1), 14–17.

Schwartz, S., Qin, X. and Zeevaart, J. (2001). Characterization of a novel carotenoid cleavage dioxygenase. *Journal of Biological Chemistry* 276(27), 25208–25211.

Speranza, G., Dada, G., Manitto, P. et al. (1984). 13-*cis*-Crocin: a new crocinoid of saffron. *Gazzetta Chimica Italiana* 114, 189–192.

Tan, H., Chen, X., Liang, N. et al. (2019). Transcriptome analysis reveals novel enzymes for apo-carotenoid biosynthesis in saffron and allows construction of a pathway for crocetin synthesis in yeast. *Journal of Experimental Botany* 70(18), 4819–4834.

von Lintig, J. and Vogt, K. (2000). Filling the gap in vitamin A research: molecular identification of an enzyme cleaving β-carotene to retinal. *Journal of Biological Chemistry* 275(16), 11915–11920.

Winterhalter, P. and Straubinger, M. (2000). Saffron—renewed interest in an ancient spice. *Food Reviews International* 16(1), 39–59.

Yang, B., Liu, X. and Gao, Y. (2009). Extraction optimization of bioactive compounds (crocin, geniposide and total phenolic compounds) from Gardenia (*Gardenia jasminoides* Ellis) fruits with response surface methodology. *Innovative Food Science & Emerging Technologies* 10(4), 610–615.

Zheng, X., Mi, J., Balakrishna, A. et al. (2022). *Gardenia* carotenoid cleavage dioxygenase 4a is an efficient tool for biotechnological production of crocins in green and non-green plant tissues. *Plant Biotechnology Journal* 20(11), 2202–2216.

Zinati, Z., Shamloo-Dashtpagerdi, R. and Behpouri, A. (2016). In silico identification of miRNAs and their target genes and analysis of gene co-expression network in saffron (*Crocus sativus* L.) stigma. *Molecular Biology Research Communications* 5(4), 233–246.

12

Saffron Apocarotenoids: Insights into Evolution, Gene Regulation, and Genetic Engineering Approaches for Crop Improvement

Aamir Hussain Pala[2], Adnan Shakeel[1], Jewel Jameeta Noor[2], Aabida Gul[2], Uzma Jan[2], Mehnaz Shakeel[2], and Nasheeman Ashraf[2,]*

[1] *Department of Botany, Central University of Kashmir, Ganderbal*
[2] *Department of Plant Molecular Biology & Biotechnology, CSIR-Indian Institute of Integrative Medicine, Srinagar, Jammu and Kashmir, India*

*Corresponding author: nashraf@iiim.ac.in

12.1 Introduction

Saffron (*Crocus sativus* L.) owes its high value both economically and medicinally to its rich content of apocarotenoids, which are unique compounds derived from the oxidative cleavage of carotenoids to produce crocin, picrocrocin, and safranal (Baba and Ashraf 2016). The process of making apocarotenoids in saffron starts in plastids with the carotenoid pathway. Here, phytoene synthase (PSY) helps to combine geranylgeranyl diphosphate (GGPP) to make phytoene. This colorless carotenoid goes through desaturation and cyclization processes to make zeaxanthin, which is the main ingredient in saffron-specific apocarotenoids (Baba et al. 2015; Ahrazem et al. 2019). Carotenoid cleavage dioxygenase 2 (CCD2) is an important enzyme in this system that cuts zeaxanthin at the 7,8 and 7′,8′ double bonds to make crocetin dialdehyde, which is a key intermediate (Tan et al. 2019). This enzyme is only found in saffron, and its expression is specific to stigmatic tissue, which shows how it has evolved to be more specialized (Ahrazem et al. 2022). A tightly co-expressed genetic module functioning during stigma development is indicated by the close parallels between its expression and that of upstream carotenoid biosynthesis genes, including β-carotene hydroxylase (BCH), PSY, and lycopene β-cyclase (LCYB) (Dhar et al. 2017; Ahrazem et al. 2019; Sharma et al. 2019). Transcriptional regulation is essential for apocarotenoid production. CsULT1, an *ultrapetala* transcription factor (TF), is a key regulator of *CCD2* and related genes, as shown in transcriptome studies (Tan et al. 2019). Research also highlights the involvement of TF families like MYB, NAC, and bZIP, which help coordinate gene expression in response to developmental and environmental changes

Plant Apocarotenoids: Biosynthesis, Signaling and Agricultural Applications, First Edition. Edited by Kaiser Iqbal Wani and Tariq Aftab.
Published 2026 by John Wiley & Sons Ltd.

(Eshaghi and Rashidi-Monfared 2024). Additionally, cis-regulatory regions in apocarotenoid gene promoters suggest a complex combinatorial control, acting as binding sites for TFs that respond to developmental signals and stress (Gómez-Gómez et al. 2010; Zhou et al. 2020). The studies discovered that two rounds of lineage-specific whole-genome triplication (WGT) took place with the chromosome-level *Crocus* genome assembly, and that these rounds played significant roles in the synthesis of carotenoids and apocarotenoids. The biosynthesis of crocin is caused by the duplication, site positive selection, and neofunctionalization of *Crocus*-specific *CCD2* from *CCD1* members, based on the kingdom-wide identification, phylogenetic analysis, and functional assays of carotenoid cleavage dioxygenases (CCDs). The KS value distribution studies in *Crocus* show that gene duplication events had a crucial role in the evolution of carotenoid biosynthesis (Xu et al. 2024). Apocarotenoids such as crocin, picrocrocin, and safranal with their various biological activities and extensive commercial applications, particularly as food colorants, have enormous promise for both health and industrial use (Hartati and Paramita 2023). Crocin is water-soluble due to its high sugar content, making it unique among carotenoids and suited for aqueous medical and food uses (Rameshrad et al. 2018). Picrocrocin is one of the most bitter chemicals discovered in saffron. It is a critical precursor in the biosynthesis of safranal, the primary fragrance ingredient of saffron, which arises during saffron drying or storage (Shahi et al. 2016). Advances in global profiling methodologies, such as genomics, transcriptomics, and metabolomics, as well as multi-omics approaches, have proven extremely beneficial in understanding the variance in apocarotenoid levels and the underlying genetic pathways (Zheng et al. 2021). Transcriptomic analysis of cultivated and wild *Crocus* species (e.g., *Crocus cartwrightianus*) revealed that high apocarotenoid accumulation consistently correlates with increased expression of *DXS*, *PDS*, *ZDS*, *Z-ISO*, *CrtISO*, *LCYB*, *CCD2*, and *UGT74AD2*, and also highlighted variations in the regulatory mechanisms (Hassanlou et al. 2023). Structural genes *CstALDH3*, *CstLcyE* (*C. sativus lycopene ε-cyclase*), as well as their transcriptional factor families like MYB, MADS, and SANT/MYB protein CstULT1 are intimately involved in the upregulation of key pathway genes (Morote et al. 2023; Thakuri et al. 2024; Khurshaid et al. 2025). Thus, this study focuses on the evolution of saffron apocarotenoids, including their biosynthesis and control: economically significant saffron apocarotenoids like crocin, picrocrocin, safranal, and genomic techniques of increasing apocarotenoids in saffron.

12.2 Biosynthesis and Regulation of Saffron Apocarotenoids

Apocarotenoid biosynthesis in saffron is a plastid-localized process beginning with the carotenoid pathway, where PSY catalyzes the condensation of GGPP to form phytoene. A defining enzymatic step in this pathway is catalyzed by *CCD2*, which cleaves zeaxanthin at the 7,8/7′,8′ double bonds to produce crocetin dialdehyde, a pivotal intermediate (Tan et al. 2019). This enzyme is unique to saffron and its expression is stigmatic tissue-specific, underlining its evolutionary specialization (Ahrazem et al. 2022). Following cleavage, crocetin dialdehyde is oxidized by aldehyde dehydrogenases (ALDHs) to yield crocetin. This aglycone is then glycosylated by UDP-glucosyltransferases (UGTs) such as *CsUGT2*, converting it into crocins—water-soluble glycosylated derivatives that are ultimately

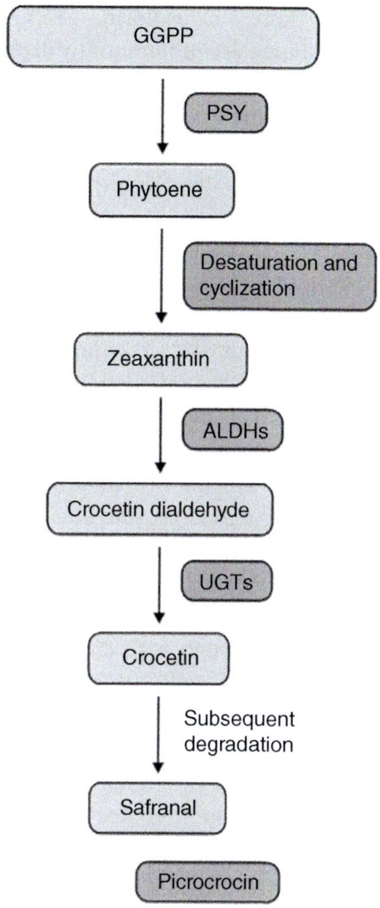

Figure 12.1 Illustrating apocarotenoid biosynthesis pathway.

transported and stored in vacuoles (Dhar et al. 2017; Tan et al. 2019). Additionally, picrocrocin, the precursor to safranal, is produced from the same cleavage event involving zeaxanthin, indicating a shared biosynthetic origin for multiple apocarotenoids (Figure 12.1) (Gómez-Gómez et al. 2010).

The biosynthesis of saffron apocarotenoids is developmentally regulated and tightly confined to specific floral tissues. Transcript profiling during stigma maturation shows that expression of *CCD2*, *UGTs*, and *ALDHs* peaks during pre-anthesis and anthesis, coinciding with maximum crocin biosynthesis and deposition (Sharma et al. 2019). This correlation highlights the developmental regulation that ensures synchronized pigment accumulation with stigma readiness. Interestingly, early transcriptional activation of carotenoid biosynthetic genes—*PSY*, *PDS*, and *BCH*—has been observed in apical and axillary buds and corm tissues, suggesting that saffron undergoes metabolic pre-programming before visible floral development (Sharma et al. 2019). These findings indicate that biosynthetic commitment precedes anatomical differentiation, possibly governed by floral meristem identity genes and early hormonal cues. Among the key enzymes, PSY gene family members exhibit specialization that contributes to flux control. Saffron harbors at least three PSY isoforms—*PSY1–PSY3*—with tissue- and stage-specific expression patterns. Some are preferentially expressed in stigmatic tissues, suggesting their role in funneling carotenoid precursors toward apocarotenoid production (Ahrazem et al. 2019). This functional partitioning of isoforms likely contributes to the regulation of metabolic branch points and minimizes competition between primary and secondary carotenoid-derived metabolites. Such specialization reflects adaptive evolution, particularly in domesticated *C. sativus*, which has been selected for enhanced stigma pigmentation and aroma.

Transcriptional regulation plays a vital role in orchestrating apocarotenoid biosynthesis. Transcriptome analyses and co-expression networks have identified CsULT1, an *ultrapetala* TF, as a critical upstream regulator of *CCD2* and related genes (Tan et al. 2019). Moreover, co-regulatory network analysis has revealed the involvement of major TF families like MYB, NAC, and bZIP, which likely coordinate expression under specific developmental and environmental conditions (Eshaghi and Rashidi-Monfared 2024). Additionally, cis-regulatory elements in the promoters of apocarotenoid pathway genes suggest complex combinatorial control. These promoter motifs provide binding sites for TFs responsive to developmental signals or stress cues, enabling dynamic modulation of gene expression (Gómez-Gómez et al. 2010; Zhou et al. 2020).

Environmental stimuli, including light, temperature, salinity, and cultivation practices, modulate apocarotenoid biosynthesis through both direct gene regulation and systemic physiological responses. For instance, Zhou et al. (2020) demonstrated that different flower cultivation regimes significantly influenced crocin content and expression levels of *BCH1* and *CCD2*, underscoring the impact of agronomic factors on metabolic outcomes. Under salt stress, *CsUGT2* expression is notably downregulated, leading to reduced crocin synthesis. This downregulation aligns with the presence of salt-responsive cis-elements in its promoter and confirms a strong link between abiotic stress and secondary metabolism (Moslemi et al. 2021). Such environmental responsiveness not only adds a layer of regulation but also raises the possibility of modulating crocin content through stress priming, controlled light exposure, or hydroponic adjustments.

Apocarotenoid biosynthesis does not operate in isolation but is part of a broader network of secondary metabolic pathways. Co-expression studies indicate that crocin-related genes show tight coordination with genes involved in flavonoid and terpenoid biosynthesis, suggesting possible transcriptional crosstalk and shared regulatory hubs (Eshaghi and Rashidi-Monfared 2024). Moreover, network pharmacology approaches propose that apocarotenoids, especially crocins, may influence their own production via retrograde regulation, that is, metabolite-mediated feedback that modulates upstream gene activity (Tiwari et al. 2023). This concept parallels plastid–nucleus signaling observed in other specialized metabolic pathways and adds a metabolite-sensing dimension to regulation. To overcome the limited natural yield of saffron apocarotenoids, metabolic engineering has emerged as a promising strategy. Key biosynthetic genes from *C. sativus* have been successfully expressed in heterologous systems like yeast and tomato, resulting in the production of crocins and related metabolites (Ahrazem et al. 2022). These synthetic platforms serve dual roles: enabling industrial-scale production and functioning as model systems to probe regulatory dynamics. For example, crocin-producing tomato lines offer insight into promoter activity, TF binding specificity, and the minimal genetic elements required for pathway reconstruction. These studies confirm that spatial and temporal regulations are preserved

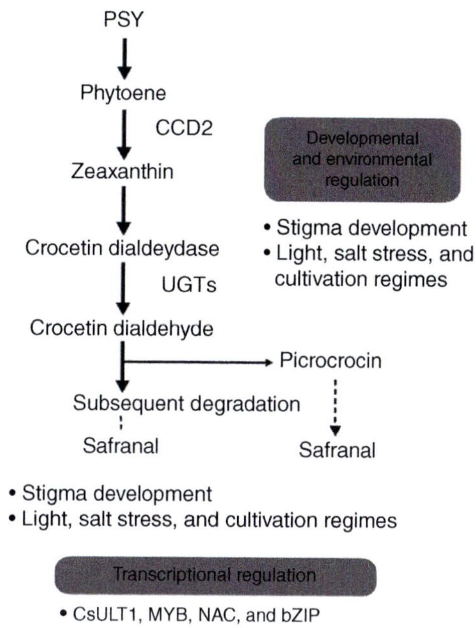

Figure 12.2 Regulation of saffron apocarotenoids.

when saffron genes are placed under appropriate promoters, validating the modular architecture of the regulatory system (Figure 12.2) (Tan et al. 2019).

12.3 Evolution of the Carotenoid Biosynthesis Pathway Genes

The apocarotenoid pathways start with PSY, the core rate-limiting enzyme (Zhou et al. 2022). This PSY enzyme has expanded into a PSY gene family ranging from 1 to 13 copies in various monocots, compared to *Arabidopsis thaliana*, which has one copy. In the *Crocus* genome, we have three different *PSY* genes; *CsPSY5* and *CsPSY6* are specifically expressed mainly in stigmas, and *CsPSY3* is uniquely transcribed in leaves. The expansion of *CsPDS*, *CsZISO*, *CsZDS*, and *CsCRTISO* genes in the *Crocus* genome has been identified at multiple positions (Xu et al. 2024). After lycopene is produced, the pathway splits into two branches. One branch involves cyclization to δ-carotene, catalyzed by lycopene ε-cyclase (LCYE), while the other leads to γ-carotene through cyclization by LCYB (Li et al. 2009). Three *LCYE*, two *LCYB*, and eight *CYP97* members were identified in the *Crocus* genome. The 9-*cis* epoxycarotenoid cleavage dioxygenase (NCED) family member NCED3 cleaves neoxanthin or violaxanthin to produce ABA. From the *Crocus* genome, six *NCED3* genes were identified; however, all *NCED3* genes showed silent expression in stigmas. The complete carotenoid biosynthesis pathway from GGPP, derived from the *Crocus* genome, was illustrated and compared with the paths in 12 other monocot species and two eudicot species. It was further concluded that there are no tandem duplication events for apocarotenoid biosynthetic genes in the *Crocus* genome. These results thus indicated that tandem duplication might not be the driving force of gene expansion related to carotenoid biosynthesis in *Crocus*. The KS distributions for all paralogous carotenoid genes in *Crocus* revealed that the average KS values for three distinct groups align with interspersed repetitive sequences and two lineage-specific whole-genome duplication (WGD) events. This suggests that these gene duplication events were significant in the evolution of carotenoid biosynthesis. Furthermore, processes like PD, TRD, and WGD have played crucial roles in the expansion of carotenoid biosynthetic genes (Xu et al. 2024).

12.3.1 Evolutionary Insights and Catalytic Activities

Crocins are the primary valuable compounds found in the stigmas of *Crocus* and the mature fruits of *Gardenia*, and the biosynthetic pathways for crocin production in both plants have been nearly fully understood. In general, the enzymes CCD, ALDH, and UGT are key players in the conversion of carotenoids into crocins (Lee et al. 2025). In a study, 20 *CCD* genes were identified in the *Crocus* genome, consisting of six *CCD1*, one *CCD1L*, two *CCD2*, three *CCD4*, two *CCD7*, and six *NCED* genes. Notably, two *CCD4* genes were highly expressed in flower tissues, including petals, stamens, and stigmas, while showing minimal expression in roots, leaves, and peduncles. In general, gene duplication has played a significant role in the development of new gene functions that aid in adaptation. When genes are duplicated, the extra copies can evolve distinct functions over time, often providing the organism with new abilities to respond to environmental challenges or selective pressures. This process can lead to the emergence of traits that enhance an organism's survival and fitness in specific conditions (Schmidt et al. 2019). Members of the CCD family play a key role in breaking down carotenoids into apocarotenoids. Notably, CCD1 and CCD4 enzymes exhibit a wide range of substrate preferences and specific cleavage sites (Hou et al. 2016; Daruwalla and Kiser 2020; Dhar et al. 2020). Compared to other monocots, the *CCD* genes in *Crocus* species exhibit a unique evolutionary pattern. Phylogenetic analysis revealed that all *Crocus CCD2* genes clustered together in a single branch, indicating a lineage-specific evolution of the *CCD2* subfamily. This *CCD2* branch was found to be sister to all *CCD1* genes from the tested species, while the *Crocus CCD1* genes were closely related to *CCD1* genes from other monocots. This suggests that *Crocus CCD1* genes have evolved rapidly. Generally, gene duplication plays a significant role in developing novel gene functions related to adaptation. This study showed strong genome synteny between *Crocus CCD2* and *CCD1* genes. The average KS value of paralogous genes in syntenic regions was around 0.17, which aligns with the KS value of WGT-2. These findings support the idea that the *CCD2* subfamily evolved from *Crocus CCD1* genes after a species-specific WGT event, with rapid evolution of the *Crocus CCD2* genes leading to neofunctionalization and the production of crocetin dialdehyde. The nonsynonymous-to-synonymous mutation ratios (ω = KA/KS) for *CCD* genes, analyzed using the two-ratio model in PAML, showed that ω values for both *CCD1* and *CCD2* branches were less than 1 ($P < 0.01$), indicating that *Crocus CCD* genes underwent purifying selection. Additionally, the testing of selection pressure at specific sites in *Crocus CCD2* using a branch-site model was performed. Thirteen sites were identified as under positive selection, including L174, T326, and N335 ($P > 99\%$), as well as S232, S257, E283, S292, H295, F341, S389, Q405, R450, and S508 ($P > 95\%$). These positively selected amino acids are likely associated with the functional divergence of the *Crocus CCD2* genes (Xu et al. 2024). In *C. sativus* (saffron), the enzyme CsCCD2 breaks down zeaxanthin at specific positions (7,8 and 7′,8′) to produce crocetin dialdehyde (Xu et al. 2020). In *Gardenia*, a different enzyme, GjCCD4a, performs the same reaction, even though it only shares 31% similarity with CsCCD2. Interestingly, GjCCD4a can act on multiple carotenoids—zeaxanthin, β-carotene, and lycopene—while CsCCD2 was previously thought to work only on zeaxanthin. GjCCD4a originated from a duplicated *CCD4* gene unique to *Gardenia*. The original CCD4 enzyme cleaved carotenoids at different positions (9,10), but the duplicated version (GjCCD4a) evolved to produce crocetin dialdehyde. In contrast, the CsCCD4 enzyme from *C. sativus* cannot break down β-carotene, lycopene, or zeaxanthin to produce crocetin dialdehyde. These findings suggest that CsCCD2 in saffron and GjCCD4a in *Gardenia* evolved independently (convergent evolution) to perform a similar function, while *CCD4* genes in *Gardenia* and *Coffea* evolved in different ways (divergent evolution). This helps explain why crocin is only found in a few, distantly related plants. Interestingly, this study shows that both CsCCD2-1 and CsCCD2-2 in saffron can actually cleave β-carotene, lycopene, and zeaxanthin—contrary to earlier findings by Frusciante and co-workers who reported that CsCCD2 could not use β-carotene or lycopene as substrates (Frusciante et al. 2014). Fang et al. (2020) cloned *FhCCD2* from *Freesia hybrida*, which shares 77.46% identity with *CsCCD2* and performs the same cleavage of zeaxanthin at the 7,8 (7′,8′) positions, key for crocin biosynthesis. Our study suggests that *CCD2* originated before the divergence of *Freesia* and *Crocus* (~16.1 MYA), after the shared WGT-2 genome triplication event. In *Crocus*, *CsCCD2* evolved from duplicated *CCD1* genes post-WGT-2, likely under strong positive selection for new 7,8 cleavage activity. Further research using advanced modeling is needed to fully understand the catalytic differences among *CCD1*, *CCD2*, and *CCD4* (Xu et al. 2024).

12.4 Structure, Function, and Potential Uses of Crocin and Picrocrocin

12.4.1 Crocin

Crocin is the major bioactive pigment in *C. sativus* (saffron), responsible for its bright red color and many medicinal properties. The intense reddish orange color of crocin comes from the conjugated double bonds in the crocetin core. These absorb light in the blue–green spectrum (~440–470 nm), reflecting orangish red. Chemically, crocin is a digentiobiosyl ester of crocetin with a molecular formula of $C_{44}H_{64}O_{24}$. Crocetin is the aglycone (non-sugar) part—a C20 dicarboxylic carotenoid. Each of the two carboxyl groups of crocetin is esterified with a gentiobiose sugar (a disaccharide of two glucose units linked β [1 → 6]). Unlike most carotenoids (which are fat-soluble), crocin is water-soluble, due to its high sugar content (hydrophilic) that makes it unique among carotenoids and ideal for aqueous medicinal and food applications (Rameshrad et al. 2018). The rigid linear structure of the crocetin backbone affects its interaction with cellular membranes and enzymes. The bulky sugar groups increase solubility and influence transport in biological systems (Wikipedia 2025). Depending on the number, type of sugar units, position, and extent of esterification, crocins are of the following types:

Crocin-1: bis-gentiobiosyl crocetin (main form);
Crocin-2: mono-gentiobiosyl mono-glucosyl crocetin;
Crocin-3: bis-glucosyl crocetin.

Crocin, a key active constituent of saffron (*C. sativus* L.), has attracted extensive scientific interest due to its diverse biological functions and therapeutic potential. Below is a comprehensive breakdown of its functions and potential applications:

Antioxidant activity: crocin exhibits a strong free radical scavenging ability and protects cells against oxidative damage (Ali et al. 2022).

Anti-inflammatory activity: it can modulate inflammatory pathways by inhibiting the expression of pro-inflammatory cytokines and suppressing various anti-inflammatory signaling pathways that in turn reduce inflammation (Hashemzaei et al. 2020).

Potential neuroprotective effects: crocin protects neurons against ischemia, neurodegeneration, and toxicity, thus suggesting its role in protecting against neurodegenerative diseases (Khazdair et al. 2015).

Anticancer: crocin has shown antiproliferative activity in a variety of cancer cell lines, including breast, colorectal, and leukemia cells. It can induce apoptosis, arrest the cell cycle, and inhibit angiogenesis (Hoshyar and Mollaei 2017).

Antidepressant: crocin has shown antidepressant effects by increasing serotonin and dopamine levels in the brain (Siddiqui et al. 2022).

Cardiovascular: it can improve heart health by reducing LDL cholesterol and triglycerides, that in turn improves blood flow (Bathaie et al. 2014).

Antidiabetic: crocin can help in managing diabetes by improving insulin sensitivity and reducing glucose (Sepahi et al. 2022).

Structural, functional, and application summary of crocin.

Property	Functions	Application areas
Color pigment	Absorbs visible light due to conjugated double bonds	Food, cosmetics
Water solubility	Glycosylation with gentiobiose increases solubility	Pharmaceutical formulations
Antioxidant	Scavenges ROS, protects lipids and DNA	Antiaging, neuroprotection
Neuroprotective	Reduces inflammation, improves cognition	Alzheimer's, ischemia
Anticancer	Induces apoptosis, inhibits angiogenesis	Cancer therapy
Biotechnological potential	Engineered production in tomato/cell cultures	Industrial biosynthesis

12.4.2 Picrocrocin

Picrocrocin is a monoterpene glycoside with a molecular formula of $C_{16}H_{26}O_7$ and is derived from oxidative cleavage of zeaxanthin by an enzyme namely CCD. It is one of the primary bitter compounds found in saffron (*C. sativus* L.). It is a key precursor in the biosynthesis of safranal, the major aroma compound of saffron that forms during drying or storage of saffron when picrocrocin undergoes enzymatic or acid-catalyzed hydrolysis, releasing safranal and glucose (Shahi et al. 2016). Picrocrocin is composed of two main parts, aglycone part (non-sugar component) that contains two double bonds (positions 2 and 6 in the chain—both E-configuration), a terminal aldehyde group at C-1, methyl groups at C-2 and C-6, glycosidic linkage at C-8 (hydroxyl group replaced by glucose), and a sugar moiety with a single β-D-glucose unit linked via a glycosidic bond at the C-8 hydroxyl group of the aglycone. Picrocrocin itself contributes to the bitter taste of saffron, a critical quality parameter, while also acting as the biochemical reservoir for safranal, thereby shaping the spice's aroma during post-harvest processing (Wikipedia 2025). Beyond organoleptic value, its ecological roles, such as UV protection or pest deterrence, remain speculative and merit further agronomic exploration. Research indicates picrocrocin possesses various biological activities, including antioxidant and anti-inflammatory properties, with potential applications in treating conditions like cancer, diabetes, and neurodegenerative diseases.

Anticancer potential: although crocin is more extensively studied in oncology, it is reported that picrocrocin inhibits HeLa cell proliferation at an IC_{50} of ≈ 3 mM, supporting its modest direct cytotoxic potential and underscoring the merit of further mechanistic and combinatorial analysis (Cayman Chemical 2022).

Antioxidant activity: It can scavenge free radicals and protect against oxidative damage (Hosseinzadeh et al. 2010).

Anti-inflammatory activity: It can modulate inflammatory pathways and reduce inflammation (Frattaruolo et al. 2023).

Potential neuroprotective effects: Studies suggest it may have a role in protecting against neurodegenerative diseases like Alzheimer's and Parkinson's (Maggi et al. 2020).

Plant defense: Picrocrocin is believed to play a protective role in *C. sativus* by deterring herbivores and pathogens through its bitter taste and potential antimicrobial properties (Mir et al. 2022).

Antiviral insight: *In vitro* investigations from 2022 highlight picrocrocin's capacity to impede HSV-1 and HIV-1 entry and replication, demonstrating antiviral potential, while *in-silico* docking suggests strong gp120 binding by crocin rather than picrocrocin. These findings support a foundation for further mechanistic studies on picrocrocin's antiviral action (Sabzian-Molaei et al. 2022).

Summary of picrocrocin functions and applications.

Function	Mechanism/effect	Application areas
Bitter taste compound	Activates bitter taste receptors	Spice quality and culinary usage
Aroma precursor	Converts to safranal on drying	Perfume and food aroma industries
Lipid regulation	Modulates LDLR expression; non-statin mode	Cardiometabolic therapy
Antiviral activity	Blocks viral replication (HSV, HIV)	Antiviral drug development
Anticancer potential	Inhibits HeLa cell growth (IC_{50} ≈ 3 mM)	Cancer therapeutics
Engineering target	Synthesized in tomato via metabolic engineering	Biofactory and nutraceutical production

12.4.3 Safranal

Among the 40 aromatic compounds identified in saffron, safranal (2,6,6-trimethyl-1,3-cyclohexadiene-1-carboxaldehyde) is the primary contributor to its distinctive aroma. In premium saffron varieties, such as *Azafrán de La Mancha*, safranal makes up more than 65% of the total aromatic compounds. Safranal is produced through the hydrolysis and dehydration of picrocrocin during the post-harvest processes, including dehydration, handling, and storage. To measure its concentration, the ISO 3632 standard recommends evaluating the absorbance at 330 nm from an aqueous saffron extract. However, this method has its drawbacks, as it can also detect cis-isomers of crocetin, which can interfere with accurate readings. To overcome this, more reliable techniques like high-performance liquid chromatography (HPLC) are used to quantify safranal, along with other components such as crocetin esters and picrocrocin. Additionally, gas chromatography (GC) is employed for the comprehensive analysis of all volatile compounds, offering enhanced precision in identifying safranal amidst other aroma compounds. These analytical techniques are vital for maintaining saffron's quality, authenticity, and consistency in both commercial and research settings (José Bagur et al. 2018).

12.4.3.1 Pharmacological Roles of Safranal

Safranal, a monoterpene aldehyde, is the key volatile compound responsible for saffron's distinctive odor. Several studies have highlighted its pharmacological properties, including antioxidant, anti-inflammatory, cardioprotective, neuroprotective, nephroprotective, and gastrointestinal protective effects. Research predominantly focuses on its CNS effects, such as antianxiety, analgesic, anticonvulsant, anti-ischemic, anti-tremor, and memory-enhancing activities. It also shows protective potential against neurodegenerative diseases like Alzheimer's, Parkinson's, and Huntington's (Esmaealzadeh et al. 2023).

a. Antioxidant and anti-inflammatory

   Several studies show safranal's antioxidant and anti-inflammatory effects. In old rats, a 0.5-mg/kg/day intraperitoneal injection of safranal for 1 month reduced age-induced oxidative damage and enhanced the antioxidant defense system. Safranal likely acts as an indirect antioxidant by inducing mild oxidative stress, which triggers the production of antioxidant enzymes (Farahmand et al. 2013).

b. Cardioprotective effects

   Extensive evidence supports safranal's antihypertensive, anti-ischemic, and relaxant effects, mediated through antioxidants, Ca^{2+} regulation, smooth muscle cell contraction inhibition, and antiapoptotic actions. In rats, safranal (1, 2, and 4 mg/kg/day) reduced systolic blood pressure in DOCA salt-induced hypertension, but its antihypertensive effects did not persist in normotensive animals with renal protective effects (Fatehi et al. 2003).

   Safranal exhibits nephroprotective effects through its antioxidant and anti-inflammatory properties. In diabetic nephropathy and chemotherapy-induced nephrotoxicity models, a 4-week safranal treatment improved renal function (reducing BUN and creatinine) and histopathological damage. This was achieved by enhancing antioxidant levels (TAS, GSH) and reducing oxidative stress (NO, TOS), along with decreasing inflammatory markers (TNF-α, IL-1β, and IFN-Υ) (Hazman and Bozkurt 2015).

c. Effects on respiratory tract

Besides its antioxidative and anti-inflammatory effects, safranal has antitussive and smooth muscle relaxant effects mediated by inhibiting the histamine (H_1) receptors and stimulating β-adrenoceptors (Boskabady et al. 2014).

12.5 Genomic Approaches to Enhance Apocarotenoids in Saffron

Saffron (*C. sativus* L.), the world's most expensive spice by weight, derives its exceptional commercial and medicinal value from a group of specialized metabolites called apocarotenoids—primarily crocin, picrocrocin, and safranal. These compounds not only impart the deep golden hue, distinctive flavor, and aroma to saffron but also exhibit potent bioactivities, including antioxidant, neuroprotective, and anti-inflammatory effects. From a biosynthetic perspective, these apocarotenoids originate from carotenoids, mainly zeaxanthin, via a series of enzymatically regulated cleavage, oxidation, and glycosylation reactions. The manipulation of this apocarotenoid pathway through modern genomic technologies has become an increasingly promising route to improve both the quality and yield of saffron's bioactive profile. At the center of this biosynthetic pathway is the CCD CstCCD2, which specifically cleaves zeaxanthin to produce crocetin dialdehyde, the immediate precursor of crocin. This reaction marks the first committed step in apocarotenoid production. Transcriptomic and molecular analyses have consistently highlighted *CstCCD2L*, a paralog of *CCD2*, as a key stigma-specific, developmentally regulated gene. Its expression closely parallels that of upstream carotenoid biosynthetic genes, such as *PSY*, *LCYB*, and *BCH*, indicating a tightly co-expressed genetic module operating during stigma development (Dhar et al. 2017; Ahrazem et al. 2019; Sharma et al. 2019). Recent genomic analyses have shed light on the evolutionary emergence of *CstCCD2* through WGD events. Using full-length PacBio SMRT transcriptome sequencing, Zhou et al. (2025) discovered that *CstCCD2* originated from a duplication of *CCD1*-like genes during a relatively recent β-WGD event. This event not only allowed for sub-functionalization of the gene in stigma tissue but also coincided with the expansion of genes associated with secondary metabolism, suggesting that genome duplication played a critical role in shaping saffron's unique apocarotenoid traits. A multilayered regulatory framework appears to coordinate this complex metabolic output. Temporal gene expression profiling across stigma developmental stages revealed co-expression of over 32 biosynthetic genes and 28 putative regulators, providing a systems-level snapshot of how apocarotenoid flux is controlled. Notably, CstALDH3, a key enzyme that oxidizes crocetin dialdehyde to crocetin, was functionally validated via heterologous expression in yeast, demonstrating its indispensability for crocin biosynthesis (Morote et al. 2023). Comparative transcriptomics across cultivated and wild *Crocus* species (e.g., *C. cartwrightianus*) showed that high apocarotenoid accumulation consistently correlates with the elevated expression of genes like *DXS*, *PDS*, *ZDS*, *Z-ISO*, *CrtISO*, *LCYB*, *CCD2*, and *UGT74AD2*, while also revealing divergence in regulatory dynamics (Hassanlou et al. 2023). Natural variation in the *CstLcyE* gene modulates carotenoid/apocarotenoid profiles and influences stress tolerance (Javid et al. 2023). In addition to structural genes, TFs have emerged as central regulators of apocarotenoid biosynthesis. Studies have shown that MYB family TFs (CstMYB1, CstMYB1R2), MADS-box factors (CstMADS19, CstMADS26), and the SANT/MYB domain protein CstULT1 are intimately involved in the activation of key pathway genes (Thakuri et al. 2024; Khurshaid et al. 2025). Overexpression of *CstULT1* led to significant increases in the expression of *PSY*, *BCH*, and *CCD2*, ultimately enhancing crocin accumulation in engineered lines (Vahedi et al. 2019). Other transcriptional regulators such as HY5, PIF1, and ORANGE (OR) further suggest a broader transcriptional circuitry responsive to light, developmental cues, and possibly stress signals (Thakuri et al. 2024). Interestingly, elicitor-based modulation has emerged as a complementary strategy to stimulate apocarotenoid biosynthesis. Moradi et al. (2022) demonstrated that treatment of saffron cell suspension cultures with salicylic acid (SA) significantly elevated the transcript levels of *CCD2* and *BCH1*, with a 3.3-fold increase observed for *CCD2* within 24 hours. This points to the possibility of enhancing secondary metabolite production via stress- or hormone-induced regulatory mechanisms. While genetic improvement in *C. sativus* is constrained by its triploid, sterile nature, significant progress has been made using heterologous expression systems. In *Nicotiana benthamiana*, transient expression of *CstCCD2*, *UGTs*, and *ALDHs* successfully reconstituted the crocin biosynthetic pathway, yielding crocin levels of up to 2.18 mg/g dry weight (Demurtas et al. 2023). Similarly, *Saccharomyces cerevisiae* strains engineered with these genes produced crocetin up to 16 mg/L, demonstrating the feasibility of pathway transfer and scale-up for industrial production (Hao et al. 2025). In parallel, the development of *in vitro* regeneration systems, such as stigma-derived callus and somatic embryogenesis protocols, has opened up possibilities for functional gene validation and future genome editing (Salami and Husaini 2022). Expression of ZCD (a functional synonym of *CCD2*) in cultured tissues has provided a useful tool for studying pathway regulation in a controlled environment. Although stable transformation remains a challenge, these advancements lay the groundwork for CRISPR/Cas9-based editing of both structural and regulatory genes. The integration of multi-omics platforms—including genomics, transcriptomics, proteomics, and metabolomics—has revolutionized our ability to dissect the apocarotenoid biosynthetic network. Platforms such as SaffronOMICS have enabled gene co-expression mapping, identification of rate-limiting nodes, and visualization of genotype-by-environment interactions affecting metabolite accumulation (Haq et al. 2022; Salami and Husaini 2022). These tools are not only advancing the understanding of saffron biology but also accelerating precision breeding and synthetic biology interventions. Although a complete reference genome for *C. sativus* is still unavailable, high-quality transcriptome assemblies, SNP datasets, and detailed metabolic maps are rapidly closing this gap. The combined power of functional genomics, metabolic engineering, elicitor-based modulation, and systems biology offers an integrated framework for enhancing the production of saffron's signature apocarotenoids. These efforts are expected to lead to the next generation of saffron cultivars—engineered for higher productivity, enriched metabolite content, and broader applicability in food, cosmetics, and pharmaceutical industries.

12.6 Conclusion

Comprehensive overview of the evolution, biosynthesis, and regulation of saffron apocarotenoids is provided, focusing particularly on their economic importance and the molecular mechanisms that govern their accumulation. Among these, crocin, picrocrocin, and safranal stand out as the most valuable compounds due to their potent biological activities and wide application in food, pharmaceutical, and cosmetic industries. The biosynthetic pathway of these apocarotenoids, which originates from carotenoid precursors, is tightly regulated by specific enzymes and gene expression patterns within *Crocus*. With increasing commercial demand and limitations posed by traditional cultivation, recent advancements in genomic techniques offer promising strategies for enhancing apocarotenoid content in saffron. Tools such as transcriptomic analysis, gene editing, and metabolic engineering now allow for a deeper understanding and targeted manipulation of key genes involved in apocarotenoid biosynthesis. These innovations not only improve yield and quality but also pave the way for sustainable and large-scale production of high-value saffron compounds.

References

Ahrazem, O., Diretto, G., Argandoña Picazo, J. et al. (2019). The specialized roles in carotenogenesis and apocarotenogenesis of the phytoene synthase gene family in saffron. *Frontiers in Plant Science* 10, 249.

Ahrazem, O., Diretto, G., Rambla, J.L. et al. (2022). Engineering high levels of saffron apocarotenoids in tomato. *Horticulture Research* 9, uhab065.

Ali, A., Yu, L., Kousar, S. et al. (2022). Crocin: functional characteristics, extraction, food applications and efficacy against brain related disorders. *Frontiers in Nutrition* 9, 1009807. https://doi.org/10.3389/fnut.2022.1009807.

Baba, S.A. and Ashraf, N. (2016). *SpringerBriefs in Plant Science: Apocarotenoids of Crocus sativus L: From Biosynthesis to Pharmacology*. Springer.

Baba, S.A., Mohiuddin, T., Basu, S. et al. (2015). Comprehensive transcriptome analysis of *Crocus sativus* for discovery and expression of genes involved in apocarotenoid biosynthesis. *BMC Genomics* 16, 1015.

Bathaie, S.Z., Farajzadeh, A. and Hoshyar, R. (2014). A review of the chemistry and uses of crocins and crocetin, the carotenoid natural dyes in saffron, with particular emphasis on applications as colorants including their use as biological stains. *Biotechnic & Histochemistry* 89, 401–411. https://doi.org/10.3109/10520295.2014.890741.

Boskabady, M.H., Byrami, G. and Feizpour, A. (2014). The effect of safranal, a constituent of *Crocus sativus* (saffron), on tracheal responsiveness, serum levels of cytokines, total NO and nitrite in sensitized guinea pigs. *Pharmacological Reports* 66, 56–61.

Cayman Chemical (2022). Product information: picrocrocin anticancer activity. https://cdn.caymanchem.com/cdn/insert/39914.pdf.

Daruwalla, A. and Kiser, P.D. (2020). Structural and mechanistic aspects of carotenoid cleavage dioxygenases (CCDs). *Biochimica et Biophysica Acta (BBA)—Molecular and Cell Biology of Lipids* 1865(11), 158590.

Demurtas, O.C., Sulli, M., Ferrante, P. et al. (2023). Production of saffron apocarotenoids in *Nicotiana benthamiana* plants genome-edited to accumulate zeaxanthin precursor. *Metabolites* 13(6), 729.

Dhar, M.K., Mishra, S., Bhat, A. et al. (2020). Plant carotenoid cleavage oxygenases: structure–function relationships and role in development and metabolism. *Briefings in Functional Genomics* 19(1), 1–9.

Dhar, M.K., Sharma, M., Bhat, A. et al. (2017). Functional genomics of apocarotenoids in saffron: insights from chemistry, molecular biology and therapeutic applications. *Briefings in Functional Genomics* 16(6), 336–347.

Eshaghi, M. and Rashidi-Monfared, S. (2024). Co-regulatory network analysis of the main secondary metabolite (SM) biosynthesis in *Crocus sativus* L. *Scientific Reports* 14(1), 2153.

Esmaealzadeh, D., Ghalibaf, A.M., Rad, M.S. et al. (2023). Pharmacological effects of safranal: an updated review. *Iranian Journal of Basic Medical Sciences* 26(10), 1131–1143. https://doi.org/10.22038/IJBMS.2023.69824.15197.

Fang, Q., Li, Y., Liu, B. et al. (2020). Cloning and functional characterization of a carotenoid cleavage dioxygenase 2 gene in safranal and crocin biosynthesis from *Freesia hybrida*. *Plant Physiology and Biochemistry* 154, 439–450.

Farahmand, S.K., Samini, F., Samini, M. et al. (2013). Safranal ameliorates antioxidant enzymes and suppresses lipid peroxidation and nitric oxide formation in aged male rat liver. *Biogerontology* 14, 63–71.

Fatehi, M., Rashidabady, T. and Fatehi-Hassanabad, Z. (2003). Effects of *Crocus sativus* petals' extract on rat blood pressure and on responses induced by electrical field stimulation in the rat isolated vas deferens and guinea-pig ileum. *Journal of Ethnopharmacology* 84 (2–3), 199–203.

Frattaruolo, L., Marra, F., Lauria, G. et al. (2023). A picrocrocin-enriched fraction from a saffron extract affects lipid homeostasis in HepG2 cells through a non-statin-like mode. *International Journal of Molecular Sciences* 24(4), 3060. https://doi.org/10.3390/ijms24043060.

Frusciante, S., Diretto, G. and Bruno, M. (2014). Novel carotenoid cleavage dioxygenase catalyzes the first dedicated step in saffron crocin biosynthesis. *Proceedings of the National Academy of Sciences of the United States of America* 111(33), 12246–12251. https://doi.org/10.1073/pnas.1404629111.

Gómez-Gómez, L., Rubio-Moraga, Á. and Ahrazem, O. (2010). Understanding carotenoid metabolism in saffron stigmas: unravelling aroma and colour formation. *Functional Plant Science and Biotechnology* 4(S2), 56–63.

Hao, C., Hu, K., Xie, J. et al. (2025). Recent advancements in the biomanufacturing of crocetin and crocins: key enzymes and metabolic engineering. *Journal of Agricultural and Food Chemistry* 73(11), 6400–6415.

Haq, S.A., Salami, S.A. and Husaini, A.M. (2022). Omics in saffron (*Crocus sativus* L.): a spice of immense medicinal value. In: *Omics in Horticultural Crops* (eds. G.R. Rout and K.V. Peter), 573–587. Academic Press.

Hartati, I. and Paramita, V. (2023). Apocarotenoids: sources, classification and their potential application as food additives. *Research in Chemical Engineering (RiCE)* 2, 44–50. https://doi.org/10.30595/rice.v2i2.112.

Hoshyar, R. and Mollaei, H. (2017). A comprehensive review on anticancer mechanisms of the main carotenoid of saffron, crocin. *Journal of Pharmacy and Pharmacology* 69(11), 1419–1427.

Hashemzaei, M., Mamoulakis, C., Tsarouhas, K. et al. (2020). Crocin: a fighter against inflammation and pain. *Food and Chemical Toxicology* 143, 111521. https://doi.org/10.1016/j.fct.2020.111521.

Hassanlou, M., Azimi Moghaddam, M.R., Salamii, S.A. et al. (2023). Investigating apocarotenoids and the structure of genes related to their synthesis in cultivated and wild saffron species using HPLC and bioinformatics tools. *Saffron Agronomy and Technology* 11(2), 183–197.

Hazman, Ö. and Bozkurt, M.F. (2015). Anti-inflammatory and antioxidative activities of safranal in the reduction of renal dysfunction and damage that occur in diabetic nephropathy. *Inflammation* 38, 1537–1545.

Hosseinzadeh, H., Shamsaie, F. and Mehri, S. (2010). Antioxidant activity of aqueous and ethanolic extracts of *Crocus sativus* L. stigma and its bioactive constituents crocin and safranal. *Pharmacognosy Magazine* 5, 419–424.

Hou, X., Rivers, J. and León, P. (2016). Synthesis and function of apocarotenoid signals in plants. *Trends in Plant Science* 21(9), 792–803.

José Bagur, M., Alonso Salinas, G.L., Jiménez-Monreal, A.M. et al. (2018). Saffron: an old medicinal plant and a potential novel functional food. *Molecules* 23(1), 30. https://doi.org/10.3390/molecules23010030.

Khazdair, M.R., Boskabady, M.H., Hosseini, M. et al. (2015). The effects of *Crocus sativus* (saffron) and its constituents on nervous system: a review. *Avicenna Journal of Phytomedicine* 5(5), 376–391.

Khurshaid, N., Shabir, N., Pala, A.H. et al. (2025). Transcriptome wide analysis of MADS box genes in *Crocus sativus* and interplay of *CstMADS19-CstMADS26* in orchestrating apocarotenoid biosynthesis. *Gene* 932, 148893.

Lee, Y., Hwang, C.Y., Cho, E.S. et al. (2025). Water-soluble carotenoid: focused on natural carotenoid crocin. *Food Science and Biotechnology* 34, 1119–1138.

Li, Z., Ahn, T.K., Avenson, T.J. et al. (2009). Lutein accumulation in the absence of zeaxanthin restores nonphotochemical quenching in the *Arabidopsis thaliana npq1* mutant. *The Plant Cell* 21(6), 1798–1812. https://doi.org/10.1105/tpc.109.066571.

Maggi, M.A., Bisti, S. and Picco, C. (2020). Saffron: chemical composition and neuroprotective activity. *Molecules* 25(23), 5618. https://doi.org/10.3390/molecules25235618.

Mir, T.U.G., Wani, A.K., Singh, J. et al. (2022). Therapeutic application and toxicity associated with *Crocus sativus* (saffron) and its phytochemicals. *Pharmacological Research—Modern Chinese Medicine* 4, 100136. https://doi.org/10.1016/j.prmcm.2022.100136.

Moradi, A., Zarinkamar, F., Mita, G. et al. (2022). Enhanced production of apocarotenoids by salicylic acid elicitation in cell suspension cultures of saffron (*Crocus sativus* L.). *Horticulturae* 8(12), 1176.

Morote, M.I., Rubio Moraga, Á., López Jiménez, A.J. et al. (2023). *Verbascum* species as a new source of saffron apocarotenoids and molecular tools for the biotechnological production of crocins and picrocrocin. *The Plant Journal* 118, 58–72.

Moslemi, F.S., Vaziri, A., Sharifi, G. et al. (2021). The effect of salt stress on the production of apocarotenoids and the expression of genes related to their biosynthesis in saffron. *Molecular Biology Reports* 48(5), 4561–4571.

Sabzian-Molaei, F., Hosseini, S., Bolhassani, A. et al. (2022). Antiviral effect of saffron compounds on the GP120 of HIV-1: an in silico study. *ChemistrySelect* 7(47), e202203471.

Salami, S.A. and Husaini, A.M. (2022). SaffronOMICS: novel approaches toward putting saffron data at work. In: *Compendium of Plant Genomes: The Saffron Genome* (eds. J. Vakhlu, S. Ambardar and S.A. Salami), 43–62. Springer International Publishing.

Schmidt, T., Heitkam, T., Liedtke, S. et al. (2019). Adding color to a century-old enigma: multicolor chromosome identification unravels the autotriploid nature of saffron (*Crocus sativus*) as a hybrid of wild *Crocus cartwrightianus* cytotypes. *New Phytologist* 222(4), 1965–1980. https://doi.org/10.1111/nph.15715.

Sepahi, S., Golfakhrabadi, M., Bonakdaran, S. et al. (2022). Effect of crocin on diabetic patients: A placebo-controlled, triple-blinded clinical trial. *Clinical Nutrition ESPEN* 50, 255–263. https://doi.org/10.1016/j.clnesp.2022.05.006.

Shahi, T., Assadpour, E. and Jafari, S.M. (2016). Main chemical compounds and pharmacological activities of stigmas and tepals of "red gold"; saffron. *Trends in Food Science & Technology* 58, 69–78. https://doi.org/10.1016/j.tifs.2016.10.010.

Sharma, M., Kaul, S. and Dhar, M.K. (2019). Transcript profiling of carotenoid/apocarotenoid biosynthesis genes during corm development of saffron (*Crocus sativus* L.). *Protoplasma* 256, 249–260.

Siddiqui, S.A., Ali Redha, A., Snoeck, E.R. et al. (2022). Anti-depressant properties of crocin molecules in saffron. *Molecules* 27(7), 2076. https://doi.org/10.3390/molecules27072076.

Tan, H., Chen, X., Liang, N. et al. (2019). Transcriptome analysis reveals novel enzymes for apo-carotenoid biosynthesis in saffron and allows construction of a pathway for crocetin synthesis in yeast. *Journal of Experimental Botany* 70(1), 39–51.

Thakuri, M.H., Chettry, U., Pathak, J. et al. (2024). Integration of apocarotenoid and transcriptome profiles of developing stigmas of saffron (*Crocus sativus* L.) reveals a multifaceted control involving PSY2, OR-a, PIF1 and HY5 genes in regulation of apocarotenoid biosynthesis. *South African Journal of Botany* 167, 467–475.

Tiwari, A., Modi, S.J., Girme, A. et al. (2023). Network pharmacology-based strategic prediction and target identification of apocarotenoids and carotenoids from standardized Kashmir saffron (*Crocus sativus* L.). *Medicine* 102(38), e34884.

Vahedi, M., Karimi, R., Panigrahi, J. et al. (2019). Salient biotechnological interventions in saffron (*Crocus sativus* L.): a major source of bio-active apocarotenoids. In: *Natural Bio-active Compounds:* Volume 3: *Biotechnology, Bioengineering, and Molecular Approaches* (eds. M. Akhtar and Swamy), 205–223. Springer.

Wikipedia (2025). Crocin. In: *Wikipedia, The Free Encyclopedia*. Retrieved 14 June 2025, https://en.wikipedia.org/w/index.php?title=Crocin&oldid=1292037069.

Xu, Z., Chen, S., Wang, Y. et al. (2024). Crocus genome reveals the evolutionary origin of crocin biosynthesis. *Acta Pharmaceutica Sinica B* 14(4), 1878–1891. https://doi.org/10.1016/j.apsb.2023.12.013.

Xu, Z., Pu, X., Gao, R. et al. (2020). Tandem gene duplications drive divergent evolution of caffeine and crocin biosynthetic pathways in plants. *BMC Biology* 18, 1–14.

Rameshrad, M., Razavi, B.M. and Hosseinzadeh, H. (2018). Saffron and its derivatives, crocin, crocetin and safranal: a patent review. *Expert Opinion on Therapeutic Patents* 28(2), 147–165.

Zheng, X., Yang, Y. and Al-Babili, S. (2021) Exploring the diversity and regulation of apocarotenoid metabolic pathways in plants. *Frontiers in Plant Science* 12, 787049. https://doi.org/10.3389/fpls.2021.787049.

Zhou, G., Li, L., Lu, J. et al. (2020). Flower cultivation regimes affect apocarotenoid accumulation and gene expression during the development of saffron stigma. *Horticulture Research* 7(1), 44.

Zhou, T., Park, Y.K., Fu, J. et al. (2025). Metabolic engineering of *Yarrowia lipolytica* for the production and secretion of the saffron ingredient crocetin. *Biotechnology for Biofuels and Bioproducts* 18(1), 1.

Zhou, X., Rao, S., Wrightstone, E. et al. (2022). Phytoene synthase: the key rate-limiting enzyme of carotenoid biosynthesis in plants. *Frontiers in Plant Science* 13, 884720.

13

Cell Biofactories for the Sustainable Production of Saffron Apocarotenoids

Tingan Zhou[1] *and Rodrigo Ledesma-Amaro*[2*]

[1] *Department of Natural Product Biosynthesis, Max Planck Institute for Chemical Ecology, Jena, Germany*
[2] *Department of Bioengineering, Imperial College Centre for Synthetic Biology, Bezos Centre for Sustainable Protein, UKRI Engineering Biology Mission Hub on Microbial Food Centre for Synthetic Biology, Imperial College London, London, UK*

*Corresponding author: r.ledesma-amaro@imperial.ac.uk

13.1 Introduction

Saffron, derived from the dried stigmas of *Crocus sativus* L., is esteemed as "red gold" (Leone et al. 2018), due to its unique color, aroma, and pharmacological properties, making it one of the world's most valuable spices (Cardone et al. 2020). With a documented history of over 4,000 years, saffron has been widely utilized in traditional healing systems as a general tonic to promote overall well-being and as a natural remedy for over 90 medical indications (Bathaie and Mousavi 2011). In recent years, the use of spices and functional foods has increased because of their alleged safety and possible nutritional and medicinal effects for the prevention of a range of degenerative diseases (Bansal et al. 2023). Notably, consumer interest in saffron has been steadily increasing, largely due to its reported therapeutic properties and potential role in supporting human health (Leone et al. 2018).

As a recurrent herb of the Iridaceae family, saffron is native to Eastern Asia and widely cultivated in the Mediterranean region (Iran, Israel, Iraq, Spain, Greece, Turkey, Egypt, France, Italy, and Switzerland), Africa, India, China, Pakistan, and Australia (Pandey et al. 2020). However, for the dry regions like Iran (Koocheki et al. 2020) or Spain (Berbel et al. 2019), substantial irrigation is needed to achieve a satisfactory yield for saffron (Koocheki et al. 2020), making saffron cultivation somehow limited by climate conditions. Additionally, each flower produces only three delicate stigmas, which must be carefully hand-harvested and dried, making saffron one of the most labor-intensive and expensive spices in the world (Gresta et al. 2008, 2009).

Considering saffron's vast application and commercial values, people started to analyze its functional components for future sustainable production via cell biofactories. Saffron

Plant Apocarotenoids: Biosynthesis, Signaling and Agricultural Applications, First Edition. Edited by Kaiser Iqbal Wani and Tariq Aftab.
Published 2026 by John Wiley & Sons Ltd.

contains three key secondary metabolites—crocin, picrocrocin, and safranal—which are primarily responsible for its characteristic color, bitter taste, and distinct aroma, respectively (Shokrpour 2019). Those compounds are also responsible for saffron's diverse applications in medicine (Leone et al. 2018). All the above three metabolites belong to the apocarotenoid family, which is a diverse group of bioactive compounds that result from the oxidative cleavage of carotenoids, such as β-carotene, zeaxanthin, and other related pigments (Ilyas et al. 2021). This degradation process typically involves the enzymatic loss of carbon atoms from the linear side chains of carotenoids, catalyzed by carotenoid cleavage dioxygenases (CCDs), producing smaller molecules with distinct chemical and biological properties (Liang et al. 2021).

13.2 Saffron Apocarotenoids

Among the apocarotenoids present in saffron, crocetin, crocin, picrocrocin, and safranal have attracted considerable scientific interest due to their notable pharmacological properties, therapeutic potential, and commercial application as a colorant (Figure 13.1).

13.2.1 Crocetin and Crocin

Crocin is a water-soluble reddish orange pigment derived from its aglycone precursor, crocetin (Abdullaev 2002; Zhao et al. 2021a). It comprises a group of structurally related

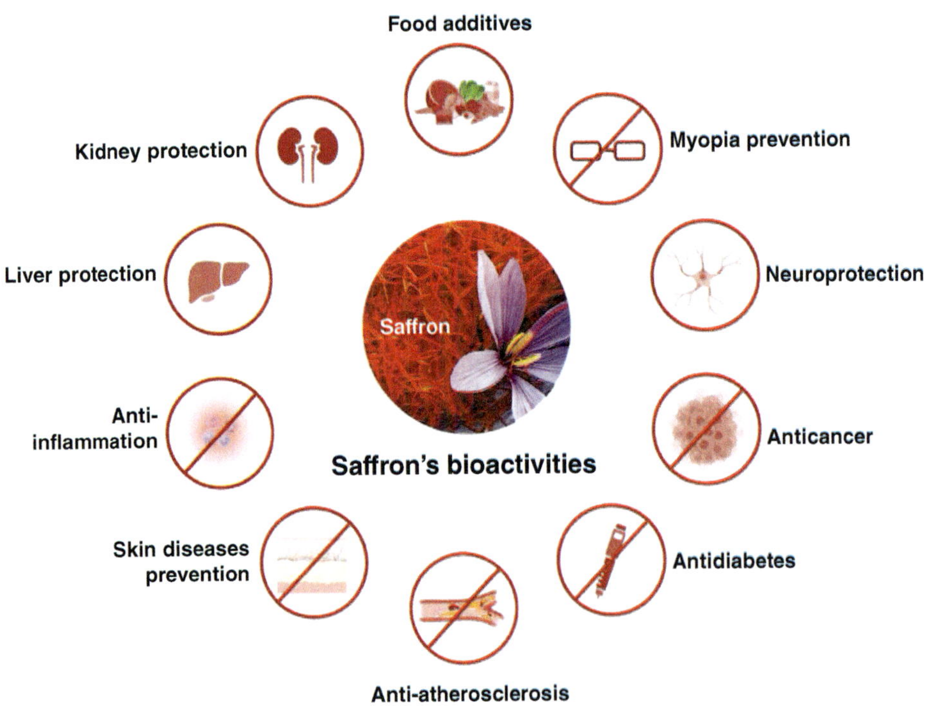

Figure 13.1 Bioactivities and applications of saffron.

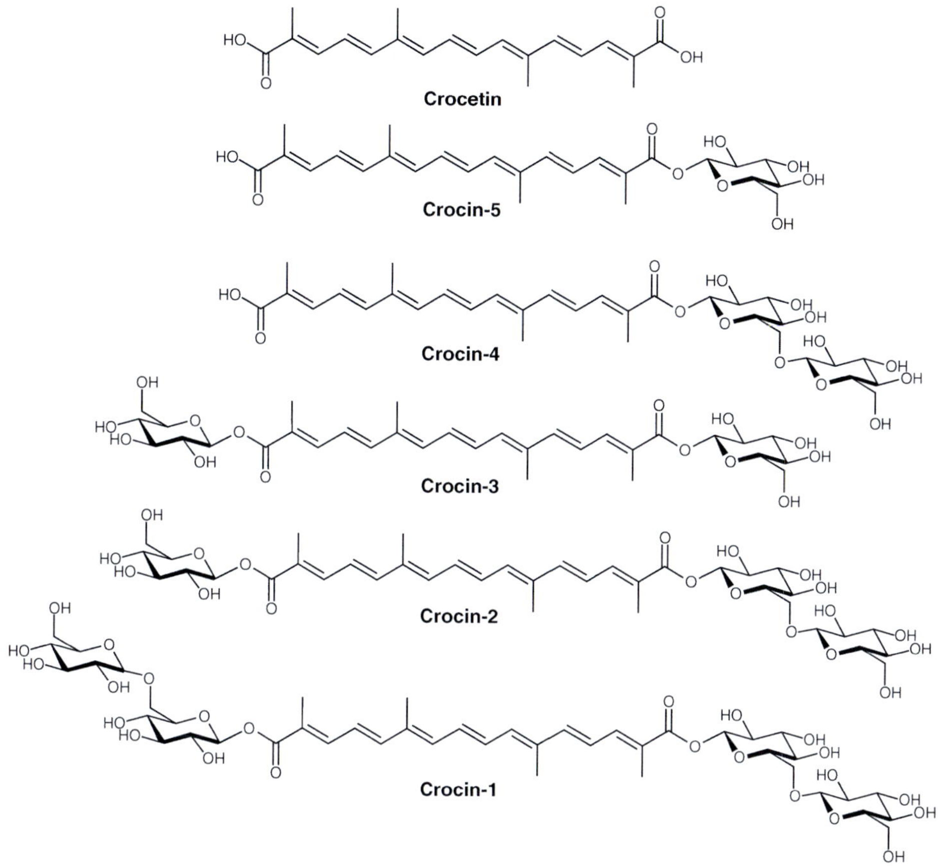

Figure 13.2 Structures of crocetin and crocins.

hydrophilic carotenoids that are formed by the esterification of crocetin with one–four sugar moieties, primarily glucose (Li et al. 1999; Abdullaev 2002). These glycosylated forms differ in the number and position of attached sugar moieties (Figure 13.2) (Li et al. 1999). Among the various crocin compounds, crocin-1 is the most abundant, accounting for over 70% of the total crocin content (Li et al. 1999; Chen et al. 2010; Song et al. 2021). More detailed classification and structural analysis of crocins, including distinctions between their cis- and trans-isomeric forms, have been comprehensively reviewed by Song et al. (2021). Unless otherwise specified, crocins referred to in this chapter are predominantly in the trans-configuration, which is the naturally predominant and pharmacologically relevant form (Shahi et al. 2016).

Crocin has been extensively studied for its broad spectrum of biological activities. Reported pharmacological properties include antigenotoxic (Koul and Abraham 2018), antioxidant (Bastani et al. 2022), anticancer (Ali et al. 2021, 2022a; Jia et al. 2023), anti-inflammatory (Ali et al. 2022a), antidiabetic (Sepahi et al. 2022; Asdaq et al. 2024), hypotensive (Razavi et al. 2013), hypoglycemic (Qiu et al. 2020), antihyperlipidemic (Lee et al. 2005), antidegenerative, and antidepressant properties (Kermanshahi et al. 2020). Its aglycone precursor, crocetin, has also demonstrated significant cardioprotective and

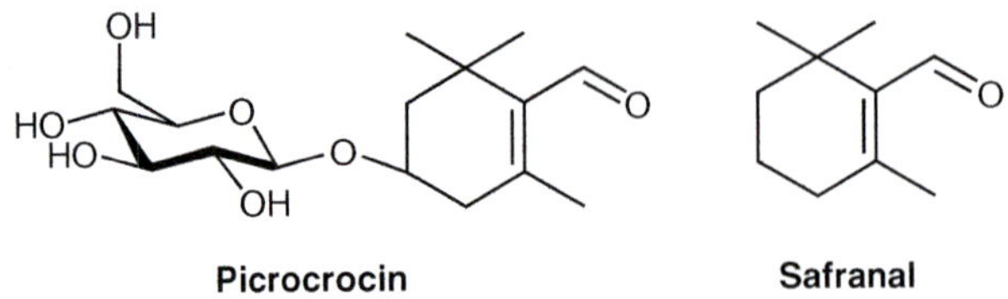

Figure 13.3 Structures of picrocrocin and safranal.

anti-atherosclerotic effects (Zheng et al. 2005; Yu et al. 2021). In addition to its medicinal potential, crocin has been utilized in the formulation of functional foods, not only to enhance their health benefits but also to improve product stability due to its antioxidant properties (Ali et al. 2022b).

13.2.2 Picrocrocin and Safranal

Safranal is derived from its monoterpene glycoside precursor, picrocrocin (Figure 13.3). Picrocrocin is primarily responsible for saffron's unique bitter taste (Chrysanthou et al. 2016). Upon enzymatic or thermal degradation (Püntener and Schlesinger 2000) leading to the loss of its glucose moiety, picrocrocin is converted into safranal, the principal volatile compound that imparts saffron's distinctive aroma (Esmaealzadeh et al. 2023).

The majority of studies on safranal have focused on its effects on the central nervous system (CNS), including its antianxiety, analgesic, anticonvulsant, anti-ischemic, and anti-tremor properties (Esmaealzadeh et al. 2023). Additionally, safranal has been reported to enhance memory function and exhibit neuroprotective effects against neurodegenerative disorders such as Alzheimer's, Parkinson's, and Huntington's diseases (Esmaealzadeh et al. 2023). Apart from its neuroprotection and treatment functions, it has also been reported to have some other beneficial effects, such as antioxidant (Nanda and Madan 2021), anti-inflammatory (Hazman and Bozkurt 2015), cardioprotective (Mehdizadeh et al. 2013), antitussive (Hosseinzadeh and Ghenaati 2006), and anticonvulsant effects (Hosseinzadeh and Talebzadeh 2005). While safranal's pharmacological properties have been extensively studied, reports on the bioactivities of picrocrocin are still limited.

13.3 Biosynthetic Pathway of Saffron Apocarotenoids

Saffron apocarotenoids, including crocetin, crocin, picrocrocin, and safranal, are derived from carotenoids, which are a class of vital pigments present in plants, algae, bacteria, and fungi. Their degradation products, known as apocarotenoids, are generated through oxidative cleavage, which is catalyzed by a family of enzymes called CCDs, specific to the cleavage of double bonds in carotenoid structures (Harrison and Bugg 2014). The biosynthesis of saffron apocarotenoids involves a highly coordinated interplay of multiple pathways, including the upstream synthesis of isopentenyl diphosphate (IPP) and dimethylallyl diphosphate (DMAPP) precursors, the midstream carotenoid biosynthetic pathway, and the downstream oxidative cleavage and subsequent modifications (Figure 13.4) (Liu et al. 2020b).

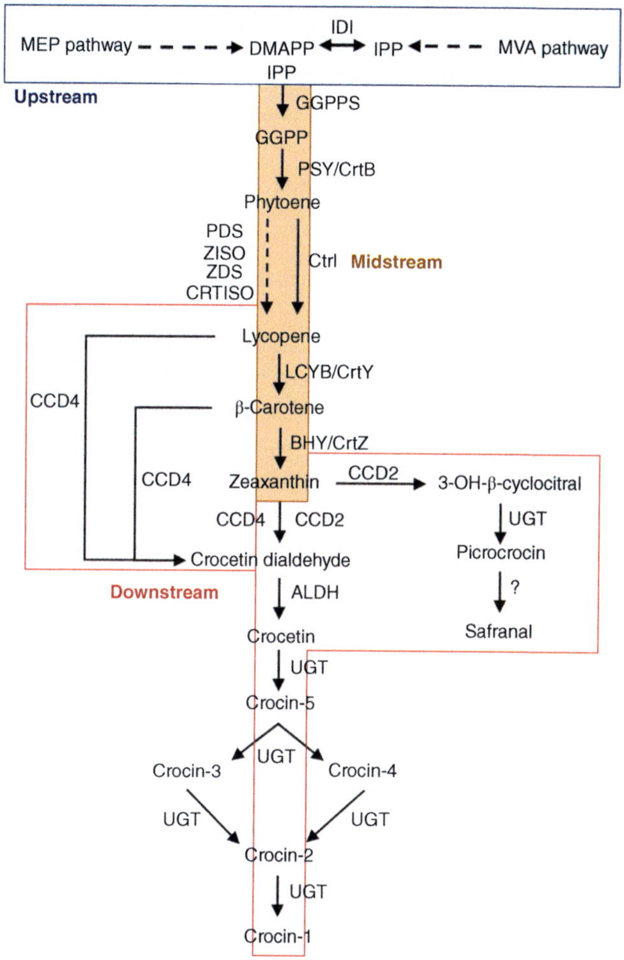

Figure 13.4 Biosynthetic pathway of saffron apocarotenoids: GGPP, geranylgeranyl diphosphate; GGPPS, geranylgeranyl diphosphate synthase; PSY/CrtB, phytoene synthase; PDS/CrtI, phytoene desaturase; ZISO, ζ-carotene isomerase; ZDS, ζ-carotene desaturase; CRTISO, carotenoid isomerase; LCYB/CrtY, lycopene β-cyclase; BHY/CrtZ, β-carotene hydrolase; CCD, carotenoid cleavage dioxygenase; ALDH, aldehyde dehydrogenase; and UGT, UDP-glycosyltransferase.

For the upstream pathway, in plants such as *C. sativus* and *Gardenia jasminoides*, IPP and DMAPP precursors are typically provided by the methylerythritol phosphate (MEP) pathway (Ji et al. 2017). However, in heterologous microbial cell factories, MEP serves as the source of IPP and DMAPP only in prokaryotic organisms like *Escherichia coli* for later production of saffron apocarotenoid (Wang et al. 2019). In contrast, in yeasts such as *Saccharomyces cerevisiae* and *Yarrowia lipolytica*, the mevalonate (MVA) pathway is utilized to provide the IPP and DMAPP pool, which is subsequently used for saffron apocarotenoid biosynthesis (Chai et al. 2017; Zhou et al. 2025).

To synthesize carotenoids, microorganisms typically initiate the pathway by producing geranylgeranyl diphosphate (GGPP), catalyzed by GGPP synthase (CrtE). Two molecules of GGPP are then condensed into phytoene under the action of phytoene synthase (CrtB).

Phytoene is subsequently converted into lycopene through a series of desaturation reactions catalyzed by phytoene dehydrogenase (CrtI). Lycopene then undergoes two cyclization reactions catalyzed by lycopene cyclase (CrtY), resulting in the formation of β-carotene. Finally, β-carotene is hydroxylated by β-carotene hydroxylase (CrtZ) to yield zeaxanthin. In plants, the overall biosynthetic process is similar but involves a more complex set of enzymes for the desaturation and isomerization steps (Zhang et al. 2018). Specifically, phytoene is converted to lycopene through the sequential action of phytoene desaturase (PDS), ζ-carotene desaturase (ZDS), ζ-carotene isomerase (ZISO), and carotenoid isomerase (CRTISO). The resulting all-*trans*-lycopene is then cyclized to form β-carotene by lycopene β-cyclase (LCYB).

In the downstream pathway, carotenoid cleavage dioxygenase 2 (CCD2) catalyzes the cleavage of the 7,8 and 7′,8′ double bonds of zeaxanthin, yielding crocetin dialdehyde and 3-hydroxy-β-cyclocitral as products. In addition to zeaxanthin, CCD4 has been reported to cleave other carotenoids such as lycopene and β-carotene (Liu et al. 2020b). Subsequently, several UDP-glycosyltransferases (UGTs) participate in the glycosylation of crocetin dialdehyde to form various crocin derivatives, as well as the glycosylation of 3-hydroxy-β-cyclocitral to produce picrocrocin. The latter is eventually converted into safranal, probably through the combined action of thermal processing and β-glucosidase activity during the drying or storage of saffron (Jain et al. 2016).

13.4 Microbial Cell Factories

Although microorganisms cannot naturally synthesize saffron apocarotenoids, many have been genetically engineered as heterologous hosts for the sustainable and rapid production of various carotenoids (Table 13.1). The biosynthesis of carotenoids such as lycopene (Ma et al. 2019, 2022; Liu et al. 2020a; Jing et al. 2021), β-carotene (Wu et al. 2020; Wang et al. 2021; Zhao et al. 2021b; Xu et al. 2023), and zeaxanthin (Shao et al. 2012; Li et al. 2015; Zhang et al. 2018, 2023) has been well established in model microorganisms, including *E. coli*, *S. cerevisiae*, and *Y. lipolytica*, indicating their great potential for biosynthesizing saffron apocarotenoids.

Escherichia coli has been fundamental to the advancement of genetics and molecular biology, laying the groundwork for the emergence of synthetic biology (Adams 2016). As a result, it is unsurprising that *E. coli* remains a dominant chassis organism in the field (Adams 2016) and is chosen to be one of the ideal strains for saffron apocarotenoids biosynthesis. In 2018, two microbial glycosyltransferases, Bs-GT and Bc-GTA, were successfully identified and introduced into *E. coli* to enable the biosynthesis of crocins from crocetin (Ding et al. 2018). Under optimized conditions, efficient glycosylation of crocetin was achieved, reaching a high conversion rate of 81.9% and yielding 476.8 mg/L of crocins (Ding et al. 2018). However, only crocin-5 and crocin-3 were produced in this system. The first *de novo* biosynthesis in *E. coli* was achieved by Wang and co-workers with only 4.42-mg/L crocetin and a detectable amount of crocin-5 (Wang et al. 2019). Using two plant-derived glycosyltransferases, GjUGT94E13 and GjUGT74F8 from *G. jasminoides*, Pu and co-workers successfully achieved the biosynthesis of all five major crocins in *E. coli* (Pu et al. 2020). By enhancing the intracellular supply of UDP-glucose, the conversion rate from crocetin reached 59.6% (Pu et al. 2020). More recently, efficient crocetin production (34.77 mg/L) was obtained in

Table 13.1 Microbial cell factories for saffron apocarotenoids biosynthesis.

Apocarotenoid	Microorganism	Performance	Reference
Crocin	*Escherichia coli*	81.9% Conversion rate from crocetin; 476.8 mg/L (mainly crocin-5 and crocin-3)	Ding et al. (2018)
Crocetin	*Escherichia coli*	4.42 mg/L	Wang et al. (2019)
Crocin	*Escherichia coli*	Detectable crocin-5	Wang et al. (2019)
Crocin	*Escherichia coli*	59.6% Conversion rate from crocetin	Pu et al. (2020)
Crocetin	*Escherichia coli*	34.77 mg/L	Lee et al. (2005)
Crocin	*Escherichia coli*	Crocin-1 (6.29 mg/L), crocin-2 (5.29 mg/L), crocin-3 (1.48 mg/L), crocin-4 (2.72 mg/L)	Lee et al. (2005)
Crocetin	*Saccharomyces cerevisiae*	1.22 mg/L, 6.28 mg/L in bioreactors	Chai et al. (2017)
Crocetin	*Saccharomyces cerevisiae*	139.67 μg/g DCW	Liu et al. (2020b)
Crocetin	*Saccharomyces cerevisiae*	1.95 mg/L, 12.43 mg/L in bioreactors	Song et al. (2021)
Crocetin	*Saccharomyces cerevisiae*	107 mg/L in bioreactors	Liang et al. (2021)
Crocetin	*Yarrowia lipolytica*	30.17 mg/L	Zhou et al. (2025)

E. coli by overexpressing a zeaxanthin cleavage dioxygenase and a novel variant of crocetin dialdehyde dehydrogenase (Lee et al. 2005). Based on this crocetin high-producing strain, the downstream biosynthesis of crocin-1 (6.29 mg/L), crocin-2 (5.29 mg/L), crocin-3 (1.48 mg/L), and crocin-4 (2.72 mg/L) was successfully achieved (Lee et al. 2005).

Considering that IPP and DMAPP serve as essential precursors for saffron apocarotenoids, yeast—with its abundant IPP/DMAPP pool—has been selected as a suitable chassis for their bioproduction. The first successful heterologous biosynthesis of crocetin in yeast was reported in 2017, following the systematic screening of CrtZ, CCD, and ALD enzymes from various sources in *S. cerevisiae*. Ultimately, a combination of CrtZ from *Erwinia uredovora*, CCD2 from *Crocus*, and ALD from *Synechocystis* sp. PCC6803 enabled crocetin production at 1.22 mg/L in shake flasks and 6.28 mg/L in bioreactors under low-temperature conditions (20 °C) (Chai et al. 2017). To address the low-temperature preference of the enzyme, Liu and co-workers engineered a temperature-responsive *S. cerevisiae* chassis, where the biosynthesis was only active at 24 °C or lower temperature, achieving crocetin production of approximately 139.67 μg/g DCW (Liu et al. 2020b). In another study, by enhancing precursor supply and disrupting the glyoxylate cycle, crocetin production reached 1.95 mg/L in shake flasks and 12.43 mg/L in 5-L bioreactors in *S. cerevisiae* (Song et al. 2021). Recently, the S323A mutant of CCD2 was employed in *S. cerevisiae*, leading to an increased crocetin production of 20 mg/L in shake flasks and 107 mg/L after fed-batch fermentation (Liang et al. 2021). Besides *S. cerevisiae*, *Y. lipolytica* has also been employed for the biosynthesis of the saffron apocarotenoid crocetin as a novel chassis, achieving yeasts' highest reported titer of 30.17 mg/L in shake flasks (Zhou et al. 2025).

13.5 Plant Cell Factories

Various plants have been tested for the heterologous production of saffron apocarotenoids (Table 13.2). As the native producer of saffron apocarotenoids, *C. sativus* cell suspension cultures have been explored for the biosynthesis of crocin and safranal (Taherkhani et al. 2021). On the 12th day of cultivation, the production of crocin and safranal peaked at 1.255 and 0.0107 mg/g of dried callus, respectively, as reported by Taherkhani et al. (2021).

When it comes to heterologous biosynthesis, *Nicotiana* species has long been considered an ideal platform for the production of plant-derived natural products (Golubova et al. 2024). In 2019, a novel uridine diphosphate glycosyltransferase, UGT709G1, was identified as being involved in the biosynthesis of picrocrocin, thereby completing the elucidation of its entire biosynthetic pathway (Diretto et al. 2019). Building upon this, efficient production of both crocins and picrocrocin was achieved in *Nicotiana benthamiana* through a virus-driven expression system (Martí et al. 2020). Using recombinant viral clones derived from tobacco etch potyvirus expressing heterologous enzymes, 0.2% crocins and 0.8% picrocrocin (based on leaf dry weight [DW]) were detected within just two weeks (Martí et al. 2020).

In further studies, *Nicotiana tabacum* and *Nicotiana glauca* were evaluated as heterologous hosts for crocin biosynthesis, yielding approximately 36 and 400 μg/g DW in leaves, respectively (Ahrazem et al. 2022b). However, the crocin profile in these transgenic leaves, dominated by *trans*-crocin-3, differed from that of saffron, which predominantly contains *trans*-crocin-4 (Ahrazem et al. 2022b). More recently, expression of *GjCCD4a*, *GjALDH2C3*, *GjUGT74F8*, and *GjUGT94E13* from *G. jasminoides* fruits in *N. benthamiana* led to

Table 13.2 Plant cell factories for saffron apocarotenoid biosynthesis.

Apocarotenoid	Plant	Performance	Reference
Crocin	*Crocus sativus* cell suspension culture	1.255 mg/g of dried callus	Taherkhani et al. (2021)
Safranal	*Crocus sativus* cell suspension culture	0.0107 mg/g of dried callus	Taherkhani et al. (2021)
Crocin	*Nicotiana benthamiana*	2-mg/g leaf DW	Martí et al. (2020)
Picrocrocin	*Nicotiana benthamiana*	8-mg/g leaf DW	Martí et al. (2020)
Crocin	*Nicotiana tabacum*	36-μg/g leaf DW	Ahrazem et al. (2022b)
Crocin	*Nicotiana glauca*	400-μg/g leaf DW	Ahrazem et al. (2022b)
Crocin	*Nicotiana benthamiana*	1-mg/g leaf DW	Xie et al. (2023)
Crocin	*Solanum lycopersicum*	Around 110 μg/g DCW	Frusciante et al. (2021)
Picrocrocin	*Solanum lycopersicum*	2.92 mg/g DW	Ahrazem et al. (2022a)
Crocin	*Solanum lycopersicum*	14.48 mg/g DW	Ahrazem et al. (2022a)

approximately 1-mg/g DW crocin production (Xie et al. 2023). Although the overall production was modest, nearly 99% of total crocins were crocin-1 and crocin-2, demonstrating the system's specificity (Xie et al. 2023).

Beyond *Nicotiana*, saffron apocarotenoid biosynthesis has also been engineered in tomato (*Solanum lycopersicum*). Expression of *Bixa orellana* CCD4-3 along with a saffron-derived glycosyl transferase, led to crocin accumulation exceeding 100 mg/g DW, closely mirroring the crocin profile of natural saffron (Frusciante et al. 2021). Similarly, introducing crocin and picrocrocin biosynthetic pathways in tomato under the control of fruit-specific and constitutive promoters resulted in 14.48 mg/g of crocins and 2.92 mg/g of picrocrocin in the tomato DW, without compromising plant growth (Ahrazem et al. 2022a).

References

Abdullaev, F.I. (2002). Cancer chemopreventive and tumoricidal properties of saffron (*Crocus sativus* L.). *Experimental Biology and Medicine (Maywood)* 227(1), 20–25.

Adams, B.L. (2016). The next generation of synthetic biology chassis: Moving synthetic biology from the laboratory to the field. *ACS Synthetic Biology* 5(12), 1328–1330.

Ahrazem, O., Diretto, G., Rambla, J.L. et al. (2022a). Engineering high levels of saffron apocarotenoids in tomato. *Horticulture Research* 9, uhac074.

Ahrazem, O., Zhu, C., Huang, X. et al. (2022b). Metabolic engineering of crocin biosynthesis in *Nicotiana* species. *Frontiers in Plant Science* 13, 861140.

Ali, A., Manzoor, M.F., Ahmad, N. et al. (2022a). The burden of cancer, government strategic policies, and challenges in Pakistan: a comprehensive review. *Frontiers in Nutrition* 9, 940514.

Ali, A., Mughal, H., Ahmad, N. et al. (2021). Novel therapeutic drug strategies to tackle immune-oncological challenges faced by cancer patients during COVID-19. *Expert Review of Anticancer Therapy* 21(12), 1371–1383.

Ali, A., Yu, L., Kousar, S. et al. (2022b). Crocin: functional characteristics, extraction, food applications and efficacy against brain related disorders. *Frontiers in Nutrition* 9, 1009807.

Asdaq, S.M.B., Mannasaheb, B.A., Orfali, R. et al. (2024). Antidiabetic and antioxidant potential of crocin in high-fat diet plus streptozotocin-induced type-2 diabetic rats. *International Journal of Immunopathology and Pharmacology* 38, 3946320231220178.

Bansal, P., Maithani, M., Gupta, V. et al. (2023). Future prospective of nutraceutical and functional food with herbs and spices. In: *Herbs, Spices and Their Roles in Nutraceuticals and Functional Foods* (eds. A. Amalraj, S. Kuttappan, K. Varma et al.), 361–381. Academic Press.

Bastani, S., Vahedian, V., Rashidi, M. et al. (2022). An evaluation on potential anti-oxidant and anti-inflammatory effects of crocin. *Biomedicine and Pharmacotherapy* 153, 113297.

Bathaie, S.Z. and Mousavi, S.Z. (2011). Historical uses of saffron: identifying potential new avenues for modern research. *Avicenna Journal of Phytomedicine* 1, 57–66.

Berbel, J., Expósito, A., Gutiérrez-Martín, C. et al. (2019). Effects of the irrigation modernization in Spain 2002–2015. *Water Resources Management* 33(5), 1835–1849.

Cardone, L., Castronuovo, D., Perniola, M. et al. (2020). Saffron (*Crocus sativus* L.), the king of spices: an overview. *Scientia Horticulturae* 272, 109560.

Chai, F., Wang, Y., Mei, X. et al. (2017). Heterologous biosynthesis and manipulation of crocetin in *Saccharomyces cerevisiae. Microbial Cell Factories* 16(1), 54.

Chen, Y., Zhang, H., Li, Y.-X. et al. (2010). Crocin and geniposide profiles and radical scavenging activity of gardenia fruits (*Gardenia jasminoides* Ellis) from different cultivars and at the various stages of maturation. *Fitoterapia* 81(4), 269–273.

Chrysanthou, A., Pouliou, E., Kyriakoudi, A. et al. (2016). Sensory threshold studies of picrocrocin, the major bitter compound of saffron. *Journal of Food Science* 81(1), S189–S198.

Ding, F., Liu, F., Shao, W. et al. (2018). Efficient synthesis of crocins from crocetin by a microbial glycosyltransferase from *Bacillus subtilis* 168. *Journal of Agricultural and Food Chemistry* 66(44), 11701–11708.

Diretto, G., Ahrazem, O., Rubio-Moraga, Á. et al. (2019). UGT709G1: a novel uridine diphosphate glycosyltransferase involved in the biosynthesis of picrocrocin, the precursor of safranal in saffron (*Crocus sativus*). *The New Phytologist* 224(2), 725–740.

Esmaealzadeh, D., Moodi Ghalibaf, A., Shariati Rad, M. et al. (2023). Pharmacological effects of safranal: an updated review. *Iranian Journal of Basic Medical Sciences* 26(10), 1131–1143.

Frusciante, S., Demurtas, O.C., Sulli, M. et al. (2021). Heterologous expression of *Bixa orellana* cleavage dioxygenase 4–3 drives crocin but not bixin biosynthesis. *Plant Physiology* 188(3), 1469–1482.

Golubova, D., Tansley, C., Su, H. et al. (2024). Engineering *Nicotiana benthamiana* as a platform for natural product biosynthesis. *Current Opinion in Plant Biology* 81, 102611.

Gresta, F., Avola, G., Lombardo, G.M. et al. (2009). Analysis of flowering, stigmas yield and qualitative traits of saffron (*Crocus sativus* L.) as affected by environmental conditions. *Scientia Horticulturae* 119(3), 320–324.

Gresta, F., Lombardo, G.M., Siracusa, L. et al. (2008). Saffron, an alternative crop for sustainable agricultural systems. A review. *Agronomy for Sustainable Development* 28(1), 95–112.

Harrison, P.J. and Bugg, T.D.H. (2014). Enzymology of the carotenoid cleavage dioxygenases: reaction mechanisms, inhibition and biochemical roles. *Archives of Biochemistry and Biophysics* 544, 105–111.

Hazman, Ö. and Bozkurt, M.F. (2015). Anti-inflammatory and antioxidative activities of safranal in the reduction of renal dysfunction and damage that occur in diabetic nephropathy. *Inflammation* 38(4), 1537–1545.

Hosseinzadeh, H. and Ghenaati, J. (2006). Evaluation of the antitussive effect of stigma and petals of saffron (*Crocus sativus*) and its components, safranal and crocin in guinea pigs. *Fitoterapia* 77(6), 446–448.

Hosseinzadeh, H. and Talebzadeh, F. (2005). Anticonvulsant evaluation of safranal and crocin from *Crocus sativus* in mice. *Fitoterapia* 76(7), 722–724.

Ilyas, M., Rana, F.A. and Riaz, M. (2021). Apocarotenoids. In: *Carotenoids: Structure and Function in the Human Body* (eds. M. Zia-Ul-Haq, S. Dewanjee and M. Riaz), 125–146. Springer International Publishing.

Jain, M., Srivastava, P.L., Verma, M. et al. (2016). *De novo* transcriptome assembly and comprehensive expression profiling in *Crocus sativus* to gain insights into apocarotenoid biosynthesis. *Scientific Reports* 6, 22456.

Ji, A., Jia, J., Xu, Z. et al. (2017). Transcriptome-guided mining of genes involved in crocin biosynthesis. *Frontiers in Plant Science* 8, 518.

Jia, Y., Yang, H., Yu, J. et al. (2023). Crocin suppresses breast cancer cell proliferation by down-regulating tumor promoter miR-122-5p and up-regulating tumor suppressors FOXP2 and SPRY2. *Environmental Toxicology* 38(7), 1597–1608.

Jing, Y., Guo, F., Zhang, S. et al. (2021). Recent advances on biological synthesis of lycopene by using industrial yeast. *Industrial & Engineering Chemistry Research* 60(9), 3485–3494.

Kermanshahi, S., Ghanavati, G., Abbasi-Mesrabadi, M. et al. (2020). Novel neuroprotective potential of crocin in neurodegenerative disorders: an illustrated mechanistic review. *Neurochemical Research* 45(11), 2573–2585.

Koocheki, A., Fallahi, H.-R. and Jami-Al-Ahmadi, M. (2020). Saffron water requirements. In: *Saffron* (eds. A. Koocheki and M. Khajeh-Hosseini), 67–92. Woodhead Publishing.

Koul, A. and Abraham, S.K. (2018). Efficacy of crocin and safranal as protective agents against genotoxic stress induced by gamma radiation, urethane and procarbazine in mice. *Human & Experimental Toxicology* 37(1), 13–20.

Lee, I.A., Lee, J.H., Baek, N.I. et al. (2005). Antihyperlipidemic effect of crocin isolated from the fructus of *Gardenia jasminoides* and its metabolite crocetin. *Biological & Pharmaceutical Bulletin* 28(11), 2106–2110.

Leone, S., Recinella, L., Chiavaroli, A. et al. (2018). Phytotherapic use of the *Crocus sativus* L. (saffron) and its potential applications: a brief overview. *Phytotherapy Research* 32(12), 2364–2375.

Li, N., Lin, G., Kwan, Y.-W. et al. (1999). Simultaneous quantification of five major biologically active ingredients of saffron by high-performance liquid chromatography. *Journal of Chromatography A* 849(2), 349–355.

Li, X.R., Tian, G.Q., Shen, H.J. et al. (2015). Metabolic engineering of *Escherichia coli* to produce zeaxanthin. *Journal of Industrial Microbiology and Biotechnology* 42(4), 627–636.

Liang, M.-H., He, Y.-J., Liu, D.-M. et al. (2021). Regulation of carotenoid degradation and production of apocarotenoids in natural and engineered organisms. *Critical Reviews in Biotechnology* 41(4), 513–534.

Liu, N., Liu, B., Wang, G. et al. (2020a). Lycopene production from glucose, fatty acid and waste cooking oil by metabolically engineered *Escherichia coli*. *Biochemical Engineering Journal* 155, 107488.

Liu, T., Yu, S., Xu, Z. et al. (2020b). Prospects and progress on crocin biosynthetic pathway and metabolic engineering. *Computational and Structural Biotechnology Journal* 18, 3278–3286.

Ma, Y., Liu, N., Greisen, P. et al. (2022). Removal of lycopene substrate inhibition enables high carotenoid productivity in *Yarrowia lipolytica*. *Nature Communications* 13(1), 572.

Ma, T., Shi, B., Ye, Z. et al. (2019). Lipid engineering combined with systematic metabolic engineering of *Saccharomyces cerevisiae* for high-yield production of lycopene. *Metabolic Engineering* 52, 134–142.

Martí, M., Diretto, G., Aragonés, V. et al. (2020). Efficient production of saffron crocins and picrocrocin in *Nicotiana benthamiana* using a virus-driven system. *Metabolic Engineering* 61, 238–250.

Mehdizadeh, R., Parizadeh, M.R., Khooei, A.R. et al. (2013). Cardioprotective effect of saffron extract and safranal in isoproterenol-induced myocardial infarction in wistar rats. *Iranian Journal of Basic Medical Sciences* 16(1), 56–63.

Nanda, S. and Madan, K. (2021). The role of safranal and saffron stigma extracts in oxidative stress, diseases and photoaging: a systematic review. *Heliyon* 7(2), e06117.

Pandey, D.K., Nandy, S., Mukherjee, A. et al. (2020). Advances in bioactive compounds from *Crocus sativus* (saffron): structure, bioactivity and biotechnology. In: *Studies in Natural Products Chemistry* (ed. Atta-ur-Rahman), 273–304. Elsevier.

Pu, X., He, C., Yang, Y. et al. (2020). *In vivo* production of five crocins in the engineered *Escherichia coli*. *ACS Synthetic Biology* 9(5), 1160–1168.

Püntener, A.G. and Schlesinger, U. (2000). Natural dyes. In: *Colorants for Non-Textile Applications* (eds. H.S. Freeman and A.T. Peters), 382–455. Elsevier Science.

Qiu, Y., Jiang, X., Liu, D. et al. (2020). The hypoglycemic and renal protection properties of crocin via oxidative stress-regulated NF-κB signaling in db/db mice. *Frontiers in Pharmacology* 11, 541.

Razavi, M., Hosseinzadeh, H., Abnous, K. et al. (2013). Crocin restores hypotensive effect of subchronic administration of diazinon in rats. *Iranian Journal of Basic Medical Sciences* 16(1), 64–72.

Sepahi, S., Golfakhrabadi, M., Bonakdaran, S. et al. (2022). Effect of crocin on diabetic patients: a placebo-controlled, triple-blinded clinical trial. *Clinical Nutrition ESPEN* 50, 255–263.

Shahi, T., Assadpour, E. and Jafari, S.M. (2016). Main chemical compounds and pharmacological activities of stigmas and tepals of 'red gold'; saffron. *Trends in Food Science & Technology* 58, 69–78.

Shao, Z., Luo, Y. and Zhao, H. (2012). DNA assembler method for construction of zeaxanthin-producing strains of *Saccharomyces cerevisiae*. *Methods in Molecular Biology* 898, 251–262.

Shokrpour, M. (2019). Saffron (*Crocus sativus* L.) breeding: opportunities and challenges. In: *Advances in Plant Breeding Strategies: Industrial and Food Crops* (eds. J.M. Al-Khayri, S.M. Jain and D.V. Johnson), 675–706. Springer.

Song, Y.-n., Wang, Y., Zheng, Y.-h. et al. (2021). Crocins: a comprehensive review of structural characteristics, pharmacokinetics and therapeutic effects. *Fitoterapia* 153, 104969.

Taherkhani, T., Zakaria, R.A., Omidi, M. et al. (2021). Expression of genes involved in the biosynthesis of crocin and safranal in cell suspension culture of saffron (*Crocus sativus*). *Pharmaceutical Chemistry Journal* 54(10), 1045–1050.

Wang, W., He, P., Zhao, D. et al. (2019). Construction of *Escherichia coli* cell factories for crocin biosynthesis. *Microbial Cell Factories* 18(1), 120.

Wang, L., Liu, Z., Jiang, H. et al. (2021). Biotechnology advances in β-carotene production by microorganisms. *Trends in Food Science & Technology* 111, 322–332.

Wu, Y., Yan, P., Li, Y. et al. (2020). Enhancing β-carotene production in *Escherichia coli* by perturbing central carbon metabolism and improving the NADPH supply. *Frontiers in Bioengineering and Biotechnology* 8, 585.

Xie, L., Luo, Z., Jia, X. et al. (2023). Synthesis of crocin I and crocin II by multigene stacking in *Nicotiana benthamiana*. *International Journal of Molecular Sciences* 24(18), 14139.

Xu, S., Zhang, X., Zhang, Y. et al. (2023). Concomitant production of erythritol and β-carotene by engineered *Yarrowia lipolytica*. *Journal of Agricultural and Food Chemistry* 71(30), 11567–11578.

Yu, L., Gao, R., Song, X. et al. (2021). Cardio-protective and anti-atherosclerosis effect of crocetin on vitamin D3 and HFD-induced atherosclerosis in rats. *Journal of Oleo Science* 70(10), 1447–1459.

Zhang, G., Chen, J., Wang, Y. et al. (2023). Metabolic engineering of *Yarrowia lipolytica* for zeaxanthin production. *Journal of Agricultural and Food Chemistry* 71(37), 13828–13837.

Zhang, Y., Liu, Z., Sun, J. et al. (2018). Biotechnological production of zeaxanthin by microorganisms. *Trends in Food Science & Technology* 71, 225–234.

Zhao, C., Kam, H.-T., Chen, Y. et al. (2021a). Crocetin and its glycoside crocin, two bioactive constituents from *Crocus sativus* L. (saffron), differentially inhibit angiogenesis by inhibiting endothelial cytoskeleton organization and cell migration through VEGFR2/SRC/FAK and VEGFR2/MEK/ERK signaling pathways. *Frontiers in Pharmacology* 12, 675359.

Zhao, Y., Zhang, Y., Nielsen, J. et al. (2021b). Production of β-carotene in *Saccharomyces cerevisiae* through altering yeast lipid metabolism. *Biotechnology and Bioengineering* 118(5), 2043–2052.

Zheng, S., Qian, Z., Tang, F. et al. (2005). Suppression of vascular cell adhesion molecule-1 expression by crocetin contributes to attenuation of atherosclerosis in hypercholesterolemic rabbits. *Biochemical Pharmacology* 70(8), 1192–1199.

Zhou, T., Park, Y.-K., Fu, J. et al. (2025). Metabolic engineering of *Yarrowia lipolytica* for the production and secretion of the saffron ingredient crocetin. *Biotechnology for Biofuels and Bioproducts* 18(1), 1.

14

Engineering Crocin Production in Heterologous Host Plants

Anu Pandita[1] *and Deepu Pandita*[2,*]

[1] *Vatsalya Clinic, New Delhi, India*
[2] *Government Department of School Education, Jammu, Jammu and Kashmir, India*

*Corresponding author: deepupandita@gmail.com

14.1 Introduction

Crocins are glycosylated crocetin derivatives, water-soluble carotenoid pigments, and very inadequate in nature, and were initially discovered in the dried saffron stigmas. Saffron has distinct pharmacological roles such as a free radical quencher, a blood activator, and resolves stagnation for tranquilization (Chinese Pharmacopoeia Commission 2000). The inimitable aroma and bright color of saffron make it a natural food colorant, a spice, and to a lesser extent a supplement (Basker and Negbi 1983; Nam et al. 2010; Georgiadou et al. 2012; Ahrazem et al. 2015, 2022a; Christodoulou et al. 2015; Bukhari et al. 2018). Crocins have properties like analgesic, sedative (Amin and Hosseinzadeh 2012), antihypertensive, anti-atherosclerotic (Hong and Yang 2013; Hashemzaei et al. 2020), anti-HSV-1 (Herpes Simplex Virus type 1), and anti-HIV-1 (Human Immunodeficiency Virus type 1) (Soleymani et al. 2018), and are useful in the Alzheimer's and neurological diseases (Ghahghaei et al. 2012), depression, dementia (Lopresti and Drummond 2014; Mazidi et al. 2016; Siddiqui et al. 2022), and various liver ailments, including chemical liver injury, fibrosis, cancer, and so forth (Ghahghaei et al. 2012; Finley and Gao 2017; Gedik et al. 2017; Skladnev and Johnstone 2017; Coll et al. 2018; Kim and Park 2018).

The fruits of *Gardenia jasminoides* (gardenia) also contain crocins but lack picrocrocin (Pfister et al. 1996; Moras et al. 2018), and *Buddleja davidii* also produces low quantities of crocins (Figure 14.1) (Liao et al. 1999). In addition to crocin and picrocrocin, safranal is also present in *B. davidii* flowers (Diretto et al. 2021).

Crocins (Figure 14.2) are divided into five types on the basis of diverse glucosyl numbers, which are crocin I–crocin V (Table 14.1) (Pu et al. 2020).

Plant Apocarotenoids: Biosynthesis, Signaling and Agricultural Applications, First Edition. Edited by Kaiser Iqbal Wani and Tariq Aftab.
Published 2026 by John Wiley & Sons Ltd.

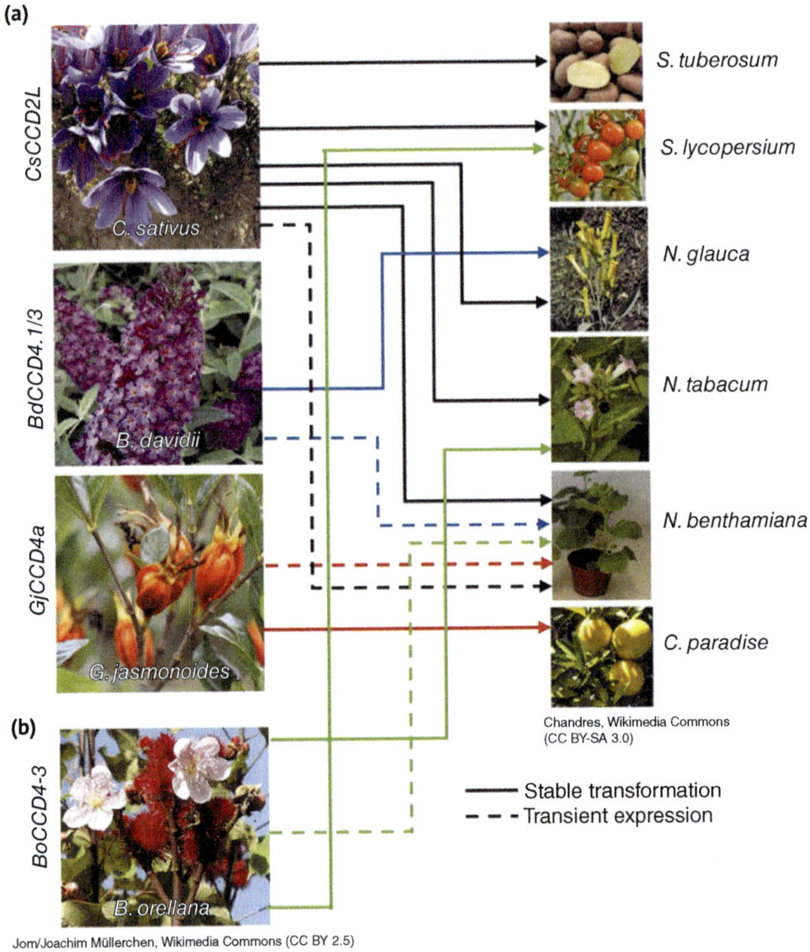

Figure 14.1 Plant sources of CCDs: (a) *C. sativus*, *B. davidii*, and *G. jasminoides* produce crocin, picrocrocin, and the approaches for their production in other plants are also given. (b) *Bixa orellana* does not produce crocins; however, BoCCD4-3 produces crocins on introduction in other plants. *Source:* Gómez-Gómez, L., et.al., (2023)/Elsevier/CC BY-NC-ND 4.0.

Crocins I and II are key components and active ingredients, which serve as quality indicators for the evaluation of the quality of saffron. The *Chinese Pharmacopoeia* requires that the content of crocins I and II in medicinal materials of saffron must be at least 10% (Chinese Pharmacopoeia Commission 2020). Crocins I and II have been shown to improve memory, provide protection against radiation, possess antidepressant properties, exhibit anticancer effects, and have antioxidant benefits (Bathaie et al. 2011; Rajabi et al. 2019; Xiao et al. 2020; Wang et al. 2022a,b), thus signifying boundless potential for clinical applications and rewards in the development of new drugs, food, and cosmetic trades. Crocins are mainly extracted from saffron, but limited sources, low yields, and high labor costs associated with harvesting and processing, as well as an absence of environmental

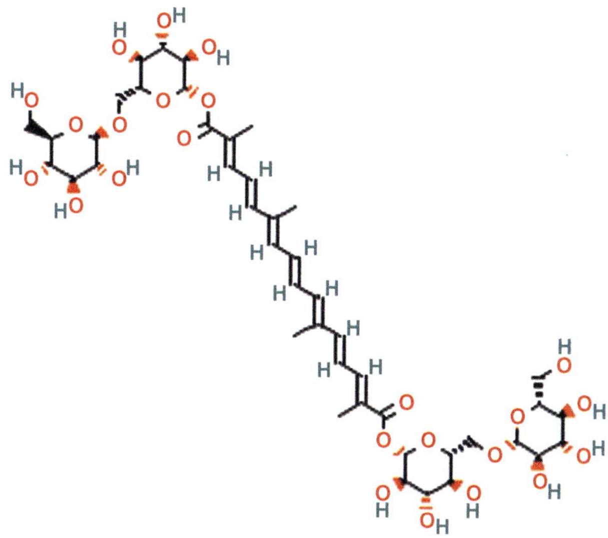

Figure 14.2 Structure of crocin.

Table 14.1 Crocin types on the basis of diverse glucosyl numbers.

Crocin type (Pu et al. 2020)	Number of glucosyls
Crocin I	Four glucosyls
Crocin II	Three glucosyls
Crocin III	Two glucosyls
Crocin IV	Two glucosyls
Crocin V	One glucosyl

Yaroslav/Adobe Stock Photos

sustainability, significantly limit their applications in the market. Crocins' high cost hinders their use by pharmaceutical and cosmetic industries. There is an urgent need to design an efficient and sustainable strategy for the production of crocin (Ahrazem et al. 2015). Saffron is expensive to produce and is affected by a variety of environmental factors (Cardone et al. 2020; Ahrazem et al. 2022b; Xie et al. 2023).

In species of *Crocus*, *Gardenia*, and *Buddleja*, zeaxanthin serves as the precursor to crocetin (Ahrazem et al. 2022a). The safranal is the precursor of picrocrocin (Tarantilis et al. 1995; Winterhalter and Straubinger 2000). The color and bitter taste of saffron are provided by crocin and picrocrocin, respectively (Cardone et al. 2020). Picrocrocin inhibits viability of the cells of melanoma and leukemia (Bukhari et al. 2018; Yu et al. 2018). Figure 14.3 shows the crocin biosynthetic pathway in saffron.

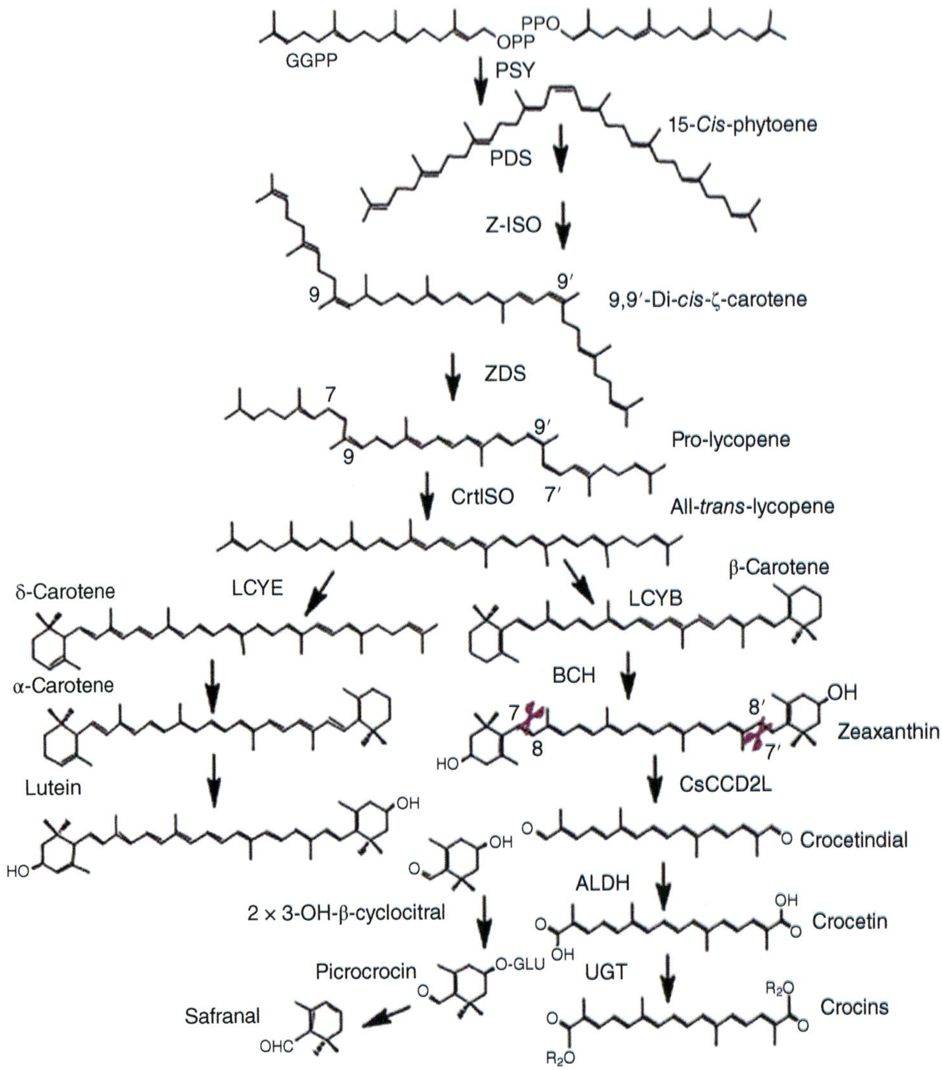

Figure 14.3 Crocin biosynthetic pathway in saffron: CrtISO, carotene isomerase; PSY, phytoene synthase; PDS, phytoene desaturase; ZDS, ζ-carotene desaturase; LCYB, β-lycopene cyclase; ALDH, aldehyde dehydrogenase; LCYE, ε-lycopene cyclase; Z-ISO, ζ-carotene isomerase; BCH, β-carotene hydroxylase; and 3-OH-β-cyclocitral or HTCC, hydroxy-2,6,6-trimethyl-1-cyclohexene-1-carboxaldehyde. Cleavage positions recognized by CsCCD2L are shown in purple. *Source:* Ahrazem et al., 2022/Frontiers Media S.A/CC BY 4.0.

14.2 Biofortified Plants for Saffron Apocarotenoid Production

14.2.1 *Nicotiana* Species as a Heterologous Host

The production of metabolites in a heterologous host has opened up opportunities for significant and maintainable manufacturing of crocins, particularly the primary bioactive

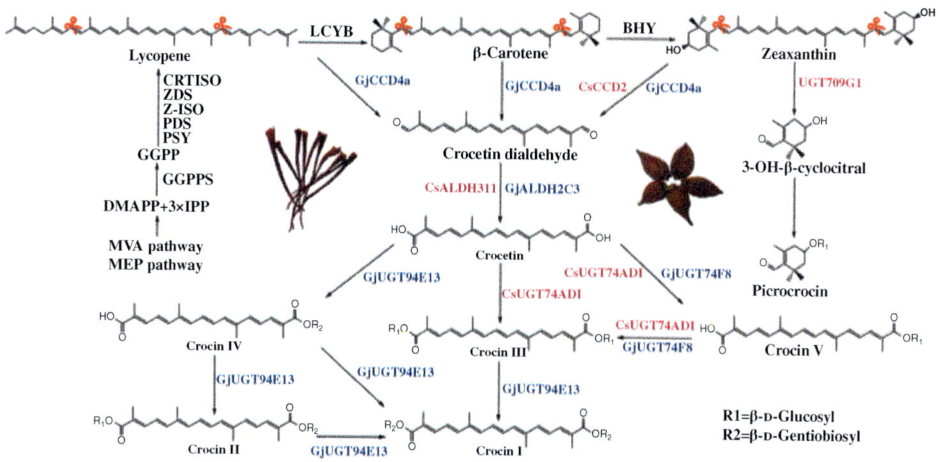

Figure 14.4 Biosynthetic pathway of crocin in *G. jasminoides* (blue) and *C. sativus* (pink): PSY, phytoene synthase; BHY, β-carotene hydrolase; CRTISO, carotenoid isomerase; Z-ISO, ζ-carotene isomerase; GjUGT74F8, GjUGT94E13, CsUGT74AD1, and UGT709G1, UDP-glucosyltransferases; PDS, phytoene desaturase; LCYB, lycopene β-cyclase; ZDS, ζ-carotene desaturase; GjALDH2C3 and CsALDH3l1, aldehyde dehydrogenases; and GjCCD4a and CsCCD2, CCDs. *Source:* Xie et al., 2023/ MDPI/CC BY 4.0.

compounds crocins I and II. The genes *GjCCD4a* (*carotenoid cleavage dioxygenase* gene), *GjALDH2C3* (*aldehyde dehydrogenase* gene), *GjUGT94E13*, and *GjUGT74F8* (two *UDP-glucosyltransferase* genes), which were isolated from the fruits of *G. jasminoides*, were expressed in the plants of *Nicotiana benthamiana*. Total crocin content in the T1 generation of *Nicotiana* can reach up to 78,362 ng/g fresh weight, while dry weight (DW) is predicted to be 1,058,945 ng/g. Interestingly, the main active components, crocins I and II, make up 99% of total crocins found in the transgenic plants. The strategy presented here provides an alternate approach for the large-scale production of crocins I and II using *Nicotiana*. The crocin biosynthetic pathway in *Crocus sativus* and *G. jasminoides* has been elucidated (Figure 14.4) (Xie et al. 2023). The *CsCCD2L* gene was transiently expressed on its own, resulting in an adequate accumulation of crocin in *Nicotiana* (Diretto et al. 2019; Martí et al. 2020; López et al. 2021).

In the transgenic plants of *Nicotiana tabacum* and *Nicotiana glauca*, crocins were produced through various transformation combinations involving carotenoid cleavage dioxygenase 2L (CsCCD2L), a key enzyme in the crocin biosynthetic pathway found in *C. sativus*. Additionally, genes related to the precursor of carotenoid biosynthesis were utilized. As a result, *N. glauca* leaves attained a crocin amount of nearly 400 μg/g DW, while *N. tabacum* had a content of 36 μg/g DW (Figure 14.5) (Martí et al. 2020; Ahrazem et al. 2022a).

While crocins I and II are the main active ingredients, they only comprise about 40% of the total, highlighting the need for further improvement. The overexpression of *GjCCD4a* in *Nicotiana* resulted primarily in the production of crocin II (Zheng et al. 2022). The viral vectors temporarily expressed CsCCD2 in genome-edited plants, which were engineered to have high levels of carotene and zeaxanthin. This resulted in the production of a significant amount of crocins I and II, primarily crocin II, which led to a total crocin amount of up to 800 g/g DW (Demurtas et al. 2023).

Figure 14.5 *Nicotiana* phenotypes: (a) plants and sixth leaves from *N. tabacum* wild-type control and T2 homozygous lines expressing *CsCCD2L* or co-expressing *CsCCD2L* + *BrCrtZ* + *AtOrMut* genes; and (b) plants and sixth leaves from *N. glauca* wild-type control and T1 hygromycin-resistant plants expressing *CsCCD2L* gene. *Source:* Ahrazem et al/Frontiers Media S.A./CC BY 4.0.

Currently, the primary component found in transgenic *Nicotiana* is crocin II, while the levels of crocin I are relatively low. Furthermore, it is important to consider both biomass and production costs if *Nicotiana* is to be utilized as a source for the extraction of crocin. There is still considerable potential for improvement in the current research. To address this issue, we implemented a strategy using in-fusion and 2A polypeptides to create a polygenic vector that contains four genes: *GjCCD4a*, *GjALDH2C3*, and two glucosyltransferase genes, *GjUGT74F8* and *GjUGT94E13*, facilitating the production of crocins in *N. benthamiana* (Demurtas et al. 2023).

Demurtas et al. (2023) synthesized four genes that are involved in crocin biosynthesis: *GjALDH2C3*, *GjCCD4a*, *GjUGT74F8*, and *GjUGT94E13*. The authors then constructed a multigene expression vector to enable the production of crocin in *Nicotiana*. They initially constructed two double-gene expression vectors. The genes *GjCCD4a* and *GjUGT74F8* were linked by 2A peptides and inserted into the pBI121 vector, utilizing the AtUBQ10 promoter and E9 terminator. Additionally, *GjALDH2C3* and *GjUGT94E13* were also connected with 2A peptides and incorporated into pBI121, this time with the CaMV35S promoter and heat-shock protein (HSP) 18.2 terminators. Subsequently, these two double-gene vectors served as templates for cloning and fusing the promoters, target genes, and terminators into the pCAMBIA1300 vector. This process resulted in the creation of the multigene expression vector named AU-CU.

The *BdCCD4.1* from *B. davidii* was expressed constitutively in *N. glauca* and the crocin levels were measured in the flowers and leaves (Huang et al. 2022). *Buddleja BdCCD4.1* was constitutively expressed in *N. glauca*. The transgenic lines produced substantial amounts of crocins in both their leaves and petals. The crocin accumulation was notably higher in the leaves than in the petals, reaching nearly 302 g/g DW. The crocin production is associated with reduced levels of endogenous carotenoids, especially β-carotene (Huang et al. 2022).

The stability of crocins in the leaves and petals was assessed after 3 years of storage, showing an average reduction of 58.06±2.20% in petals and 78.37±5.08% in leaves demonstrating the effectiveness of *BdCCD4.1* in promoting crocin production in *N. glauca*, highlighting the significant role of tissue in maintaining the stability of high-value metabolites during storage (Huang et al. 2022).

The *GjCCD4* from *G. jasminoides* showed transient expression in *N. benthamiana*, and the levels of picrocrocin and crocin were equated with those obtained under the same conditions using *CsCCD2L* from saffron (Zheng et al. 2022). In *GjCCD4a* transient expression, an accumulation of 1.630 mg/g DW was observed, while the leaves infiltrated with *CsCCD2L* gene showed only 0.67 mg/g DW (Zheng et al. 2022).

Frusciante et al. (2022) investigated six plastidial *BoCCD4-3* genes to produce crocetin dialdehyde from lycopene. Additionally, β-carotene and zeaxanthin served as substrates for *BoCCD4-3*, leading to the production of crocin. The *BoCCD4-3* expression in the leaves of *N. benthamiana* led to the accumulation of 0.026 mg/g DW of crocins (Frusciante et al. 2022).

Martí et al. (2020) have developed a virus-driven system that enables rapid production of significant amounts of crocins and picrocrocin in *N. benthamiana* within just 14 days. This system utilizes viral clones derived from the tobacco etch potyvirus, which express specific carotenoid cleavage dioxygenase (CCD) enzymes from *C. sativus* and *B. davidii.*

Metabolomics of the infected tissues revealed that expressing either *C. sativus CsCCD2L* or *B. davidii BdCCD4.1* solely through virus resulted in the production of crocin, picrocrocin, and safranal. When using the recombinant virus that expressed *CsCCD2L*, Martí et al. (2020) achieved accumulations of 0.2% crocin and 0.8% picrocrocin in DW of leaves within only 14 days.

To enhance the production of apocarotenoids in *N. benthamiana*, Martí et al. (2020) co-expressed CsCCD2L with other carotenogenic enzymes, such as *Pantoea ananatis* phytoene synthase (PaCrtB) and saffron β-carotene hydroxylase 2 (BCH2), using the same viral system. This combinatorial approach resulted in an additional increase in crocin levels, reaching up to 0.35% in the leaves where CsCCD2L and PaCrtB were co-expressed (Martí et al. 2020).

Given that saffron apocarotenoids are typically harvested at a high cost from flower stigmas only per annum and that flowers of *Buddleja* spp. accumulate lesser content, this system presents an attractive substitute for the maintainable production of these valuable metabolites (Martí et al. 2020).

14.2.2 *Solanum tuberosum* as a Heterologous Host

Gómez Gómez et al. (2022) developed a unique construct referred to as O6, which includes *UGT74AD1*, *CsCCD2L*, and *UGT709G1*. These genes are responsible for the synthesis of saffron apocarotenoids and are driven by a patatin promoter, aimed at producing potato tubers that generate crocins and picrocrocin. The diverse potato tubers accumulated crocins in amounts ranging from 19.41 to 360 mg/g DW and picrocrocin from 105 to 800 mg/g DW. The main compounds detected included crocetin, crocin I, and crocin II.

The carotenoid and apocarotenoid patterns differed between wild-type and transgenic tubers, which was linked to changes in the expression of specific pathway genes, particularly *PSY2*, *CCD1*, and *CCD4*. Additionally, modified tubers exhibited a higher antioxidant capacity, nearly four times greater than that of the wild type. This finding is encouraging regarding the potential health benefits of these products (Gómez Gómez et al. 2022).

14.2.3 *Solanum lycopersicum* as a Heterologous Host

Tomato, with a small genome size of 950 Mb, is a first-rate model for both basic and functional research (Gerszberg et al. 2015). *Solanum lycopersicum* fruit serves as an ideal platform for crocin production due to its high potential for carotenoid accumulation. *BoCCD4-3* and *CsUGT74AD1* expressed constitutively in *S. lycopersicum* fruits, resulting in the generation of 0.110-mg/g crocin (Frusciante et al. 2022). Ahrazem et al. (2022b) introduced a biosynthetic pathway for crocin and picrocrocin in *S. lycopersicum* by utilizing genes from saffron, specifically *CsUGT2* (*CsUGT74AD1*), *CsCCD2L*, and *CsUGT709G1*. They combined pE8 and p2A11 promoters, which are fruit-specific, with CaMV p35S, which is a constitutive promoter. The resulting *S. lycopersicum* fruits, with the *CsCCD2L* gene placed under the control of fruit-specific promoters, amassed up to 14,800 mg/g of crocins and 2,920 mg/g of picrocrocin DW. This represents the highest levels of these specific apocarotenoids reported in heterologous systems to date, achieved through transient or stable metabolic engineering, while not negotiating the normal development of plants.

Morote et al. (2023) combined transgenic tomato fruits engineered to produce crocins with high-pigmented 3/BETA (hp3/BSh) *S. lycopersicum* mutants, which have alterations in carotenoid metabolism, to rise the crocin levels in tomatoes. The fruits of *S. lycopersicum* from F4 plants displayed a higher antioxidant activity, and crocin-rich juice derived from these fruits protected neuroblastoma cells against oxidative damage by activating the nuclear factor erythroid 2 (Nrf2). The bio-accessibility of crocins in juice was found to be comparable to that of saffron. Complete results indicated that crocin-fortified *S. lycopersicum* could enhance the availability of crocin, offer added health benefits, and deliver better protection against chronic diseases related to oxidative stress in humans (Morote et al. 2023).

Transgenic *S. lycopersicum*, which express *CsCCD2* in their fruits, referred to as Tomaffron, can amass higher levels of saffron apocarotenoids, even with low substrate availability for *CsCCD2*. *CsCCD2* has been introduced into the Xantomato variety, which already amasses higher levels of zeaxanthin and β-carotene in ripe fruits due to a combination of four mutant alleles. By combining the genotypes of Xantomato and Tomaffron, researchers aimed to optimize the production of the apocarotenoids (Lobato-Gómez et al. 2024).

The best transgenic lines produced 15 and 14 times more crocins and picrocrocin than Tomaffron, respectively, along with a fourfold increase in β-carotene in contrast to Xantomato, though this came at the expense of the yield of fruits. The segregation of four mutations was performed to identify the optimal combination for achieving higher levels of saffron apocarotenoids without negatively impacting the yield of the fruits (Lobato-Gómez et al. 2024).

Plants featuring hp3 and BSh mutations amassed 6 and 15 times more crocin and picrocrocin than Tomaffron, without any observable pleiotropic effects. Notably, these elevated saffron apocarotenoid levels were achieved in fruits that also contained high amounts of both lycopene and β-carotene, either independently or in combination. This suggests that the apocarotenoids produced may play a regulatory role and indicates the potential to simultaneously increase the levels of both types of health-promoting compounds (Lobato-Gómez et al. 2024).

The crocin and picrocrocin biosynthesis pathway was introduced into *S. lycopersicum* plants by both fruit-specific and constitutive promoters. This approach led to 14.48-mg/g crocin and 2.92-mg/g picrocrocin in DW of tomatoes, all without negotiating the growth of the plant. The approach expressed *CsCCD2L* to produce crocetin dialdehyde and 2,6,6-trimethyl-4-hydroxy-1-carboxaldehyde-1-cyclohexene. Additionally, glycosyltransferases UGT709G1 and CsUGT2 were utilized for the production of picrocrocin and crocin, respectively (Ahrazem et al. 2022b).

Metabolic analyses of transgenic fruits showed that picrocrocin and crocetin-(β-D-gentiobiosyl)-(β-D-glucosyl)-ester were major compounds, along with safranal, while the usual carotenoids found in tomatoes were reduced. This study demonstrated the highest levels of crocins ever achieved through metabolic engineering in heterologous systems. Furthermore, modified *S. lycopersicum* exhibited enhanced antioxidant activity and provided protection against neurological disorders. Consequently, these newly designed *S. lycopersicum* could serve as an innovative platform for producing economic apocarotenoids in saffron (Ahrazem et al. 2022b). The carotenoid pathways in *S. lycopersicum* and crocin and picrocrocin pathways from saffron introduced into *S. lycopersicum* fruits are shown in Figure 14.6.

14.2.4 *Citrus* Species as a Heterologous Host

Citrus fruits containing carotenoids are part of the diet (Lado et al. 2016). The origin and evolution of citrus fruits at the genomic level have been established (Wu et al. 2018), which reveal complex taxonomic relationships. Biotechnological tools are available (Pena and Navarro 2000). Besides *Nicotiana* spp. and tomatoes, citrus has also been used as a model system for producing crocin. Specifically, white callus derived from *Citrus paradise* Macf,

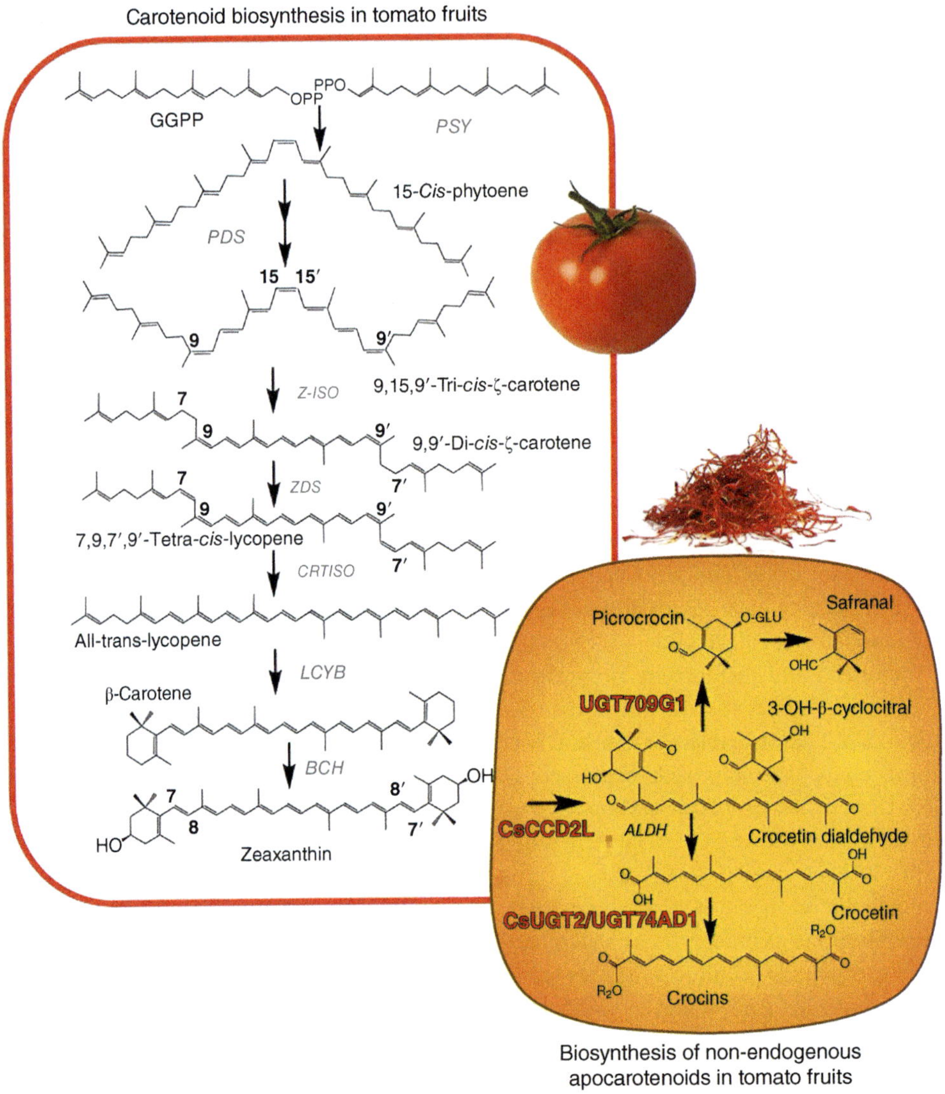

Figure 14.6 Carotenoid pathways in *S. lycopersicum* and crocin and picrocrocin pathways from *C. sativus* introduced into *S. lycopersicum* fruits. Introgressed enzymes are emphasized in red. CRTISO, carotene isomerase; ZDS, ζ-carotene desaturase; HTCC, hydroxy-2,6,6-trimethyl-1-cyclohexene-1-carboxaldehyde; Z-ISO, ζ-carotene isomerase; and ALDH, aldehyde dehydrogenase. *Source:* Ahrazem et al/Oxford University Press/CC BY 4.0.

naturally containing low levels of carotenoids, was modified to create various lines which express diverse combinations of carotenoid and apocarotenoid biosynthetic genes: *tpCrtB* (a bacterial PSY with a transit peptide), *OsBCH* (rice β-carotene hydroxylase), and either *GjCCD4a* or *CsCCD2L* (Zheng et al. 2022).

The initial step involved agrobacterium-mediated transformations using the bacterial *PSY* gene fused with transit peptide, TpCrtB. This resulted in orangish red callus due to high levels of β-carotene and α-carotene. The second step focused on modifying the content of xanthophyll by utilizing *OsBCH* in a second round of transformation. The resulting PSY/BCH yellow callus lines demonstrated higher levels of β-xanthophylls while showing a lower content of cyclic carotenoids (Zheng et al. 2022).

These two newly produced transgenic lines were subsequently super-transformed with *CsCCD2L* and *GjCCD4a*, respectively, resulting in the creation of four additional cell lines: CCD2/PSY (CsCCD2L/OsBCH/TpCrtB), CCD2/BCH/PSY (CsCCD2L/TpCrtB), CCD4/PSY (GjCCD4a/TpCrtB), and CCD4/BCH/PSY (GjCCD4a/OsBCH/TpCrtB). Excluding CCD2/PSY line, which seemed slightly orange, all other lines were pale yellow, indicating good substrate consumption (Zheng et al. 2022).

14.3 Conclusion

In addition to *Nicotiana* spp., potato, and tomatoes, citrus white callus has been utilized as a model system for the production of crocins. The biosynthetic pathway for crocin and picrocrocin was introduced into *S. lycopersicum* plants using both fruit-specific and constitutive promoters. In potato, a unique construct was developed, referred to as O6, which includes the *CsCCD2L*, *UGT74AD1*, and *UGT709G1* genes. These genes are responsible for the biosynthesis of saffron apocarotenoids and are driven by a patatin promoter, aimed at producing potato tubers that generate crocins and picrocrocin.

References

Ahrazem, O., Diretto, G., Rambla, J.L. et al. (2022b). Engineering high levels of saffron apocarotenoids in tomato. *Horticulture Research* 9, uhac074. https://doi.org/10.1093/hr/uhac074.

Ahrazem, O., Rubio-Moraga, A., Nebauer, S.G. et al. (2015). Saffron: its phytochemistry, developmental processes, and biotechnological prospects. *Journal of Agricultural and Food Chemistry* 63, 8751–8764. https://doi.org/10.1021/acs.jafc.5b03194.

Ahrazem, O., Zhu, C., Huang, X. et al. (2022a). Metabolic engineering of crocin biosynthesis in *Nicotiana* species. *Frontiers in Plant Science* 13, 861140. https://doi.org/10.3389/fpls.2022.861140.

Amin, B. and Hosseinzadeh, H. (2012). Evaluation of aqueous and ethanolic extracts of saffron, *Crocus sativus* L., and its constituents, safranal and crocin in allodynia and hyperalgesia induced by chronic constriction injury model of neuropathic pain in rats. *Fitoterapia* 83, 888–895. https://doi.org/10.1016/j.fitote.2012.03.022.

Basker, D. and Negbi, M. (1983). Uses of saffron. *Economic Botany* 37, 228–236.

Bathaie, S.Z., Shams, A. and Kermani, F.M.Z. (2011). Crocin bleaching assay using purified di-gentiobiosyl crocin (alpha-crocin) from Iranian saffron. *Iranian Journal of Basic Medical Sciences* 14, 399–406.

Bukhari, S.I., Manzoor, M. and Dhar, M.K. (2018). A comprehensive review of the pharmacological potential of *Crocus sativus* and its bioactive apocarotenoids. *Biomedicine and Pharmacotherapy* 98, 733–745. https://doi.org/10.1016/j.biopha.2017.12.090.

Cardone, L., Castronuovo, D., Perniola, M. et al. (2020). The influence of soil physical and chemical properties on saffron (*Crocus sativus* L.) growth, yield and quality. *Agronomy* 10, 1154.

Chinese Pharmacopoeia Commission (2020). *Chinese Pharmacopoeia: Volume 1*. China Medical Science Press.

Christodoulou, E., Kadoglou, N.P., Kostomitsopoulos, N. et al. (2015). Saffron: a natural product with potential pharmaceutical applications. *Journal of Pharmacy and Pharmacology* 67, 1634–1649. https://doi.org/10.1111/jphp.12456.

Coll, M., Perea, L., Boon, R. et al. (2018). Generation of hepatic stellate cells from human pluripotent stem cells enables *in vitro* modeling of liver fibrosis. *Cell Stem Cell* 23, 101–113.

Demurtas, O.C., Sulli, M., Ferrante, P. et al. (2023). Production of saffron apocarotenoids in *Nicotiana benthamiana* plants genome-edited to accumulate zeaxanthin precursor. *Metabolites* 13, 729.

Diretto, G., Ahrazem, O., Rubio-Moraga, A. et al. (2019). UGT709G1: a novel uridine diphosphate glycosyltransferase involved in the biosynthesis of picrocrocin, the precursor of safranal in saffron (*Crocus sativus*). *New Phytologist* 224, 725–740. https://doi.org/10.1111/nph.16079.

Diretto, G., López-Jiménez, A.J., Ahrazem, O. et al. (2021). Identification and characterization of apocarotenoid modifiers and carotenogenic enzymes for biosynthesis of crocins in *Buddleja davidii* flowers. *Journal of Experimental Botany* 72, 3200–3218.

Finley, J.W. and Gao, S. (2017). A perspective on *Crocus sativus* L. (saffron) constituent crocin: a potent water-soluble antioxidant and potential therapy for Alzheimer's disease. *Journal of Agricultural and Food Chemistry* 65, 1005–1020. https://doi.org/10.1021/acs.jafc.6b04398.

Frusciante, S., Demurtas, O.C., Sulli, M. et al. (2022). Heterologous expression of *Bixa orellana* cleavage dioxygenase 4-3 drives crocin but not bixin biosynthesis. *Plant Physiology* 188, 1469–1482.

Gedik, S., Erdemli, M.E., Gul, M. et al. (2017). Hepatoprotective effects of crocin on biochemical and histopathological alterations following acrylamide-induced liver injury in Wistar rats. *Biomedicine and Pharmacotherapy* 95, 764–770.

Georgiadou, G., Tarantilis, P.A. and Pitsikas, N. (2012). Effects of the active constituents of *Crocus Sativus* L., crocins, in an animal model of obsessive-compulsive disorder. *Neuroscience Letters* 528, 27–30. https://doi.org/10.1016/j.neulet.2012.08.081.

Gerszberg, A., Hnatuszko-Konka, K., Kowalczyk, T. et al. (2015). Tomato (*Solanum lycopersicum* L.) in the service of biotechnology. *Plant Cell, Tissue and Organ Culture* 120, 881–902. https://doi.org/10.1007/s11240-014-0664-4.

Ghahghaei, A., Bathaie, S.Z. and Bahraminejad, E. (2012). Mechanisms of the effects of crocin on aggregation and deposition of Aβ1–40 fibrils in Alzheimer's disease. *International Journal of Peptide Research and Therapeutics* 18, 347–351.

Gómez Gómez, L., Morote, L., Frusciante, S. et al. (2022). Fortification and bioaccessibility of saffron apocarotenoids in potato tubers. *Frontiers in Nutrition* 9, 1045979. https://doi.org/10.3389/fnut.2022.104597.

Gómez-Gómez, L., Rubio-Moraga, A. and Morote, L. et al. (2023). Engineering the production of crocins and picrocrocin in heterologous plant systems. *Industrial Crops & Products* 194, 116283.

Hashemzaei, M., Mamoulakis, C., Tsarouhas, K. et al. (2020). Crocin: a fighter against inflammation and pain. *Food and Chemical Toxicology* 143, 111521.

Hong, Y.J. and Yang, K.S. (2013). Anti-inflammatory activities of crocetin derivatives from processed *Gardenia jasminoides*. *Archives of Pharmacal Research* 36, 933–940.

Huang, X., Morote, L., Zhu, C. et al. (2022). The biosynthesis of non-endogenous apocarotenoids in transgenic *Nicotiana glauca*. *Metabolites* 12, 575.

Kim, B. and Park, B. (2018). Saffron carotenoids inhibit STAT3 activation and promote apoptotic progression in IL-6-stimulated liver cancer cells. *Oncology Reports* 39, 1883–1891.

Lado, J., Zacarias, L. and Rodrigo, M.J. (2016). Regulation of carotenoid biosynthesis during fruit development. *Subcellular Biochemistry* 79, 161–198.

Liao, Y.H., Houghton, P.J. and Hoult, J.R. (1999). Novel and known constituents from *Buddleja* species and their activity against leukocyte eicosanoid generation. *Journal of Natural Products* 62, 1241–1245. https://doi.org/10.1021/np990092+.

Lobato-Gómez, M., Drapal, M., Fernández-Muñoz, R. et al. (2024). Maximizing saffron apocarotenoid production in varied tomato fruit carotenoid contexts. *The Plant Journal* 120(3), 966–983. https://doi.org/10.1111/tpj.17030.

López, A.J., Frusciante, S., Niza, E. et al. (2021). A new glycosyltransferase enzyme from family 91, UGT91P3, is responsible for the final glucosylation step of crocins in saffron (*Crocus sativus* L.). *International Journal of Molecular Sciences* 22, 8815. https://doi.org/10.3390/ijms22168815.

Lopresti, A.L. and Drummond, P.D. (2014). Saffron (*Crocus sativus*) for depression: a systematic review of clinical studies and examination of underlying antidepressant mechanisms of action. *Human Psychopharmacology* 29, 517–527. https://doi.org/10.1002/hup.2434.

Martí, M., Diretto, G., Aragones, V. et al. (2020). Efficient production of saffron crocins and picrocrocin in *Nicotiana benthamiana* using a virus-driven system. *Metabolic Engineering* 61, 238–250. https://doi.org/10.1016/j.ymben.2020.06.009.

Mazidi, M., Shemshian, M., Mousavi, S.H. et al. (2016). A double-blind, randomized and placebo-controlled trial of saffron (*Crocus sativus* L.) in the treatment of anxiety and depression. *Journal of Integrative Medicine* 13, 195–199. https://doi.org/10.1515/jcim-2015-0043.

Moras, B., Loffredo, L. and Rey, S. (2018). Quality assessment of saffron (*Crocus sativus* L.) extracts via UHPLC-DAD-MS analysis and detection of adulteration using gardenia fruit extract (*Gardenia jasminoides* Ellis). *Food Chemistry* 257, 325–332. https://doi.org/10.1016/j.foodchem.2018.03.025.

Morote, L., Lobato-Gómez, M., Ahrazem, O. et al. (2023). Crocins-rich tomato extracts showed enhanced protective effects *in vitro*. *Journal of Functional Foods* 101, 105432.

Nam, K.N., Park, Y.M., Jung, H.J. et al. (2010). Anti-inflammatory effects of crocin and crocetin in rat brain microglial cells. *European Journal of Pharmacology* 648, 110–116. https://doi.org/10.1016/j.ejphar.2010.09.003.

Pena, L. and Navarro, L. (2000). Transgenic citrus. In: *Transgenic Trees* (ed. Y.P.S. Bajaj), 39–54. Springer.

Pfister, S., Steck, A. and Pfander, H. (1996). Isolation and structure elucidation of carotenoid glycoslyesters on gardenia fruits (*Gardenia jasminoides*) and saffron (*Crocus sativus*). *Journal of Agricultural and Food Chemistry* 44, 2612–2615. https://doi.org/10.1021/JF950713E.

Pu, X., He, C., Yang, Y. et al. (2020). *In vivo* production of five crocins in the engineered *Escherichia coli*. *ACS Synthetic Biology* 9, 1160–1168.

Rajabi, H., Jafari, S.M., Rajabzadeh, G. et al. (2019). Chitosan-gum Arabic complex nanocarriers for encapsulation of saffron bioactive components. *Colloids and Surfaces A: Physicochemical and Engineering Aspects* 578, 123644.

Siddiqui, S.A., Ali Redha, A., Snoeck, E.R. et al. (2022). Anti-depressant properties of crocin molecules in saffron. *Molecules* 27, 2076.

Skladnev, N.V. and Johnstone, D.M. (2017). Neuroprotective properties of dietary saffron: more than just a chemical scavenger? *Neural Regeneration Research* 12, 210–211. https://doi.org/10.4103/1673-5374.198976.

Soleymani, S., Zabihollahi, R., Shahbazi, S. et al. (2018). Antiviral effects of saffron and its major ingredients. *Current Drug Delivery* 15, 698–704.

Tarantilis, P.A., Tsoupras, G. and Polissiou, M. (1995). Determination of saffron (*Crocus sativus* L.) components in crude plant extract using high-performance liquid chromatography-UV-visible photodiode-array detection-mass spectrometry. *Journal of Chromatography A* 699, 107–118. https://doi.org/10.1016/0021-9673(95)00044-n.

Wang, L., Cao, Y., Zhang, X. et al. (2022a). Reactive oxygen species-responsive nanodrug of natural crocin-i with prolonged circulation for effective radioprotection. *Colloids and Surfaces B: Biointerfaces* 213, 112441.

Wang, P., Li, M., Guo, Z. et al. (2022b). Comparison of the pharmacokinetics of crocin-I in normoxic and hypoxic rats. *Toxicology and Applied Pharmacology* 447, 116088.

Winterhalter, P. and Straubinger, M. (2000). Saffron-renewed interest in an ancient spice. *Food Reviews International* 16, 39–59.

Wu, G.A., Terol, J., Ibanez, V. et al. (2018). Genomics of the origin and evolution of citrus. *Nature* 554, 311–316.

Xiao, Q., Shu, R., Wu, C. et al. (2020). Crocin-I alleviates the depression-like behaviors probably via modulating "microbiota-gut-brain" axis in mice exposed to chronic restraint stress. *Journal of Affective Disorders* 276, 476–486.

Xie, L., Luo, Z., Jia, X. et al. (2023). Synthesis of crocin I and crocin II by multigene stacking in *Nicotiana benthamiana*. *International Journal of Molecular Sciences* 24(18), 14139. https://doi.org/10.3390/ijms241814139.

Yu, L., Li, J. and Xiao, M. (2018). Picrocrocin exhibits growth inhibitory effects against SKMEL-2 human malignant melanoma cells by targeting JAK/STAT5 signaling pathway, cell cycle arrest and mitochondrial mediated apoptosis. *Journal of B.U.O.N.: Official Journal of the Balkan Union of Oncology* 23, 1163–1168.

Zheng, X., Mi, J. and Balakrishna, A. (2022). *Gardenia* carotenoid cleavage dioxygenase 4a is an efficient tool for biotechnological production of crocins in green and non-green plant tissues. *Plant Biotechnology Journal* 20, 2202–2216.

Section IV

Physiological Roles and Stress Responses

15

Apocarotenoids in Plant Biology: Defense Signaling and Lignin Interactions

Nayaab Shahir Pandit[1], *Mohsin Shafi*[2], *Ishrat Mehmmod*[3,*], *and Kaiser Iqbal Wani*[4,5]

[1] *School of Computational and Integrative Sciences (SC&IS), Jawaharlal Nehru University, New Delhi, India*
[2] *Genomic Biology Lab, Department of Biosciences, Jamia Millia Islamia, New Delhi, India*
[3] *Department of Biophysics, Panjab University, Chandigarh, India*
[4] *Key Laboratory for Quality Regulation of Tropical Horticultural Crops of Hainan Province, School of Tropical Agriculture and Forestry, Hainan University, Haikou, Hainan, China*
[5] *National Key Laboratory for Tropical Crop Breeding, School of Breeding and Multiplication (Sanya Institute of Breeding and Multiplication), Hainan University, Sanya, Hainan, China*

*Corresponding author: ishratmehmood943@gmail.com

15.1 Introduction

Plants exist in a constantly challenging environment, facing threats from both living organisms (biotic stress) and environmental factors (abiotic stress). To survive and thrive, they deploy a sophisticated network of defense mechanisms. Among the most critical components of this arsenal are secondary metabolites—diverse chemical compounds not directly involved in primary growth but essential for adaptation and protection (Nawaz et al. 2023). Within this vast chemical repertoire, apocarotenoids represent a particularly fascinating group. Derived from the oxidative breakdown of carotenoids, these molecules were initially recognized primarily for their roles in fruit and flower coloration and aroma. However, research over the past few decades has revealed their much broader significance as crucial signaling molecules involved in growth regulation, stress responses, and interactions with other metabolic pathways, including the formation of lignin (Simkin 2021).

The discovery of key apocarotenoids like abscisic acid (ABA) and strigolactones (SLs) revolutionized our understanding, demonstrating their profound impact on plant development and stress physiology. Subsequent identification of the enzymes responsible for their formation, particularly carotenoid cleavage dioxygenases (CCDs) and 9-*cis*-epoxycarotenoid dioxygenases (NCEDs), provided the molecular keys to unlock their biosynthetic pathways (Zhou et al. 2019). Modern genomics, metabolomics, and gene editing techniques continue to illuminate the intricate roles of apocarotenoids,

Plant Apocarotenoids: Biosynthesis, Signaling and Agricultural Applications, First Edition. Edited by Kaiser Iqbal Wani and Tariq Aftab.
Published 2026 by John Wiley & Sons Ltd.

confirming they are not mere degradation by-products but precisely regulated agents integral to plant life. This chapter explores the multifaceted functions of apocarotenoids, focusing specifically on their contributions to plant defense systems and their emerging connection to lignin biosynthesis, a process vital for structural integrity and pathogen resistance (Akbar et al. 2023). Understanding these connections remains a significant, yet rewarding, challenge in plant science, hindered by the multilayered regulation, cross-pathway interactions, and the dynamic nature of secondary metabolism influenced by development and environment.

15.2 Lignin: A Structural and Defensive Polymer

Integral to the discussion of plant defense is lignin, a complex phenolic polymer fortified within the secondary cell walls of vascular plants. Primarily composed of monolignols (coniferyl, sinapyl, and *p*-coumaryl alcohols) joined through oxidative polymerization, lignin provides essential mechanical strength and rigidity, enables efficient water transport, and forms a robust physical barrier against pests and pathogens (Ninkuu et al. 2022). Beyond its structural role, lignin is an active participant in defense. Plants often respond to attack or injury by rapidly increasing lignin deposition at the site of stress—a strategy known as defensive lignification. This reinforces cell walls, hinders pathogen invasion, and makes plant tissues less digestible to microbes and herbivores (Bhuiyan et al. 2009).

15.3 Biosynthesis of Apocarotenoids: From Carotenoids to Signaling Molecules

Apocarotenoid production begins with carotenoids, themselves synthesized in plastids via the methylerythritol phosphate (MEP) pathway. This pathway starts with geranylgeranyl diphosphate (GGPP) and proceeds through key intermediates like phytoene and lycopene, eventually branching to produce α- and β-carotene, lutein, and various other xanthophylls. The specific pool of carotenoids available acts as the substrate source for apocarotenoid generation (Zheng et al. 2021b). The transformation of carotenoids into apocarotenoids occurs primarily through two routes discussed in the following text.

15.3.1 Enzymatic Cleavage

This is the major, highly regulated pathway mediated by specific enzymes. NCEDs are principally responsible for cleaving 9-*cis*-epoxycarotenoids (like violaxanthin and neoxanthin) to produce xanthoxin, the direct precursor to the crucial stress hormone ABA. A broader family, the CCDs (including isoforms like CCD1, CCD4, CCD7, and CCD8), cleaves various carotenoids at different positions to generate a diverse array of molecules. These include volatile compounds like β-ionone (often associated with aroma and signaling), SLs (critical for plant architecture and symbiotic interactions), and β-cyclocitral (a stress-responsive signal). The specific type, location, and activity level of these enzymes determine the profile of apocarotenoids produced (Walter and Strack 2011).

15.3.2 Nonenzymatic Oxidation

Under conditions of high oxidative stress, such as intense light or pathogen attack, reactive oxygen species (ROS), particularly singlet oxygen, can directly attack and cleave carotenoids. This nonenzymatic process generates bioactive apocarotenoids like β-cyclocitral and β-ionone. These molecules often function as internal stress alerts, triggering protective responses within the cell. This route offers a rapid mechanism for producing signaling molecules directly in response to oxidative damage (Walter and Strack 2011).

The overall biosynthesis of apocarotenoids is tightly controlled at multiple levels. Gene expression for biosynthetic enzymes (CCDs, NCEDs) is modulated by stress signals, developmental cues, hormones, and environmental factors like light. Feedback loops involving apocarotenoids themselves or downstream signaling components add another layer of regulation. Furthermore, epigenetic modifications, such as changes in DNA methylation or histone structure, are emerging as important regulators of these pathways, particularly under stress (Zheng et al. 2021b).

15.4 Key Apocarotenoids and Their Diverse Functions

The enzymatic and nonenzymatic cleavages of carotenoids yield a wide spectrum of bioactive apocarotenoids (Figure 15.1) with distinct roles.

15.4.1 Abscisic Acid

Perhaps the most studied apocarotenoid, ABA is indispensable for adapting to abiotic stresses, especially drought and salinity. It triggers stomatal closure to conserve water, regulates seed dormancy and germination, and influences root and shoot growth. ABA signaling operates through a well-defined pathway involving PYR/PYL/RCAR receptors, PP2C phosphatases, and SnRK2 kinases, leading to widespread changes in gene expression (Vishwakarma et al. 2017).

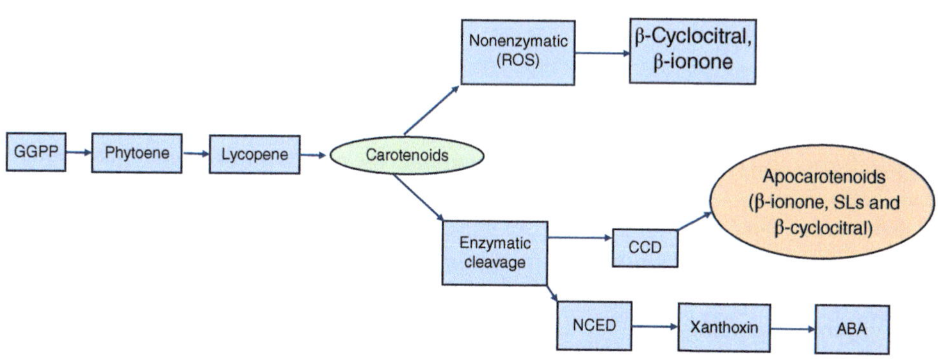

Figure 15.1 Apocarotenoid biosynthesis.

15.4.2 Strigolactones

Primarily synthesized in the roots, SLs are crucial regulators of plant form and interaction. They suppress the growth of axillary buds, thus controlling shoot branching patterns, and promote beneficial symbiotic relationships with arbuscular mycorrhizal fungi (AMFs), enhancing nutrient uptake. Intriguingly, they also act as germination cues for parasitic plants like *Striga* and *Orobanche*, which exploit this signal to locate hosts. SL perception involves the D14 receptor and downstream components that regulate gene expression related to development (Helmich 2025).

15.4.3 β-Cyclocitral

This volatile apocarotenoid, often generated nonenzymatically during photooxidative stress, functions as a potent stress signal. It acts as a reactive electrophilic species (RES), modulating cellular redox state and activating detoxification and antioxidant pathways. It has also been shown to enhance root growth under certain conditions and may act as a priming agent, preparing plants for future stress encounters (Faizan et al. 2022).

15.4.4 Anchorene and Zaxinone

These apocarotenoids highlight the ongoing discovery in this field. Anchorene appears to promote anchor root formation, potentially influencing auxin signaling. Zaxinone, found in rice, affects root and shoot development, possibly by interacting with sugar metabolism and SL levels, and shows potential as an agricultural bio-stimulant (Felemban et al. 2019).

15.4.5 Other Bioactive Forms

Beyond these major players, compounds like β-ionone (contributing to fragrance and possessing antimicrobial properties), dihydroactinidiolide (involved in stress-induced plastid-to-nucleus signaling), and various C13-norisoprenoids (influencing aroma and ecological interactions) contribute to the diverse functional landscape of apocarotenoids (Magome et al. 2023).

15.5 Apocarotenoid Signaling in Plant Defense Mechanisms

Apocarotenoids are now understood to be significant players in orchestrating plant defense responses. Their mobility and chemical reactivity allow them to function both directly against threats and indirectly as signals that modulate the plant's immune system (Moreno et al. 2021).

Under stress, certain apocarotenoids like β-cyclocitral and β-ionone, generated enzymatically or via ROS action, serve as rapid alert signals. They can initiate retrograde signaling from stressed plastids to the nucleus, prompting adjustments in metabolism and activating defense genes, including those for detoxification. Their volatility suggests potential roles in cell-to-cell or even plant-to-plant communication, possibly contributing to systemic acquired resistance or allelopathic interactions (Moreno et al. 2021).

In response to biotic agents (pathogens and herbivores), apocarotenoids interact complexly with the plant immune network. For instance, β-ionone exhibits direct antifungal/antibacterial effects and can induce defense gene expression. ABA's role is multifaceted; it often antagonizes defense pathways effective against biotrophic pathogens (reliant on live tissue) but can enhance defenses against necrotrophic pathogens (which kill tissue) and herbivores, frequently interacting with jasmonic acid (JA) signaling. Apocarotenoids can also be involved downstream of pathogen detection, contributing to physical defenses like callose deposition and cell wall reinforcement (Felemban et al. 2019).

Apocarotenoids are also central to tolerating abiotic stress. ABA's role in managing water stress through stomatal control and gene regulation is paramount. β-Cyclocitral enhances tolerance to oxidative stress arising from high light, ozone, or heavy metals by upregulating antioxidant systems (Faizan et al. 2022). Other apocarotenoids, like zaxinone, appear to modulate responses to nutrient limitations, influencing root system architecture and interactions within the rhizosphere. Collectively, these molecules enable plants to mount rapid, adaptive responses crucial for survival in fluctuating and often harsh environments (Felemban et al. 2019).

15.6 Apocarotenoid Interactions: Crosstalk with Hormones and Microbes

Apocarotenoid signaling pathways are deeply integrated within the plant's broader regulatory network, engaging in extensive crosstalk with classical defense hormones like salicylic acid (SA), JA, and ethylene (ET) as summarized in Table 15.1. The antagonism between ABA and SA, where ABA often dampens SA-mediated defenses against biotrophic pathogens, is a well-established example (Yang et al. 2015). However, this relationship is context-dependent, as ABA can sometimes work synergistically with JA and ET to bolster defenses against necrotrophic pathogens. SLs exert influence more indirectly, shaping immunity by altering plant architecture and potentially modifying root exudate profiles, affecting the composition and readiness of rhizosphere microbial communities. Volatile signals like β-cyclocitral further bridge these networks, known to upregulate genes responsive to JA as well as those involved in antioxidant defense, positioning them at a nexus between stress signaling and redox homeostasis. These intricate interactions demonstrate that apocarotenoids help orchestrate a balanced defense strategy, managing the critical trade-offs between growth, stress responses, and resource allocation (Havaux 2020).

Beyond internal hormonal networks, apocarotenoids are pivotal mediators of plant interactions with the microbial world. SLs exuded from roots serve as crucial chemical signals that initiate beneficial symbioses with AMFs by promoting fungal hyphal branching (Deshpande et al. 2021). Paradoxically, these same signals are exploited by parasitic weeds like *Striga* and *Orobanche* as cues to trigger their germination and locate a host. Apocarotenoids like zaxinone can also indirectly shape the root microbial landscape by modulating SL levels and root metabolism. Furthermore, volatile apocarotenoids, including β-ionone and dihydroactinidiolide, may function as antimicrobial compounds or interfere with microbial communication (quorum sensing) (Wang 2021). Through these diverse mechanisms, apocarotenoids significantly influence both beneficial and antagonistic relationships, thereby shaping the plant's immediate microbial environment and impacting nutrient uptake, disease susceptibility, and overall rhizosphere health.

Table 15.1 Summarizing the crosstalk of apocarotenoids with hormones and microbes, organized by interaction type, key apocarotenoids, roles, and effects.

Interaction type	Key apocarotenoid(s)	Associated hormones/ microbes	Mechanism/role	Effect/outcome
Hormonal antagonism	ABA	SA	ABA suppresses SA-mediated defenses	Reduced resistance to biotrophic pathogens
Hormonal synergy	ABA	JA, ET	ABA enhances JA/ET signaling	Strengthened defense against necrotrophic pathogens
Defense signaling	β-Cyclocitral	JA-responsive genes, antioxidant pathways	Upregulates JA-induced and redox-regulating genes	Links stress signaling to redox homeostasis
Plant architecture modulation	SLs	—	Alter shoot/root architecture	Indirectly impact immunity and resource allocation
Root–microbe communication	SLs	AMFs	Stimulate fungal hyphal branching	Promote mutualistic symbiosis
Parasitic plant activation	SLs	*Striga*, *Orobanche*	Act as germination cues	Facilitate host recognition and infection
Root microbiome modulation	Zaxinone	Root microbial community	Regulates SL biosynthesis and root metabolism	Shapes rhizosphere composition and function
Volatile antimicrobials	β-Ionone, dihydroactinidiolide	Soil pathogenic microbes	Possess antimicrobial properties; may inhibit quorum sensing	Defend against pathogenic microbes; possibly alter microbial signaling
Trade-off regulation	Multiple (e.g., ABA, β-cyclocitral)	Hormonal networks, microbial signals	Integrate hormonal and microbial signals to balance stress response, growth, and immunity	Optimize resource allocation, stress tolerance, and immune readiness

15.7 Lignin: Biosynthesis, Structure, and Regulation

Lignin stands as a cornerstone of plant structural biology and defense. This complex polyphenolic polymer, embedded within the secondary cell wall, imparts mechanical strength, rigidity, and resistance against biological and environmental challenges (Jeffri et al. 2024). Its biosynthesis originates from the phenylpropanoid pathway (Figure 15.2), a central route

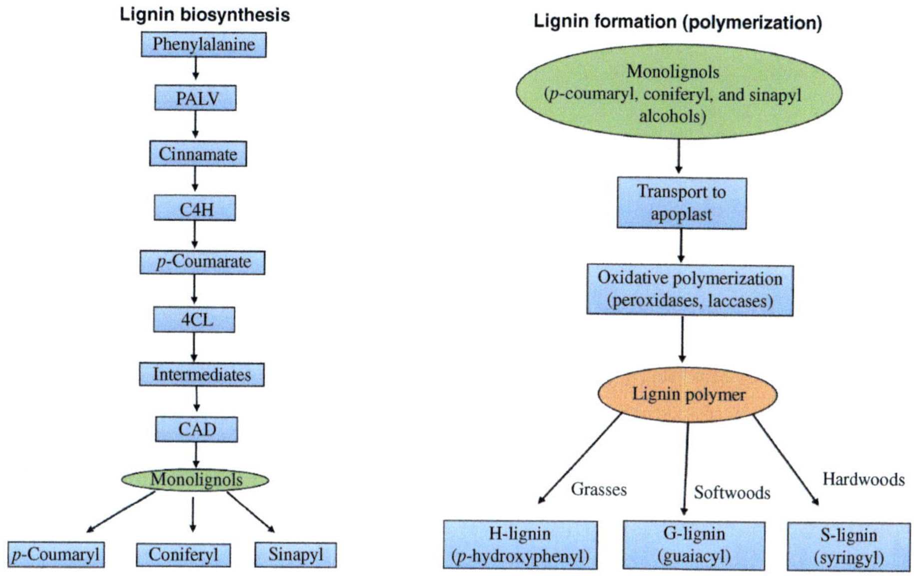

Figure 15.2 Lignin biosynthesis and formation.

in plant secondary metabolism starting with the conversion of phenylalanine to cinnamate by the enzyme phenylalanine ammonia-lyase, a critical regulatory point. Subsequent enzymatic steps involving enzymes like cinnamate-4-hydroxylase (C4H) and 4-coumarate:CoA ligase (4CL) lead to activated intermediates that are channeled toward the synthesis of the three primary monolignols: *p*-coumaryl alcohol, coniferyl alcohol, and sinapyl alcohol (Lavhale et al. 2018). These monolignol units are produced via the action of enzymes such as cinnamyl alcohol dehydrogenase (CAD) (Pan et al. 2014).

Once synthesized, these monolignols are transported to the cell wall space (apoplast), where they undergo oxidative polymerization catalyzed by peroxidases and laccases. This process links the monomers through various carbon–carbon and ether bonds, forming the intricate lignin polymer (Morel 2023). The relative proportions of the three monolignols incorporated—giving rise to *p*-hydroxyphenyl (H), guaiacyl (G), and syringyl (S) lignin units, respectively—result in significant structural diversity. Lignin structure is thus heterogeneous, varying between species, tissue types, and developmental stages. For instance, G-lignin (from coniferyl alcohol) is common in softwoods, while S-lignin (from sinapyl alcohol) is abundant in hardwoods; H-lignin (from *p*-coumaryl alcohol) is often found in grasses. This structural variation profoundly impacts lignin's physical properties, influencing wood characteristics, cell wall digestibility, and the efficacy of the cell wall as a defensive barrier (Lu et al. 2020).

Lignin deposition is tightly regulated throughout plant development and in response to environmental cues (Figure 15.3). It is particularly prominent in tissues requiring reinforcement, such as xylem and sclerenchyma fibers. Hormones like auxins, gibberellins, ET, and JA play key roles; for example, auxin promotes vascular differentiation (rich in lignin), while JA and ET often enhance lignification during stress responses like wounding. External factors, including mechanical stress (wind, touch), drought, light intensity, and nutrient availability, also strongly influence lignin production. High light and low nitrogen,

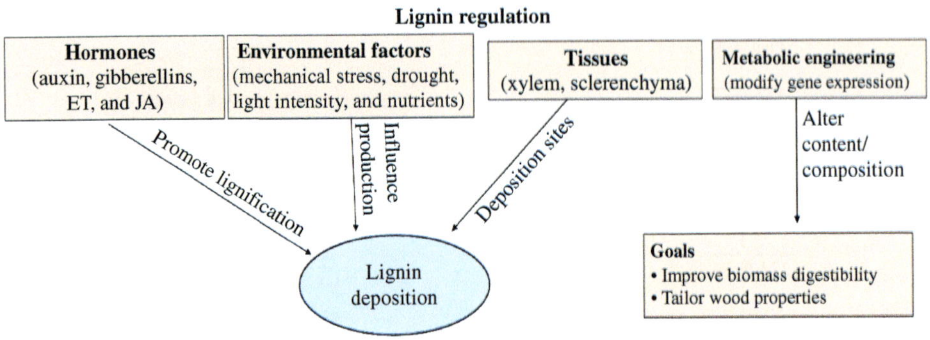

Figure 15.3 Lignin regulation.

for example, often promote increased lignification (Frei 2013). By modifying the expression of key biosynthetic genes, scientists aim to alter lignin content or composition to improve biomass digestibility or tailor wood properties for specific industrial applications (Yoon et al. 2015).

15.7.1 Interplay Between Apocarotenoid and Lignin Pathways

Despite originating from distinct metabolic pathways—carotenoid cleavage for apocarotenoids and phenylpropanoid metabolism for lignin—these two classes of compounds exhibit significant interplay, sharing regulatory elements and influencing each other's production and function, particularly in the context of stress responses, as shown in Figure 15.4 (Zheng et al. 2021b). Transcription factors, especially from the MYB family, have been identified that regulate genes in both pathways, suggesting points of coordinated control. For example, certain MYBs known to control phenylpropanoid biosynthesis can affect carotenoid metabolism, and vice versa, particularly under stress conditions like high light (Pratyusha and Sarada 2022). Signaling molecules like JA also act as common regulators, capable of inducing both apocarotenoid production (e.g., via CCD enzymes) and key enzymes in lignin biosynthesis, enabling a coordinated enhancement of defenses against pests or pathogens. Hormones like auxin and ET further integrate these pathways within broader developmental and stress response programs (Imtiaz et al. 2023).

Oxidative stress provides a critical link between apocarotenoids and lignin (Figure 15.4). ROS generated during stress can nonenzymatically produce apocarotenoids like β-cyclocitral and trigger increased lignification to reinforce cell walls. Apocarotenoids, acting as RES or signaling molecules, can modulate cellular redox balance and potentially influence the expression of lignin biosynthetic genes, possibly through redox-sensitive transcription factors. β-Cyclocitral, in particular, might act as a priming signal, enhancing the plant's capacity for subsequent defensive lignification upon further stress (Havaux 2014; Felemban et al. 2019).

Hormonal signaling provides another layer of coordination (Figure 15.4). ABA, derived from apocarotenoids, is central to drought responses and has been shown to induce genes in both its precursor pathway and the phenylpropanoid pathway, leading to both elevated ABA levels and increased lignification, which helps protect tissues from desiccation (Muhammad Aslam et al. 2022). Similarly, JA, triggered by wounding or herbivory,

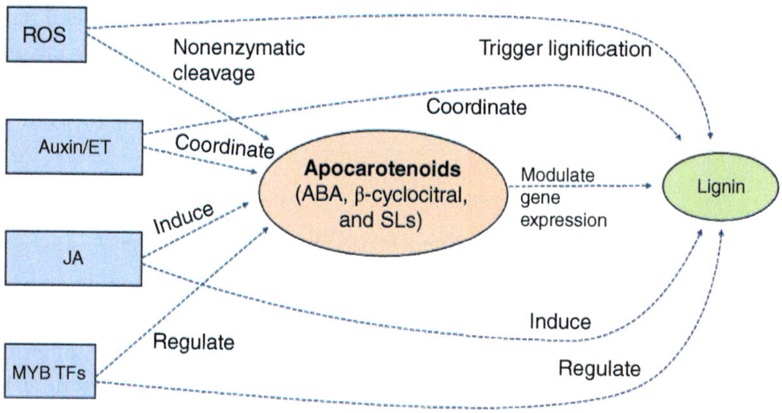

Figure 15.4 Apocarotenoid–lignin interplay.

can stimulate both apocarotenoid synthesis and lignin deposition, creating stronger physical barriers. Emerging evidence directly supports the idea that apocarotenoids influence lignin biosynthesis (Moreno et al. 2021). Studies suggest ABA, SLs, and β-cyclocitral can modulate the expression of key lignin genes, possibly mediated through downstream transcription factors or hormonal crosstalk. The precise molecular mechanisms likely involve complex signaling cascades initiated by stress cues like ROS, which simultaneously activate transcriptional programs for both apocarotenoid production and cell wall fortification via lignin (Faizan et al. 2022).

15.7.2 Coordinated Roles in Plant Development and Stress Adaptation

The interplay between apocarotenoids and lignin extends beyond immediate stress responses to shape fundamental aspects of plant development and architecture. In root systems, SLs are key regulators of architecture, influencing lateral root formation and elongation, particularly in response to nutrient availability like phosphate deficiency, while also mediating mycorrhizal associations. Concurrently, lignin provides essential structural reinforcement and waterproofing to root vascular tissues and the endodermis, crucial for anchorage and function, especially under mechanical stress from the soil. Hormonal interactions, particularly involving ABA and auxin, further integrate stress signals with developmental programs affecting both root growth patterns and lignification for structural integrity, notably under drought (Felemban et al. 2019).

In the development of vascular tissues (xylem and phloem), lignin is indispensable. Its deposition in xylem vessel walls provides the rigidity needed for water transport under tension and contributes to pathogen resistance (Yadeta and Thomma 2013). Hormones like auxin guide vascular differentiation, inherently linked to lignification. SLs also participate by influencing cell division and differentiation in the vascular cambium. ABA contributes by modulating cell wall properties and enhancing lignification, especially as an adaptation to water stress, thereby reinforcing the water transport system (Gall et al. 2015).

Plants also rely on the coordinated action of apocarotenoids and lignin to withstand mechanical stress. Physical forces trigger increased lignin biosynthesis, particularly in stems and roots, to bolster structural support (Saini et al. 2024). Simultaneously, stress signals, potentially involving apocarotenoids like β-cyclocitral or JA-mediated pathways, can trigger or enhance this defensive lignification response. This synergy ensures that plants can effectively fortify their structures against physical damage, demonstrating how these distinct metabolic pathways converge to enhance overall plant resilience and adaptation (Weng et al. 2021).

15.7.3 Fine-tuning Developmental Processes and Cellular Specificity

Beyond broad developmental roles, the interplay between apocarotenoids and lignin is fine-tuned during major developmental phase transitions and at the cellular level. The shift from juvenile to mature growth, or from vegetative to reproductive stages, often involves significant changes in the metabolism of both compound classes (Xie et al. 2018). For instance, lignification typically increases in stems and floral structures during the transition to flowering, providing the necessary support for reproductive organs and enhancing defense during this vulnerable period (Marjamaa et al. 2007). Apocarotenoids, particularly ABA and SLs, influence these transitions by mediating stress responses and resource allocation (Felemban et al. 2019). ABA, for example, plays roles in seed maturation and senescence, processes often correlated with increased lignin in protective seed coats. Conversely, appropriate timing is crucial; premature or excessive lignification can impede growth (Brookbank et al. 2021). There is speculation that certain apocarotenoids, like zaxinone, might help regulate this timing, potentially delaying lignification in actively growing tissues to maintain flexibility before structural reinforcement occurs (Felemban et al. 2019).

This regulation operates with remarkable precision down to the level of individual cells. Lignin deposition is heavily concentrated in xylem vessels to ensure water transport efficiency and mechanical strength, while occurring at lower levels in parenchyma or phloem, where its role might be more focused on localized defense (Boyce et al. 2004). Similarly, apocarotenoid biosynthesis and signaling are often highly localized. ABA production might increase specifically in vascular tissues under drought to coordinate water conservation and lignification, whereas SLs are primarily synthesized and act in the roots to influence architecture and microbial interactions (Milborrow 2001; Andreo-Jimenez et al. 2015). This cell-specific control ensures that resources are used efficiently, deploying these potent structural and signaling molecules only where and when they are most needed, thereby optimizing overall plant performance.

15.8 Analytical Approaches for Studying Apocarotenoids and Lignin

Unraveling the complex biosynthesis, regulation, and function of apocarotenoids and lignin necessitates a diverse toolkit of sophisticated analytical methods. Foundational techniques include chromatography coupled with mass spectrometry (MS). High-performance liquid chromatography (HPLC), often with UV–vis detection, is widely used to separate and quantify nonvolatile apocarotenoids like ABA and SLs (Butnariu 2016). For volatile apocarotenoids such as β-ionone and β-cyclocitral, gas chromatography–mass spectrometry (GC–MS) is the method of choice (Leroux et al. 2022). Liquid chromatography–mass spectrometry (LC–MS), with its high sensitivity and specificity, offers powerful capabilities for comprehensive profiling of both apocarotenoids and lignin precursors (like hydroxycinnamic acids) and their derivatives (Zheng et al. 2021a). For analyzing the structure of the complex lignin polymer itself, matrix-assisted laser desorption/ionization mass spectrometry (MALDI-MS) can be employed (Letourneau and Volmer 2023).

Understanding where these compounds accumulate is crucial, and imaging technologies provide spatial context. Confocal laser scanning microscopy, often combined with specific fluorescent stains (like phloroglucinol for lignin) or tagged molecules, allows high-resolution visualization within tissues and cells (Maceda and Terrazas 2022). Mass spectrometry imaging (MSI) takes this further, enabling the direct mapping of various metabolites, including apocarotenoids and lignin components, across tissue sections, offering powerful insights into metabolic localization and pathway crosstalk. Raman spectroscopy provides a noninvasive way to obtain chemical fingerprints of lignin and related phenolics directly within plant tissues (Dong and Aharoni 2022).

Genetic and molecular tools are indispensable for dissecting the underlying pathways and regulatory networks. Gene expression analysis, using techniques like quantitative real-time PCR (qRT-PCR) or RNA sequencing, reveals how the transcription of biosynthetic genes (e.g., *CCDs*, *PAL*) changes under different conditions or in different genetic backgrounds (Baba et al. 2015). Functional genomics, employing tools like CRISPR-Cas9 gene editing or RNA interference (RNAi), allows researchers to precisely modify specific genes to determine their function in apocarotenoid or lignin production and downstream physiological processes (Bhojiya and Joshi 2024). Reporter gene constructs (e.g., using GFP) fused to promoters of interest enable visualization of where and when key pathway genes are active within the living plant (Chiu et al. 1996).

Metabolomic approaches provide a broader perspective on the chemical landscape. Targeted metabolomics focuses on quantifying known apocarotenoids, lignin precursors, and related compounds, tracking their dynamics during development or stress. Untargeted metabolomics aims for a comprehensive snapshot of all detectable metabolites, facilitating the discovery of novel compounds or unexpected metabolic shifts resulting from genetic or environmental perturbations (Allwood et al. 2021). Stable isotope labeling and flux analysis can further illuminate the flow of metabolites through these interconnected pathways (Wang et al. 2020). Increasingly, integrating data from these various “omics” platforms (genomics, transcriptomics, proteomics, and metabolomics) using systems biology approaches allows for a more holistic understanding. This multi-omics integration helps map pathway interconnections, identify critical regulatory hubs, and predict how plants respond to changes, paving the way for targeted manipulation of these pathways (Pinu et al. 2019).

15.9 Agricultural and Biotechnological Applications

The growing understanding of apocarotenoid and lignin biology has significant implications for agriculture and biotechnology, offering avenues to improve crop performance and sustainability. A major focus is enhancing crop resilience. Knowledge of ABA’s role in stomatal control and stress gene regulation can be leveraged to develop drought-tolerant varieties with improved water-use efficiency (Cardoso et al. 2020). Manipulating SL biosynthesis or signaling could lead to crops with enhanced nutrient uptake efficiency, particularly beneficial in nutrient-poor soils (Naz et al. 2025). Engineering lignin content or composition can bolster resistance to pathogens, pests, and physical stresses like wind or lodging (Riaz et al. 2023).

Optimizing plant architecture and yield is another key application area. Modifying SL levels can influence root system architecture for better soil exploration or control shoot branching for ideal crop structures (Kelly et al. 2023). While lignin is essential for structure, reducing or altering its composition in specific biomass crops like poplar or switchgrass can significantly improve their processability for biofuel production or as animal feed by increasing the accessibility of cellulose (Welker et al. 2015). Furthermore, harnessing novel growth regulators identified among apocarotenoids, such as β-cyclocitral or zaxinone, might offer new ways to optimize plant development and productivity (Wang et al. 2021).

These insights also contribute to sustainable agriculture. Enhancing natural plant defenses mediated by apocarotenoids could reduce the reliance on synthetic pesticides. The recalcitrant nature of lignin means it contributes significantly to stable soil organic carbon; optimizing lignin characteristics in crop residues could potentially enhance soil carbon sequestration, benefiting soil health and mitigating climate change (Zhou et al. 2022). Developing crop varieties with inherently better nutrient efficiency and stress tolerance through targeted manipulation of these pathways aligns well with low-input farming systems. Finally, there is potential to harness specific apocarotenoids (like β-cyclocitral) or lignin-derived compounds as novel, natural bioactive agents for crop protection, serving as biofungicides or antimicrobial agents as alternatives to synthetic chemicals (Santra and Banerjee 2020; Sánchez-Hernández et al. 2022).

15.10 Illustrative Examples: From Model Systems to Crops

Our understanding of apocarotenoid and lignin functions has been significantly advanced through studies in diverse plant systems. The model organism *Arabidopsis thaliana*, with its tractable genetics and short life cycle, has been invaluable. Research using *Arabidopsis* mutants, such as those affecting CCDs or key lignin biosynthetic enzymes like phenylalanine ammonia-lyase and caffeic acid *O*-methyltransferase, has been instrumental in dissecting biosynthetic pathways, identifying regulatory genes, revealing crosstalk between pathways, and linking specific compounds (like ABA or altered lignin) to distinct roles in development (e.g., seed germination) and stress responses (e.g., drought sensitivity, pathogen resistance) (Vanholme et al. 2012).

Insights from model systems are increasingly translated to important crop plants. In rice, a global staple, apocarotenoids like SLs are recognized as key regulators of root architecture and phosphorus uptake efficiency, offering targets for improving crop performance in nutrient-poor soils (Wang 2021). Concurrently, manipulating lignin biosynthesis in rice, for instance by targeting *OsPAL*, is being explored to enhance biomass digestibility for biofuel production. The critical role of ABA in conferring drought tolerance in rice has also been well documented, guiding efforts to engineer more water-use-efficient varieties (Shad et al. 2024).

Woody plants, such as poplar, serve as prime models for studying lignocellulosic biomass. Much research focuses on modifying lignin content and composition—often through downregulating biosynthetic genes—to improve the efficiency of converting wood into biofuels or other bio-based products (Pazhany and Henry 2019). Apocarotenoids like ABA and β-cyclocitral also play roles in these long-lived plants, influencing vascular tissue development and contributing to overall structural integrity and stress adaptation. Furthermore, many medicinal plants are rich sources of bioactive apocarotenoids (e.g., antioxidants like lutein and zeaxanthin) and lignin derivatives (e.g., lignans with antimicrobial or antioxidant properties), highlighting the potential for discovering novel compounds with therapeutic value (Jaswir et al. 2011). Research is also expanding into emerging model systems, including mosses and ferns to understand the evolutionary origins of these pathways, and legumes to investigate their roles in symbiotic nitrogen fixation (Mathesius 2022).

15.11 Future Directions and Perspectives

The study of apocarotenoids and lignin is a dynamic field with exciting future prospects. Systems biology approaches, integrating multi-omics data (genomics, transcriptomics, proteomics, and metabolomics), are becoming essential for constructing comprehensive metabolic network models (Vanholme et al. 2012; Pinu et al. 2019). These models will help unravel the complex regulatory interactions within and between pathways, identify key control points, and predict metabolic responses to genetic or environmental changes. Synthetic biology offers powerful tools to engineer metabolic pathways, potentially enabling the production of novel apocarotenoids with specific functions or the development of microbial "factories" for scalable, cost-effective synthesis of high-value compounds for agriculture, medicine, or industry (Arendt et al. 2016).

Addressing the challenges of climate change is a critical future direction. Understanding how apocarotenoids (like ABA and SLs) and lignin contribute to plant adaptation to drought, extreme temperatures, and altered nutrient availability will be crucial for developing climate-resilient crops. Furthermore, exploring lignin's role in long-term carbon sequestration in soils and biomass offers potential strategies for climate change mitigation (Feng et al. 2008). Continued advancements in enabling technologies, particularly precise gene editing tools like CRISPR-Cas9 and RNAi techniques, will accelerate both fundamental research and the development of improved plant varieties.

Ultimately, translating laboratory discoveries into practical applications is paramount. This requires robust translational research, including field trials of engineered or bred varieties under realistic agricultural conditions, and strong collaborations between academic researchers, biotechnology companies, and the agricultural sector.

15.12 Conclusion

In conclusion, apocarotenoids represent a diverse class of signaling molecules derived from carotenoids, whose functions are deeply intertwined with the biosynthesis and deposition of lignin. Together, these compounds play fundamental roles in shaping plant architecture, mediating defense against biotic and abiotic stresses, and orchestrating developmental programs. Significant progress, driven by advanced analytical, genetic, and imaging techniques, has illuminated the intricacies of their biosynthesis, regulation, and interplay. This knowledge opens exciting opportunities for agricultural innovation—enhancing crop resilience to environmental challenges, optimizing yield and biomass quality for food, feed, and biofuel applications, and developing sustainable crop protection strategies. Continued interdisciplinary research into apocarotenoid and lignin biology holds immense promise for addressing global challenges related to food security, environmental sustainability, and the transition toward a renewable bioeconomy.

References

Akbar, M.U., Aqeel, M., Shah, M.S. et al. (2023). Molecular regulation of antioxidants and secondary metabolites act in conjunction to defend plants against pathogenic infection. *South African Journal of Botany* 161, 247–257.

Allwood, J.W., Williams, A., Uthe, H. et al. (2021). Unravelling plant responses to stress—the importance of targeted and untargeted metabolomics. *Metabolites* 11(8), 558.

Andreo-Jimenez, B., Ruyter-Spira, C., Bouwmeester, H.J. et al. (2015). Ecological relevance of strigolactones in nutrient uptake and other abiotic stresses, and in plant-microbe interactions below-ground. *Plant and Soil* 394, 1–19.

Arendt, P., Pollier, J., Callewaert, N. et al. (2016). Synthetic biology for production of natural and new-to-nature terpenoids in photosynthetic organisms. *The Plant Journal* 87(1), 16–37.

Baba, S.A., Mohiuddin, T., Basu, S. et al. (2015). Comprehensive transcriptome analysis of *Crocus sativus* for discovery and expression of genes involved in apocarotenoid biosynthesis. *BMC Genomics* 16, 698.

Bhojiya, A.A. and Joshi, H. (2024). Crispr gene editing for secondary metabolite production: a review. In: *Gene Editing in Plants: CRISPR-Cas and Its Applications* (eds. A. Kumar, S. Arora, S. Ogita et al.), 437–475. Springer.

Bhuiyan, N.H., Selvaraj, G., Wei, Y. et al. (2009). Role of lignification in plant defense. *Plant Signaling & Behavior* 4(2), 158–159.

Boyce, C.K., Zwieniecki, M.A., Cody, G.D. et al. (2004). Evolution of xylem lignification and hydrogel transport regulation. *Proceedings of the National Academy of Sciences of the United States of America* 101(50), 17555–17558.

Brookbank, B.P., Patel, J., Gazzarrini, S. et al. (2021). Role of basal ABA in plant growth and development. *Genes* 12(12), 1936.

Butnariu, M. (2016). Methods of analysis (extraction, separation, identification and quantification) of carotenoids from natural products. *Journal of Ecosystem & Ecography* 6(2), 193.

Cardoso, A.A., Gori, A., Da-Silva, C.J. et al. (2020). Abscisic acid biosynthesis and signaling in plants: key targets to improve water use efficiency and drought tolerance. *Applied Sciences* 10(18), 6322.

Chiu, W.-l., Niwa, Y., Zeng, W. et al. (1996). Engineered GFP as a vital reporter in plants. *Current Biology* 6(3), 325–330.

Deshpande, S., Purkar, V. and Mitra, S. (2021). β-Cyclocitral, a master regulator of multiple stress-responsive genes in *Solanum lycopersicum* L. plants. *Plants* 10(11), 2465.

Dong, Y. and Aharoni, A. (2022). Image to insight: exploring natural products through mass spectrometry imaging. *Natural Product Reports* 39(7), 1510–1530.

Faizan, M., Tonny, S.H., Afzal, S. et al. (2022). β-Cyclocitral: emerging bioactive compound in plants. *Molecules* 27(20), 6845.

Felemban, A., Braguy, J., Zurbriggen, M.D. et al. (2019). Apocarotenoids involved in plant development and stress response. *Frontiers in Plant Science* 10, 1168.

Feng, X., Simpson, A.J., Wilson, K.P. et al. (2008). Increased cuticular carbon sequestration and lignin oxidation in response to soil warming. *Nature Geoscience* 1(12), 836–839.

Frei, M. (2013). Lignin: characterization of a multifaceted crop component. *The Scientific World Journal* 2013(1), 436517.

Gall, H.L., Philippe, F., Domon, J.-M. et al. (2015). Cell wall metabolism in response to abiotic stress. *Plants* 4(1), 112–166.

Havaux, M. (2014). Carotenoid oxidation products as stress signals in plants. *The Plant Journal* 79(4), 597–606.

Havaux, M. (2020). β-Cyclocitral and derivatives: emerging molecular signals serving multiple biological functions. *Plant Physiology and Biochemistry* 155, 35–41.

Helmich, R. (2025). Unveiling the strigolactone signaling pathway: from receptors to responses. In: *Strigolactones: Emerging Plant Hormones* (eds. T. Aftab and I. Wani), 35–67. Wiley.

Imtiaz, H., Arif, Y., Alam, P. et al. (2023). Apocarotenoids biosynthesis, signaling regulation, crosstalk with phytohormone, and its role in stress tolerance. *Environmental and Experimental Botany* 210, 105337.

Jaswir, I., Noviendri, D., Hasrini, R.F. et al. (2011). Carotenoids: sources, medicinal properties and their application in food and nutraceutical industry. *Journal of Medicinal Plants Research* 5(33), 7119–7131.

Jeffri, N.I., Rawi, N.F.M., Kassim, M.H.B.M. et al. (2024). Unlocking the potential: evolving role of technical lignin in diverse applications and overcoming challenges. *International Journal of Biological Macromolecules* 274, 133506.

Kelly, J.H., Tucker, M.R. and Brewer, P.B. (2023). The strigolactone pathway is a target for modifying crop shoot architecture and yield. *Biology* 12(1), 95.

Lavhale, S.G., Kalunke, R.M. and Giri, A.P. (2018). Structural, functional and evolutionary diversity of 4-coumarate-CoA ligase in plants. *Planta* 248(5), 1063–1078.

Leroux, J., Truong, T.T., Pogson, B.J. et al. (2022). Detection and analysis of novel and known plant volatile apocarotenoids. In: *Methods in Enzymology: Volume 670* (ed. E.T. Wurtzel), 311–368. Elsevier.

Letourneau, D.R. and Volmer, D.A. (2023). Mass spectrometry-based methods for the advanced characterization and structural analysis of lignin: a review. *Mass Spectrometry Reviews* 42(1), 144–188.

Lu, Y.-C., Lu, Y. and Fan, X. (2020). Structure and characteristics of lignin. In: *Lignin: Biosynthesis and Transformation for Industrial Applications* (eds. S. Sharma and A. Kumar), 17–75. Springer.

Maceda, A. and Terrazas, T. (2022). Fluorescence microscopy methods for the analysis and characterization of lignin. *Polymers* 14(5), 961.

Magome, H., Arai, M., Oyama, K. et al. (2023). Multiple loss-of-function mutations of carotenoid cleavage dioxygenase 4 reveal its major role in both carotenoid level and apocarotenoid composition in flue-cured mature tobacco leaves. *Scientific Reports* 13(1), 12992.

Marjamaa, K., Kukkola, E.M. and Fagerstedt, K.V. (2007). Lignification in development. *International Journal of Plant Developmental Biology* 1(1), 160–169.

Mathesius, U. (2022). Are legumes different? Origins and consequences of evolving nitrogen fixing symbioses. *Journal of Plant Physiology* 276, 153765.

Milborrow, B. (2001). The pathway of biosynthesis of abscisic acid in vascular plants: a review of the present state of knowledge of ABA biosynthesis. *Journal of Experimental Botany* 52(359), 1145–1164.

Morel, O. (2023). Characterization of the spatial distribution of lignins in plant cell walls using chemical reporters and Raman. Doctoral dissertation. Université de Lille; Universität für Bodenkultur.

Moreno, J.C., Mi, J., Alagoz, Y. et al. (2021). Plant apocarotenoids: from retrograde signaling to interspecific communication. *The Plant Journal* 105(2), 351–375.

Muhammad Aslam, M., Waseem, M., Jakada, B.H. et al. (2022). Mechanisms of abscisic acid-mediated drought stress responses in plants. *International Journal of Molecular Sciences* 23(3), 1084.

Nawaz, M., Sun, J., Shabbir, S. et al. (2023). A review of plants strategies to resist biotic and abiotic environmental stressors. *Science of The Total Environment* 900, 165832.

Naz, M., Afzal, M.R., Raza, M.A. et al. (2025). The significant effects of strigolactones on plant growth and microbe interactions: a review. *Plant Growth Regulation* 105, 937–957.

Ninkuu, V., Yan, J., Fu, Z. et al. (2022). Lignin and its pathway-associated phytoalexins modulate plant defense against fungi. *Journal of Fungi* 9(1), 52.

Pan, H., Zhou, R., Louie, G.V. et al. (2014). Structural studies of cinnamoyl-CoA reductase and cinnamyl-alcohol dehydrogenase, key enzymes of monolignol biosynthesis. *The Plant Cell* 26(9), 3709–3727.

Pazhany, A.S. and Henry, R.J. (2019). Genetic modification of biomass to alter lignin content and structure. *Industrial & Engineering Chemistry Research* 58(35), 16190–16203.

Pinu, F.R., Beale, D.J., Paten, A.M. et al. (2019). Systems biology and multi-omics integration: viewpoints from the metabolomics research community. *Metabolites* 9(4), 76.

Pratyusha, D.S. and Sarada, D.V. (2022). MYB transcription factors—master regulators of phenylpropanoid biosynthesis and diverse developmental and stress responses. *Plant Cell Reports* 41(12), 2245–2260.

Riaz, M.W., Yousaf, M.I., Hussain, Q. et al. (2023). Role of lignin in wheat plant for the enhancement of resistance against lodging and biotic and abiotic stresses. *Stresses* 3(2), 434–453.

Saini, N., Anmol, A., Kumar, S. et al. (2024). Exploring phenolic compounds as natural stress alleviators in plants—a comprehensive review. *Physiological and Molecular Plant Pathology* 133, 102383.

Sánchez-Hernández, E., Langa-Lomba, N., González-García, V. et al. (2022). Lignin–chitosan nanocarriers for the delivery of bioactive natural products against wood-decay phytopathogens. *Agronomy* 12(2), 461.

Santra, H.K. and Banerjee, D. (2020). Natural products as fungicide and their role in crop protection. In: *Natural Bioactive Products in Sustainable Agriculture* (eds. J. Singh and A. Yadav), 131–219. Springer.

Shad, M.A., Li, X., Rao, M.J. et al. (2024). Exploring lignin biosynthesis genes in rice: evolution, function, and expression. *International Journal of Molecular Sciences* 25(18), 10001.

Simkin, A.J. (2021). Carotenoids and apocarotenoids in planta: their role in plant development, contribution to the flavour and aroma of fruits and flowers, and their nutraceutical benefits. *Plants* 10(11), 2321.

Vanholme, R., Storme, V., Vanholme, B. et al. (2012). A systems biology view of responses to lignin biosynthesis perturbations in *Arabidopsis*. *The Plant Cell* 24(9), 3506–3529.

Vishwakarma, K., Upadhyay, N., Kumar, N. et al. (2017). Abscisic acid signaling and abiotic stress tolerance in plants: a review on current knowledge and future prospects. *Frontiers in Plant Science* 8, 161.

Walter, M.H. and Strack, D. (2011). Carotenoids and their cleavage products: biosynthesis and functions. *Natural Product Reports* 28(4), 663–692.

Wang, J.Y. (2021). Zaxinone, a natural apocarotenoid, regulates growth and strigolactone biosynthesis in rice. Doctoral dissertation. King Abdullah University of Science and Technology.

Wang, J.Y., Lin, P.-Y. and Al-Babili, S. (2021). On the biosynthesis and evolution of apocarotenoid plant growth regulators. *Seminars in Cell & Developmental Biology* 109, 3–11.

Wang, Y., Wondisford, F.E., Song, C. et al. (2020). Metabolic flux analysis—linking isotope labeling and metabolic fluxes. *Metabolites* 10(11), 447.

Welker, C.M., Balasubramanian, V.K., Petti, C. et al. (2015). Engineering plant biomass lignin content and composition for biofuels and bioproducts. *Energies* 8(8), 7654–7676.

Weng, J.-K., Lynch, J.H., Matos, J.O. et al. (2021). Adaptive mechanisms of plant specialized metabolism connecting chemistry to function. *Nature Chemical Biology* 17(10), 1037–1045.

Xie, M., Zhang, J., Tschaplinski, T.J. et al. (2018). Regulation of lignin biosynthesis and its role in growth-defense tradeoffs. *Frontiers in Plant Science* 9, 1427.

Yadeta, K.A. and Thomma, B.P.H.J. (2013). The xylem as battleground for plant hosts and vascular wilt pathogens. *Frontiers in Plant Science* 4, 97.

Yang, Y.-X., Ahammed, G.J., Wu, C. et al. (2015). Crosstalk among jasmonate, salicylate and ethylene signaling pathways in plant disease and immune responses. *Current Protein and Peptide Science* 16(5), 450–461.

Yoon, J., Choi, H. and An, G. (2015). Roles of lignin biosynthesis and regulatory genes in plant development. *Journal of Integrative Plant Biology* 57(11), 902–912.

Zheng, X., Mi, J., Deng, X. et al. (2021a). LC–MS-based profiling provides new insights into apocarotenoid biosynthesis and modifications in citrus fruits. *Journal of Agricultural and Food Chemistry* 69(6), 1842–1851.

Zheng, X., Yang, Y. and Al-Babili, S. (2021b). Exploring the diversity and regulation of apocarotenoid metabolic pathways in plants. *Frontiers in Plant Science* 12, 787049.

Zhou, Q., Li, Q., Li, P. et al. (2019). Carotenoid cleavage dioxygenases: identification, expression, and evolutionary analysis of this gene family in tobacco. *International Journal of Molecular Sciences* 20(22), 5796.

Zhou, Y., Zhang, J., Xu, L. et al. (2022). Long-term fertilizer postponing promotes soil organic carbon sequestration in paddy soils by accelerating lignin degradation and increasing microbial necromass. *Soil Biology and Biochemistry* 175, 108839.

16

Role of Carotenoid Cleavage Dioxygenase-encoding Genes in Mitigating Stress in Plants

Najwa Shabir[1,2] *and Nasheeman Ashraf*[1,2,*]

[1] *Plant Molecular Biology and Biotechnology Division, CSIR-Indian Institute of Integrative Medicine, Srinagar, Jammu and Kashmir, India*
[2] *Academy of Scientific and Innovative Research (AcSIR), Ghaziabad, Uttar Pradesh, India*

*Corresponding author: nashraf.iiim@csir.res.in

16.1 Introduction

With the increasing global environmental challenges and a growing demand for sustainable agricultural practices in order to meet the world's ever-increasing demand for food, a thorough understanding of plants' adaptation mechanisms to environmental stresses has become a research priority. Plants routinely face a variety of abiotic and biotic stress factors, including drought, salinity, temperature extremes, and pathogen attacks, all of which can disrupt physiological processes, hinder development, and reduce overall crop yield. To mitigate these challenges, plants have evolved an extensive array of adaptive mechanisms, including hormonal reprogramming, redox-based signaling, and the accumulation of secondary metabolites with protective functions.

Amongst the vast repertoire of bioactive compounds synthesized by plants to aid them in fighting unfavorable growth conditions, carotenoids represent a class of isoprenoid compounds that perform critical roles in photosynthesis, photoprotection, and developmental regulation (Zia-Ul-Haq 2021). These C40 tetraterpenoids are not only involved in contributing toward the structural and functional integrity of photosystems but also serve as biosynthetic precursors for apocarotenoids, which influence developmental, signaling, and stress responses within plant systems (Sierra et al. 2022).

Apocarotenoids are signaling molecules that have played vital roles in biological communication and adaptation throughout evolutionary history (Carnà et al. 2025). They are a chemically diverse group of low-molecular-weight compounds with over 700 different structures found in nature and include volatile signaling molecules such as β-ionone and β-cyclocitral as well as nonvolatile hormones like abscisic acid (ABA) and strigolactones (SLs) (Leroux et al. 2022). These compounds are increasingly recognized not only for their roles in regulating plant growth and development but also for mediating biotic and abiotic stress responses, inter-plant communication, and ecological interactions (Chen et al. 2025). Their biological significance necessitates the tight regulation of their biosynthesis, ensuring that apocarotenoid production is responsive to internal cues and environmental triggers (Moreno et al. 2021). Studies have demonstrated that numerous apocarotenoids serve crucial biological roles in the growth and development of plants (Ohmiya et al. 2019). For instance, SLs, a group of apocarotenoid-derived plant hormones, modulate shoot branching, root growth, and responses to environmental stressors (Felemban et al. 2019; Imtiaz et al. 2023). Additionally, other apocarotenoids, including β-ionone and β-cyclocitral, function as volatile organic compounds that can attract pollinators or serve as defensive compounds against herbivores and pathogens (Paparella et al. 2021). The diverse array of biological functions exhibited by apocarotenoids underscores their crucial importance in various aspects of plant physiology and ecology.

In plants, the formation of apocarotenoids occurs through the oxidative cleavage of carotenoid backbones either nonenzymatically via interactions with reactive oxygen species (ROS) under stress conditions or enzymatically through the action of a highly conserved group of enzymes called the carotenoid cleavage dioxygenases (CCDs) (Liang et al. 2021). Since the nonenzymatic cleavage is nontargeted, it often results in the formation of random oxidative products (Sun et al. 2022). In comparison, the CCD enzymes ensure targeted breakdown of carotenoids and are preferred over the nonenzymatic breakdown for the generation of biologically relevant compounds (Harrison and Bugg 2014).

Given their pivotal function in stress-related apocarotenoid production, genes encoding CCDs are increasingly recognized as significant regulators of plant resilience. A comprehensive understanding of their expression patterns, catalytic characteristics, and regulatory mechanisms is essential for elucidating how plants integrate metabolic and signaling pathways to alleviate environmental stress. To appreciate the broad spectrum of biological roles attributed to CCDs, it is essential to examine their structural organization, phylogenetic classification, and catalytic behavior across plant systems. Equally important is the analysis of the regulatory mechanisms that govern *CCD* gene expression and enzyme activity, as these processes determine the spatial and temporal production of apocarotenoids. Together, such investigations will help offer critical insights into how plants dynamically modulate apocarotenoid biosynthesis in response to developmental signals and environmental cues, thereby integrating metabolic and signaling pathways to enhance stress resilience.

16.2 The CCD Family

CCDs are a nonheme iron enzyme family that requires oxygen to cleave double bonds in carotenoid skeletons (Dzib-Cauich et al. 2020). These enzymes possess strong regio- and stereospecificity in cleaving double bonds of carotenoids but show high substrate promiscuity (Zhang et al. 2023). Thus, the same enzyme can cleave multiple substrates, resulting in the formation of an extensive array of apocarotenoids and subsequently contributing to the biochemical complexity and functional diversity of this metabolite class (Moreno et al. 2021).

CCDs have been identified across various organisms, but their structural and functional diversity is particularly notable in plants due to their involvement in crucial physiological processes, including mediating plant–environment interactions. These interactions include attracting pollinators, deterring herbivores, and defending against pathogens (Gómez-Sagasti et al. 2023).

Over the course of evolution, carotenoid cleavage oxygenases (CCOs) have diversified into distinct clades with specialized substrate specificities and biochemical functions. Cyanobacterial apocarotenoid oxygenases (ACOs) specifically cleave all-*trans* apocarotenoids, while mammalian and insect CCOs target C40 carotenoids like β-carotene (Liang et al. 2018). Within the plant kingdom, two major subfamilies of CCOs have been characterized: the CCD and the NCED (9-*cis*-epoxy carotenoid dioxygenase) subfamilies (Ohmiya 2009). Through highly specific cleavage reactions, CCDs contribute to multiple physiological processes, including photosynthesis, pigmentation, photoprotection, and stress signaling (Auldridge et al. 2006b). Their ability to generate diverse bioactive metabolites with ecological and economic relevance—particularly in high-value species such as *Crocus sativus*—underscores their significance in both fundamental plant biology and applied crop biotechnology (Daruwalla et al. 2020).

16.3 Discovery and Early Functional Studies

The initial carotenoid-cleaving enzyme, viviparous14 (VP14), was first identified through the investigation of a maize mutant that was deficient in the stress hormone ABA (Schwartz et al. 1997). Enzymatic analysis of VP14 revealed that it cleaves the 11,12 double bonds of the 9-*cis* isomers of the carotenoids neoxanthin and violaxanthin, generating xanthoxin, which serves as the precursor for the stress hormone ABA (Schwartz et al. 2003). This breakthrough not only elucidated the enzymatic origin of ABA biosynthesis but also marked the first functional characterization of a *CCD* gene. The foundational work on VP14 enabled the subsequent identification of homologous enzymes across various plant species and other organisms, which highlighted that carotenoid cleavage is a conserved metabolic process fundamental to plant development and stress signaling. Following the initial discovery, CCDs have been subsequently identified and studied in a variety of plant species, including tomato (Simkin et al. 2004a), common bean (Qin and Zeevaart 1999), cowpea (Iuchi et al. 2000), avocado (Chernys and Zeevaart 2000), annatto (Rodríguez-Ávila et al. 2011), saffron (Bouvier et al. 2003), *Petunia* (Simkin et al. 2004b), and melon (Cheng et al. 2022), among others. Subsequent investigations have expanded our understanding of the CCD family, revealing a growing number of *CCD* genes, each exhibiting distinct functions in hormone biosynthesis, pigmentation, volatile compound formation, and stress adaptation. The identification of *CCD2* in *C. sativus*, for instance, linked carotenoid cleavage directly to the biosynthesis of crocins and safranal—signature apocarotenoids that define saffron's quality (Ahrazem et al. 2016b). Similarly, studies in other plants have revealed that *CCD1* contributes to the production of key volatile apocarotenoids such as β-ionone by cleaving multiple carotenoid substrates in the cytosol, while *CCD4* acts within plastids to control carotenoid content, influencing both pigmentation and aroma profiles in flowers and fruits (Dhar et al. 2020). Collectively, research studies have established CCDs as a metabolically versatile and biologically indispensable enzyme family, prompting further investigation into their structural diversity, subcellular localization, and regulatory complexity across different plant systems.

16.4 Structural Features and Catalytic Architecture of CCDs

CCDs are notable for their ability to execute reactions other than cleavage, including double bond isomerizations and structural rearrangement of the carbon skeleton. They are nonheme iron-containing enzymes which catalyze the oxidative scission of carotenoids and apocarotenoids (Sui et al. 2013). These enzymes are distinguished by the presence of four His and three Glu (or, less usually, Asp) residues that work together to coordinate an Fe II cofactor required for catalytic activity (Sui et al. 2016). The crystal structures of *Arabidopsis* CCD1 and other CCD enzymes have revealed a highly conserved β-propeller fold with seven blades as shown in Figure 16.1 (Priya et al. 2017). It is believed that the unique structural characteristics and precise arrangement of the active site residues in different CCD isoforms contribute to their distinct substrate preferences and cleavage site specificities (Kloer and Schulz 2006). Structure–function studies have been used to investigate the molecular mechanisms underlying the regioselective cleavage of carotenoids by CCDs, employing techniques such as targeted mutagenesis, biochemical approaches, and crystallographic analysis across bacterial, plant, and animal systems (Auldridge et al. 2006a).

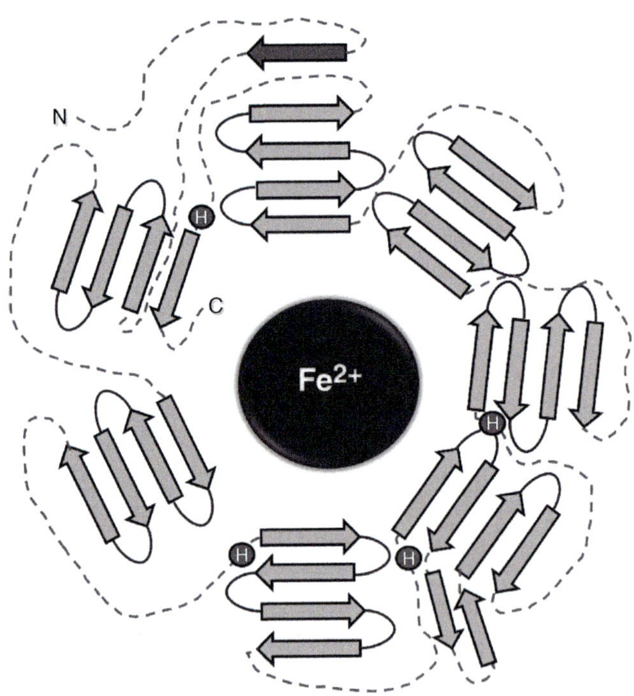

Figure 16.1 Three-dimensional structure of a CCD enzyme. The enzyme family features a conserved seven-bladed β-propeller fold centered around a nonheme Fe^{2+} ion, enabling regioselective cleavage of carotenoids to produce diverse apocarotenoids involved in plant signaling and stress responses.

The first CCD with a three-dimensional structure was *Synechocystis* sp. PCC6803's ACO revealed in 2005 (Daruwalla et al. 2020). The structure revealed a fundamental CCD fold consisting of a seven-bladed β-propeller domain capped by α-helical and loop regions, which collectively form a dome-like structure enclosing the active iron site. Initial attempts to examine substrate binding revealed a doubly bent electron density pattern in the active site, indicating substrate isomerization. However, this was eventually identified as non-carotenoid electron density due to the detergent used during crystallization (Daruwalla and Kiser 2020). Until recently, similar structural studies of VP14 with carotenoids had failed due to the difficulties in trapping CCD–carotenoid complexes due to carotenoids' water insolubility, which impedes the development of high-occupancy complexes. More recently, the structure of a maize VP14 was determined in the presence of its substrate, β-carotene (Messing et al. 2010). This has provided key insights into substrate recognition and specificity, as well as the mechanism of carotenoid cleavage. As per the study, VP14 features two domains: a conserved seven-bladed β-propeller and an α-helical domain. The β-propeller is common among CCD enzymes, while the helical domain might distinguish plant CCDs. This domain analysis, combined with previous biochemical studies, suggests a mechanism where VP14 might penetrate the thylakoid membrane and access the substrate through a tunnel leading to the catalytic center (Messing et al. 2010). Researchers have also revealed a striking similarity between the active sites of CCD and RPE65, despite the evolutionary differences, especially in the substrate-interacting residues and the iron metal center, suggesting a similar biosynthesis mechanism for apocarotenoids across these enzymes (Daruwalla et al. 2020).

16.5 Classification and Functional Roles of CCDs in Plant Stress Responses

CCDs play indispensable roles in modulating plant responses to diverse abiotic and biotic stress conditions through their enzymatic products—apocarotenoids. These apocarotenoids function either as direct defense molecules or as signaling intermediates that integrate with hormonal and transcriptional networks to regulate stress-responsive gene expression (Imtiaz et al. 2023). As such, these enzymes have been recognized as central players in multiple plant physiological and ecological processes.

16.5.1 Phylogenetic Classification and Evolutionary Divergence

Following the initial discovery, research throughout the early 2000s led to more detailed characterization and classification of the CCO family in plants (Akram et al. 2023). Plant CCOs occur in two distinct forms, CCDs and NCEDs, as shown in Figure 16.2. While both groups have been implicated in mediating tolerance to environmental challenges by modulating hormone biosynthesis, altering redox dynamics, enhancing metabolite accumulation, and facilitating cross-communication between signaling pathways (Priya et al. 2019), these subgroups differ in substrate specificity and physiological roles but are collectively involved in the production of apocarotenoids with key functions in plant growth regulation, stress adaptation, and signaling. CCDs are primarily associated with the formation of volatile compounds and SLs, whereas NCEDs are central to ABA biosynthesis (Imtiaz et al. 2023).

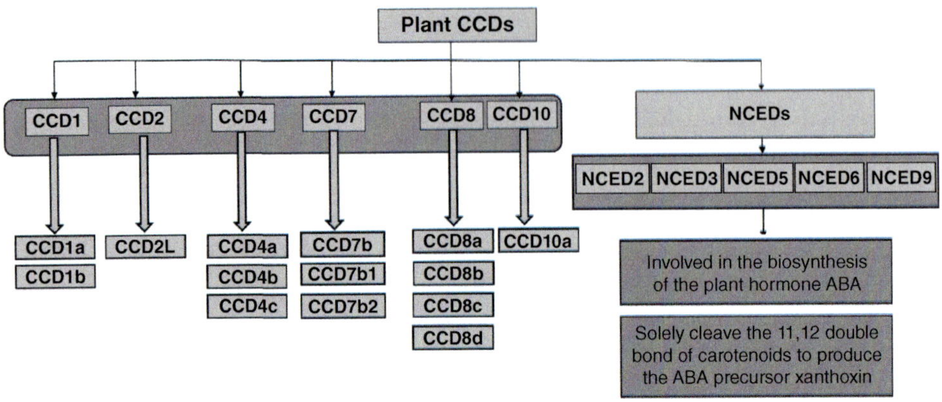

Figure 16.2 Classification of the *CCD* gene family in plants.

Thus, their functional divergence reflects the adaptive strategies plants have evolved to modulate growth, development, and stress responses in changing environments.

To date, a total of 158 *CCD* genes have been documented among multiple plant species (Yao et al. 2022). These genes exhibit varied exon–intron structures, chromosomal locations, and regulatory elements, which contribute to their differential expression patterns and biochemical specialization across taxa (Priya et al. 2019). Phylogenetic studies have categorized the plant CCD family into two main groups: Group I, which includes *CCD7* and *CCD8*, which have distinct evolutionary histories; and Group II consisting of CCD1, CCD4, CCD-like (CCDL), and *NCEDs*, which are presumed to share a common evolutionary origin (Su et al. 2021). This classification provides a foundational framework for further dissecting the molecular roles of individual *CCD* members across plant lineages.

16.5.2 Functional Roles of *NCED* Genes in ABA-mediated Development and Drought Response

In the model plant *Arabidopsis thaliana*, the *NCED* gene family comprises several members, including *NCED2*, *NCED3*, *NCED5*, *NCED6*, and *NCED9*, which have been classified based on phylogenetic analysis and their distinct functional roles (Tan et al. 2003). Among these, the *AtNCED3* isoform occupies a pivotal part in regulating the levels of the stress hormone ABA, particularly under drought stress conditions (Ren et al. 2007). Overexpression of *AtNCED3* has been observed to increase drought resilience in *Arabidopsis* through the reduction of leaf transpiration rates. Additionally, *AtNCED6* and *AtNCED9* are involved in the developmental regulation of ABA biosynthesis in seeds, while *AtNCED2*, *AtNCED3*, and *AtNCED5* contribute significantly to overall endogenous ABA accumulation (Lefebvre et al. 2006). The *NCED* gene family is also diversified in rice, with members like *OsNCED1*–*OsNCED5* displaying tissue-specific expression patterns and distinct functional roles (Haider et al. 2023). For instance, *OsNCED1* and *OsNCED3* are associated with drought resistance, while *OsNCED2* plays a part in delaying seed germination (Teng et al. 2014; Huang et al. 2018). Overexpression of *OsNCED5* and *OsNCED3* genes in the model plant *Arabidopsis* has been found to enhance tolerance to drought conditions, delay the onset of seed dormancy, and modulate plant architecture and leaf characteristics (Hwang et al. 2010; Huang et al. 2019).

In the economically important crop *C. sativus* (saffron), the *CstNCED* genes are closely linked to the accumulation of ABA under various stress conditions, including drought, salinity, and low temperatures (Ahrazem et al. 2012). These results underline the crucial role of NCED enzymes in ABA biosynthesis and their mediation of ABA-dependent signaling pathways, which are crucial for plant stress responses.

16.5.3 Multifunctional Roles of CCD1 in Volatile Apocarotenoid Biosynthesis, Defense Regulation, and Reproductive Development

Unlike NCEDs, which typically have specific cleavage sites, CCD enzymes exhibit a broader range of substrate specificities, allowing them to cleave various carotenoids and apocarotenoids (Figure 16.3). The CCD enzymes are further divided into CCD1, CCD4, CCD7, and CCD8 based on their distinct substrate preferences and cleavage sites (Dhar et al. 2020). The CCD1 enzyme possesses a high degree of versatility, enabling it to cleave a wide array of diverse carotenoids at multiple positions (9,10; 9,10,9′,10′; 5,6,5′,6′; or 7,8,7′,8′) (Vogel et al. 2008; Floss and Walter 2009). Being the only CCD that operates in the cytosol, researchers have suggested that it acts on C27 apocarotenoids derived from C40 carotenoids through a sequential cleavage mechanism where another CCD enzyme, like CCD4, carries out the initial cleavage in the plastid (Timmins et al. 2020). CCD1 has been implicated in the synthesis of volatile compounds such as β-ionone, which contributes to the scent and taste profiles of various plant species and is used in different industries (Wang et al. 2022). It has also been shown to cleave lycopene to produce 6-methyl-5-hepten-2-one, a significant volatile flavor compound in tomatoes (Ilg et al. 2014). Studies have proven that CCD1's role extends beyond flavor and aroma production to include significant functions in plant development and interaction with mycorrhizal fungi. In the mycorrhizal root system of legume *Medicago truncatula*, RNA interference (RNAi)-mediated repression of the *CCD1* gene (*MtCCD1*) has been shown to lead to the accumulation of C27 apocarotenoids (Floss et al. 2008). This finding underscores the enzyme's significant role in the metabolism of these compounds during symbiosis with arbuscular mycorrhizal fungi. CCD1 has

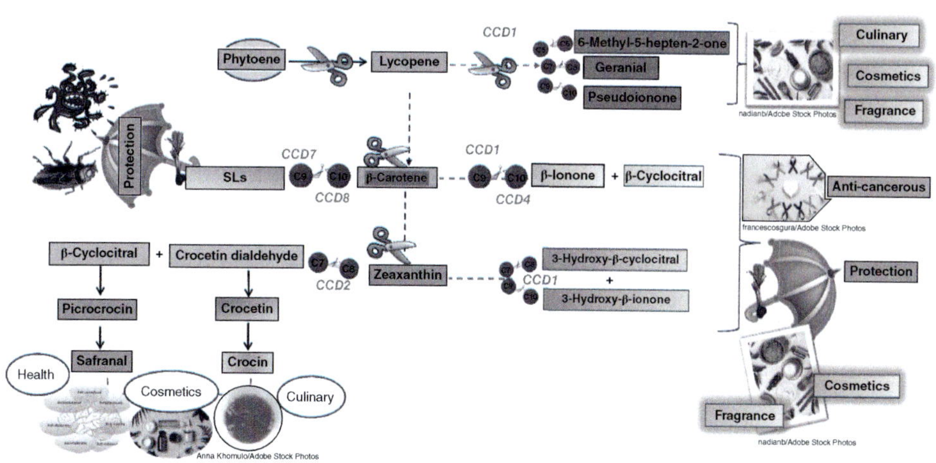

Figure 16.3 Overview of apocarotenoid biosynthesis mediated by CCDs and their downstream functional roles.

also been associated with the regulation of abiotic stress responses, with different expression patterns observed under various adverse environmental conditions. It has also been implicated in modulating plant immune responses beyond metabolic functions. A study in *Nicotiana edwardsonii* demonstrated that a *CCD1* gene introgressed from *Nicotiana bigelovii* selectively suppressed hypersensitive response (HR)-mediated cell death triggered by the P6 protein of *Cauliflower mosaic virus* (CaMV) strain W260, without compromising the overall resistance to the virus. This suggests a regulatory role of CCD1 in uncoupling localized cell death from pathogen resistance (HR) pathways (Cawly et al. 2005).

In *C. sativus*, two variants of *CstCCD1* enzyme (*CstCCD1a* and *CstCCD1b*) have been identified within the *Crocus* stigma. These isoforms have been observed to catalyze the 9,10 (9′,10′) oxidative cleavage of β-carotene, resulting in the production of β-ionone, which is regarded as an essential signaling molecule during abiotic stress (Rubio et al. 2008; Baba et al. 2015). However, as per studies, the expression of *CCD1* genes does not always correlate directly with the emission of apocarotenoid volatiles in the stigma tissue. This discrepancy may be due to differences in the subcellular localization of the enzyme, which affects substrate availability. It has been suggested that since the subcellular localization of this enzyme limits its access to carotenoids stored in plastids, CCD1's function may be more related to carotenoid turnover and response to stress (Ahrazem et al. 2016a). *CstCCD1a* has been shown to be constitutively expressed and may play a role in the rapid production of C13 cyclohexane apocarotenoids in response to local damage or stress, including the production of β-ionone, which has antifungal and antimicrobial activities (Rubio-Moraga et al. 2013). In contrast, the *CstCCD1b* isoform is specifically expressed in the stigma tissue, indicating it may have a more specialized function in governing stigma development and in the biosynthesis of apocarotenoids within this tissue (Rubio et al. 2008). Additionally, CCD1 enzymes may also contribute to attracting pollinators through the production of aromatic compounds (Shu et al. 2020).

16.5.4 CCD4 as a Key Mediator of Carotenoid Homeostasis and Environmental Stress Acclimation

In contrast to the broad substrate specificities of CCD1, the CCD4 subfamily exhibits a more restricted substrate range, characteristically cleaving carotenoids at the 9,10 and 9′,10′ double bonds or the 7′,8′ positions (Rubio-Moraga et al. 2014b). This targeted catalytic activity of CCD4 enzymes is crucial in determining the level of pigmentation in various plant tissues (Huang et al. 2009). *CCD4* genes have been shown to be well conserved across different plant species (Ahrazem et al. 2010). Different CCD4 alleles and isoforms have been identified, such as CCD4a–e, indicating evolutionary diversification and adaptation to specific physiological roles (Zheng et al. 2015). Higher expression of CCD4 has been inversely proven to correlate with β-carotene accumulation in banana, suggesting its role in carotenoid degradation (Awasthi et al. 2022). *CCD4* genes in cotton (*Gossypium* spp.) have been shown to be upregulated by various abiotic stresses such as heat, salinity, and ABA application, suggesting their role in stress acclimation (Zhang et al. 2021). In *C. sativus*, researchers have revealed that the expression of CCD4 is highly regulated by environmental factors such as light, heat, and various stresses besides being tissue-specific, with different CCD4 isoforms showing distinct patterns of regulation and functional roles in stress response and development (Baba et al. 2015). Furthermore, the overexpression of

CstCCD4b in *Arabidopsis* has been shown to enhance tolerance to abiotic stresses such as salinity, water deficiency, and oxidative damage by modulating the ROS machinery (Baba et al. 2015).

16.5.5 CCD7 and CCD8: Orchestrators of SL Production and Axillary Bud Suppression

CCD7 and CCD8 have been recognized as key players in the biosynthesis of a novel plant hormone that regulates axillary bud development and branching in plants (Rubio-Moraga et al. 2014a). Studies have revealed that these two enzymes act sequentially in the SL biosynthesis pathway, with CCD7 catalyzing the initial cleavage of carotenoids and CCD8 converting the intermediate into the SL precursor carlactone, an important signaling molecule involved in plant developmental processes (Auldridge et al. 2006b). The functions of CCD7 and CCD8 in controlling branch development are conserved across various plant species, including *Arabidopsis*, pea, *Petunia*, rice, and kiwifruit (Ledger et al. 2010). Reduction in the expression of CCD8 in *Petunia* led to increased branch development and delayed leaf senescence, indicating its role in inhibiting shoot branching (Snowden et al. 2005). Bulbous plants like *Crocus* rely heavily on vegetative propagation, and the control of dormancy is critical for improving crop yield and distribution. SLs are crucial in both bud development and inhibition of sprouting; however, the molecular and hormonal mechanisms governing bud sprouting in *C. sativus* are not fully understood (Umehara et al. 2015). However, a study by Rubio-Moraga et al. (2014a) highlights the role of these two key genes, *CstCCD7* and *CstCCD8*. They have revealed that these genes are expressed in various tissues of saffron, with *CstCCD8* showing higher levels in quiescent axillary buds, which decrease following the removal of the apical bud. In contrast, *CstCCD7* expression increases after decapitation, suggesting that *CstCCD8* primarily regulates SL production linked to bud dormancy. The study also suggests that these genes are involved in the development of vascular tissue and stigma in saffron, with SLs potentially playing a role in these processes, particularly under the control of *CstCCD8* (Rubio-Moraga et al. 2014a).

16.6 Recent Discoveries of CCD Variants and Their Lineage-specific Functional Adaptations

The classification of CCDs in plants is ever evolving, with ongoing research revealing new functions and regulatory mechanisms in various species. Some CCD members with novel functions have been discovered in other plants recently. In maize, *ZmCCD10a* has been identified (Zhong et al. 2020). *ZmCCD10a* expression has been shown to be related to phosphorus (Pi) absorption and allocation regulation, particularly under low-Pi conditions, which increases Pi accumulation in shoots and improves plant tolerance to Pi constraint (Luo et al. 2024). Research suggests that *ZmCCD10a* regulates Pi homeostasis via interactions with major Pi transporters and signaling pathways, presumably in an ARF19-dependent manner (Zhong et al. 2020). Wang et al. (2019) identified a *Zaxinone Synthase* (*ZAS*) gene from rice that acts as a regulatory metabolite, promoting the biosynthesis of SL and ABA in *Arabidopsis* roots. Votta et al. (2022) showed that rice lines overexpressing *ZAS* were generated, leading to higher endogenous zaxinone levels, improved root growth, more productive tillers, and a 30% increase in grain yield. This overexpression also resulted in lower SL levels, higher monosaccharide sugars, and enhanced nutrient uptake, particularly phosphate, improving tolerance to low-phosphate conditions (Votta et al. 2022). CCDL, a novel member of the CCD family, was also recently discovered in tomatoes, rice, watermelon, and melon (Cheng et al. 2022; Haider et al. 2023; Xue et al. 2023). In *C. sativus*, Frusciante and co-workers reported a novel *CCD2* gene responsible for converting zeaxanthin into crocetin (Frusciante et al. 2014). In spring *Crocus*, which accumulates crocins in the stigma and tepals, *CaCCD2* was identified, whose comparison with the closely related *CstCCD2* from *C. sativus* indicated that the previously reported *CstCCD2* was incomplete and hence a longer CCD2 version, named as *CstCCD2L*, was identified (Ahrazem et al. 2016b). Besides *Crocus*, researchers were able to detect safranal and crocin in *Freesia hybrida*, which prompted them to investigate the CCD family in *Freesia*. Phylogenetic investigation and biochemical analysis in *Freesia* led to the identification of *FhCCD2*, which catalyzes the cleavage of zeaxanthin at the 7,8 and 7′,8′ double bonds to generate intermediates prerequisite for the biosynthesis of safranal and crocin (Fang et al. 2020). Besides *Crocus* and *Freesia*, CCD2 enzyme has not been documented in any plant taxa to date.

More recently, another CCD subfamily member, *NtCCD10*, was characterized in *Nicotiana tabacum*, marking the first identification of CCD10 in a dicot (Li et al. 2022). *NtCCD10* was found to possess the canonical seven-bladed β-propeller structure with a central Fe^{2+} cofactor, yet it diverges structurally from CCD1 by containing a unique β-sheet in the α-helical region, non-antiparallel α1 and α3 helices, and a longer substrate access tunnel. Biochemically, *NtCCD10* exhibits the ability to cleave both C9—C10/C9′—C10′ and C5—C6/C5′—C6′ double bonds in carotenoids, producing compounds such as β-ionone and 6-methyl-5-hepten-2-one. Its higher catalytic activity compared to *PhCCD1* in heterologous expression systems suggests its potential as a more efficient enzyme for apocarotenoid biosynthesis. Notably, *NtCCD10* was localized to the cytoplasm, like CCD1, although other tobacco CCD10 paralogs showed predicted localizations in both chloroplasts and plasma membranes. This variation suggests potential sub-functionalization or diversification among CCD10 paralogs. These findings not only establish CCD10 as a functionally distinct lineage but also highlight its biotechnological promise in producing high-value volatiles.

16.7 Regulation of CCD Genes Under Stress Conditions

CCDs are homologous enzymes found in different plant species, but their expression patterns vary significantly across tissues and developmental stages. For instance, the expression of *MdCCD4b* is predominantly reported to be localized to the flowers of *Malus domestica*, however, it is not expressed in fruits or buds (Chen et al. 2018). In *Solanum lycopersicum*, the *CCD4b* gene is expressed across all tissues, with the highest levels observed in mature leaves and the lowest expression detected in fruit (Wei et al. 2016). Only the petals and fruit peel of *Citrus × sinensis* express *CsCCD4b1*. *CsCCD4b1* expression is reported to be the lowest during the green stage and reaches its peak in the late breaker stage, and then gradually declines as the fruit progresses toward ripening (Rodrigo et al. 2008). The CCD4 enzyme is predominantly expressed in leaves and flowers of potato plants while exhibiting significantly lower expression in stems, tubers, and roots (Campbell et al. 2010). The rose cultivar *Rosa × damascena* CCD4 is primarily expressed in flowers, with minimal expression in leaves, stems, and roots (Huang et al. 2009). *VvCCD4b* is extensively expressed in mature leaves, flowers, and ripe fruit and is upregulated by ABA (Meng et al. 2020). In *C. sativus*, *CstCCD1a* expression was discovered in all floral organs, with high levels in stigma and petals, while *CstCCD1b* expression was limited to the stigma. *CstCCD1a* expression was also found to be restricted to non-stressed leaves (Rubio et al. 2008). *CstCCD1* expression was also observed in corms undergoing pre-senescence and senescence in saffron (Rubio-Moraga et al. 2010). Studies revealed that *CstCCD4* was strongly expressed in the stigma and petal tissues of *C. sativus*, and cold and heat treatments led to an increase in *CstCCD4c* expression (Rubio-Moraga et al. 2014b). The expression of CCD2 was found to be restricted to stigma (Ahrazem et al. 2010).

The expression profiles of *CCD* genes indicate that they are transcriptionally regulated in response to various developmental and environmental cues. In plants, cis-regulatory and trans-regulatory elements primarily govern gene transcription. Cis-acting elements in the promoter region regulate temporal and geographical gene expression patterns, as well as how these patterns respond to stress. Trans-acting elements, particularly transcription factors, bind to cis-acting regulatory regions and control gene expression. A recent study in *C. × sinensis* discovered that *CsMADS6* increased *CsCCD1* expression (Lu et al. 2021). In *Osmanthus fragrans* Lour., *OfWRKY3* and *OfERF61* induce *OfCCD4* expression, leading to carotenoid cleavage and β-ionone production in sweet osmanthus petals (Han et al. 2019). *VvMADS4* was found to be a negative regulator of *VvCCD4b* expression (Meng et al. 2020). In sweet potato (*Ipomoea batatas*), *IbCCD4* regulation is controlled by *IbWRKY20* and *IbCBF2*, indicating a putative regulation mechanism for carotenoid accumulation via ABA signaling (Zhang et al. 2023). In *Chrysanthemum morifolium*, two MADS-box-containing transcription factors, namely APETALA3 (AP3) and ULT1 Interacting Factor 1 (UIF1), can potentially interact with the PISTILLATA (PI) protein to form the CmAP3–CmPI–CmUIF1 protein complex. This complex has been found to regulate carotenoid metabolism through direct activation of the expression of *CmCCD4a-2* (Lu et al. 2022).

16.8 Expanding Functional Horizons of CCDs

The expanding functional scope of CCDs highlights their multifaceted roles in plant development, stress adaptation, and ecological interactions. Beyond their established contributions to the biosynthesis of hormones such as ABA and SLs, CCDs catalyze the production of a wide range of volatile and nonvolatile apocarotenoids that participate in signaling, defense, and developmental regulation (Simkin 2021). Acting as metabolic gatekeepers, CCD enzymes influence diverse physiological processes including tissue pigmentation, aroma biosynthesis, root architecture, and pathogen response (Imtiaz et al. 2023). Subfamilies such as NCED, CCD1, CCD4, CCD7, CCD8, and newly described CCDL forms such as CCD2L and ZAS demonstrate considerable enzymatic and functional diversities across plant taxa, reflecting evolutionary adaptation to specific ecological niches and developmental cues (Tan et al. 2019; Ablazov et al. 2023).

In high-value crops like *C. sativus*, CCD2 and CCD4 have been functionally linked to the biosynthesis of crocins, picrocrocin, and β-ionone compounds that determine saffron's medicinal value and market quality (Baba et al. 2015; Xu et al. 2024). Advances in transcriptomics, metabolomics, and genome editing have accelerated the characterization of CCD isoforms and revealed their integration into broader metabolic and stress-response networks (Choi et al. 2025). Canonical members such as NCEDs are central to ABA-mediated drought tolerance, while CCD1 and CCD4 generate important aroma volatiles including β-ionone, α-ionone, and geranylacetone (Espasandin et al. 2014; Shi et al. 2020; Timmins et al. 2020). Recent discoveries, such as ZAS in *Oryza sativa*, further illustrate the regulatory depth of this enzyme family. ZAS cleaves apo-10′-zeaxanthinal to produce zaxinone, a C18 apocarotenoid that suppresses SL biosynthesis and enhances plant growth (Wang et al. 2019; Ablazov et al. 2023).

Interestingly, zaxinone exhibits species-specific effects. In *A. thaliana*, which lacks a ZAS ortholog, it instead promotes ABA and SL accumulation while inhibiting hypocotyl elongation, suggesting the existence of alternative biosynthetic or perception mechanisms (Ablazov et al. 2020). Likewise, *Zea mays* ZmCCD10a cleaves a broad array of carotenoids to yield volatiles such as β-ionone and geranylacetone (Zhong et al. 2020). These examples highlight the substrate promiscuity, catalytic flexibility, and functional specialization of different CCD enzymes.

Beyond developmental and metabolic roles, several CCD-derived apocarotenoids function in cellular signaling. For instance, β-cyclocitral—a product of enzymatic or oxidative carotenoid cleavage—acts as a retrograde signal during oxidative stress, modulating root growth and gene expression. Similarly, β-ionone serves as a defense-related signal, enhancing pathogen resistance independently of classical jasmonate or salicylate pathways. CCD-derived compounds also contribute to interspecific interactions, influencing microbial associations, rhizosphere composition, and parasitic plant germination. These findings establish CCDs not only as metabolic enzymes but also as integrators of environmental and developmental signaling.

Despite these advances, key gaps remain in our understanding of isoform-specific functions, subcellular targeting, and the regulation of CCD gene expression in diverse plant systems. With many CCD isoforms still uncharacterized, particularly in non-model species, further investigation is needed to resolve their roles within hormonal and metabolic networks. Integrating multi-omics platforms with synthetic biology and precision breeding offers exciting prospects for engineering crops with enhanced stress resilience and optimized apocarotenoid profiles. The discovery of zaxinone analogs such as MiZax compounds also presents new avenues for developing bio-stimulants and crop protection strategies (Jamil et al. 2023). Altogether, CCDs represent a rapidly expanding frontier in plant science, linking metabolic regulation with plant adaptability and agricultural innovation.

References

Ablazov, A., Mi, J., Jamil, M. et al. (2020). The apocarotenoid zaxinone is a positive regulator of strigolactone and abscisic acid biosynthesis in *Arabidopsis* roots. *Frontiers in Plant Science* 11, 578. https://doi.org/10.3389/fpls.2020.00578.

Ablazov, A., Votta, C., Fiorilli, V. et al. (2023). *ZAXINONE SYNTHASE 2* regulates growth and arbuscular mycorrhizal symbiosis in rice. *Plant Physiology* 191(1), 382–399. https://doi.org/10.1093/plphys/kiac472.

Ahrazem, O., Gómez-Gómez, L., Rodrigo, M.J. et al. (2016a). Carotenoid cleavage oxygenases from microbes and photosynthetic organisms: features and functions. *International Journal of Molecular Sciences* 17(11), 1781. https://doi.org/10.3390/ijms17111781.

Ahrazem, O., Rubio-Moraga, A., Berman, J. et al. (2016b). The carotenoid cleavage dioxygenase CCD2 catalysing the synthesis of crocetin in spring crocuses and saffron is a plastidial enzyme. *New Phytologist* 209(2), 650–663. https://doi.org/10.1111/nph.13609.

Ahrazem, O., Rubio-Moraga, A., Trapero, A. et al. (2012). Developmental and stress regulation of gene expression for a 9-*cis*-epoxycarotenoid dioxygenase, *CstNCED*, isolated from *Crocus sativus* stigmas. *Journal of Experimental Botany* 63(2), 681–694. https://doi.org/10.1093/jxb/err293.

Ahrazem, O., Trapero, A., Gómez, M.D. et al. (2010). Genomic analysis and gene structure of the plant carotenoid dioxygenase 4 family: a deeper study in *Crocus sativus* and its allies. *Genomics* 96(4), 239–250. https://doi.org/10.1016/j.ygeno.2010.07.003.

Akram, J., Siddique, R., Shafiq, M. et al. (2023). Genome-wide identification of *CCO* gene family in cucumber (*Cucumis sativus*) and its comparative analysis with *A. thaliana*. *BMC Plant Biology* 23(1), 640. https://doi.org/10.1186/s12870-023-04647-4.

Auldridge, M.E., Block, A., Vogel, J.T. et al. (2006a). Characterization of three members of the Arabidopsis carotenoid cleavage dioxygenase family demonstrates the divergent roles of this multifunctional enzyme family. *The Plant Journal* 45(6), 982–993. https://doi.org/10.1111/j.1365-313x.2006.02666.x.

Auldridge, M.E., McCarty, D.R. and Klee, H.J. (2006b). Plant carotenoid cleavage oxygenases and their apocarotenoid products. *Current Opinion in Plant Biology* 9(3), 315–321. https://doi.org/10.1016/j.pbi.2006.03.005.

Awasthi, P., Khan, S., Lakhani, H. et al. (2022). Transgene-free genome editing supports CCD4 role as a negative regulator of β-carotene in banana. *Journal of Experimental Botany* 73, 3401–3416. https://doi.org/10.1093/jxb/erac042.

Baba, S.A., Jain, D., Abbas, N. et al. (2015). Overexpression of *Crocus* carotenoid cleavage dioxygenase, *CsCCD4b*, in *Arabidopsis* imparts tolerance to dehydration, salt and oxidative stresses by modulating ROS machinery. *Journal of Plant Physiology* 189, 114–125. https://doi.org/10.1016/j.jplph.2015.11.001.

Bouvier, F., Suire, C., Mutterer, J. et al. (2003). Oxidative remodeling of chromoplast carotenoids: identification of the carotenoid dioxygenase *CsCCD* and *CsZCD* genes involved in *Crocus* secondary metabolite biogenesis. *The Plant Cell* 15(1), 47–62. https://doi.org/10.1105/tpc.006536.

Campbell, R., Ducreux, L.J., Morris, W.L. et al. (2010). The metabolic and developmental roles of carotenoid cleavage dioxygenase4 from potato. *Plant Physiology* 154(2), 656–664.

Carnà, M., Korwin Krukowski, P., Tosato, E. et al. (2025). Apocarotenoids as stress signaling molecules in plants. *Agriculture* 15(9), 926. https://doi.org/10.3390/agriculture15090926.

Cawly, J.D., Cole, A.B., Király, L. et al. (2005). The plant gene *CCD1* selectively blocks cell death during the hypersensitive response to *Cauliflower mosaic virus* infection. *Molecular Plant-Microbe Interactions* 18(3), 212–219. https://doi.org/10.1094/mpmi-18-0212.

Chen, H., Zuo, X., Shao, H. et al. (2018). Genome-wide analysis of carotenoid cleavage oxygenase genes and their responses to various phytohormones and abiotic stresses in apple (*Malus domestica*). *Plant Physiology and Biochemistry* 123, 81–93. https://doi.org/10.1016/j.plaphy.2017.12.001.

Chen, Y., Jiang, Z., Wu, S. et al. (2025). Structure and release function of fragrance glands. *Horticulture Research* 12(5), uhafo31. https://doi.org/10.1093/hr/uhaf031.

Cheng, D., Wang, Z., Li, S. et al. (2022). Genome-wide identification of *CCD* gene family in six cucurbitaceae species and its expression profiles in melon. *Genes* 13(2), 262. https://doi.org/10.3390/genes13020262.

Chernys, J.T., and Zeevaart, J.A.D. (2000). Characterization of the 9-cis-epoxycarotenoid dioxygenase gene family and the regulation of abscisic acid biosynthesis in avocado. *Plant Physiology* 124(1), 343–354. https://doi.org/10.1104/pp.124.1.343.

Choi, H., Yi, T.G., Gho, Y.S. et al. (2025). Augmenting carotenoid accumulation by multiplex genome editing of the redundant CCD family in rice. *Plant Physiology and Biochemistry* 225, 110008. https://doi.org/10.1016/j.plaphy.2025.110008.

Daruwalla, A. and Kiser, P.D. (2020). Structural and mechanistic aspects of carotenoid cleavage dioxygenases (CCDs). *Biochimica et Biophysica Acta (BBA)—Molecular and Cell Biology of Lipids* 1865(11), 158590. https://doi.org/10.1016/j.bbalip.2019.158590.

Daruwalla, A. Zhang, J., Lee, H.J. et al. (2020). Structural basis for carotenoid cleavage by an archaeal carotenoid dioxygenase. *Proceedings of the National Academy of Sciences of the United States of America* 117(33), 19914–19925. https://doi.org/10.1073/pnas.2004116117.

Dhar, M.K., Mishra, S., Bhat, A. et al. (2020). Plant carotenoid cleavage oxygenases: structure–function relationships and role in development and metabolism. *Briefings in Functional Genomics* 19(1), 1–9. https://doi.org/10.1093/bfgp/elz037.

Dzib-Cauich, J., Us-Camas, R. and Rivera-Madrid, R. (2020). Natural sources of apocarotenoids and their applications. In: *Biology, Chemistry and Applications of Apocarotenoids* (eds. S. Ramamoorthy, R.R. Madrid and C.G.P. Doss), 11–39. CRC Press. https://doi.org/10.1201/9780429344206-2.

Espasandin, F.D., Maiale, S.J., Calzadilla, P. et al. (2014). Transcriptional regulation of 9-*cis*-epoxycarotenoid dioxygenase (NCED) gene by putrescine accumulation positively modulates ABA synthesis and drought tolerance in *Lotus tenuis* plants. *Plant Physiology and Biochemistry* 76, 29–35. https://doi.org/10.1016/j.plaphy.2013.12.018.

Fang, Q., Li, Y., Liu, B. et al. (2020). Cloning and functional characterization of a carotenoid cleavage dioxygenase 2 gene in safranal and crocin biosynthesis from *Freesia hybrida. Plant Physiology and Biochemistry* 154, 439–450. https://doi.org/10.1016/j.plaphy.2020.06.035.

Felemban, A., Braguy, J., Zurbriggen, M.D. et al. (2019). Apocarotenoids involved in plant development and stress response. *Frontiers in Plant Science* 10, 1168. https://doi.org/10.3389/fpls.2019.01168.

Floss, D.S., Schliemann, W., Schmidt, J. et al. (2008). RNA interference-mediated repression of *MtCCD1* in mycorrhizal roots of *Medicago truncatula* causes accumulation of C_{27} apocarotenoids, shedding light on the functional role of CCD1. *Plant Physiology* 148(3), 1267–1282. https://doi.org/10.1104/pp.108.125062.

Floss, D.S. and Walter, M.H. (2009). Role of carotenoid cleavage dioxygenase 1 (CCD1) in apocarotenoid biogenesis revisited. *Plant Signaling & Behavior* 4(3), 172–175. https://doi.org/10.4161/psb.4.3.7840.

Frusciante, S., Diretto, G., Bruno, M. et al. (2014). Novel carotenoid cleavage dioxygenase catalyzes the first dedicated step in saffron crocin biosynthesis. *Proceedings of the National Academy of Sciences of the United States of America* 111(33), 12246–12251. https://doi.org/10.1073/pnas.1404629111.

Gómez-Sagasti, M.T., López-Pozo, M., Artetxe, U. et al. (2023). Carotenoids and their derivatives: a "Swiss Army knife-like" multifunctional tool for fine-tuning plant-environment interactions. *Environmental and Experimental Botany* 207, 105229. https://doi.org/10.1016/j.envexpbot.2023.105229.

Haider, M.Z., Sami, A., Shafiq, M. et al. (2023). Genome-wide identification and in-silico expression analysis of carotenoid cleavage oxygenases gene family in *Oryza sativa* (rice) in response to abiotic stress. *Frontiers in Plant Science* 14, 1269995. https://doi.org/10.3389/fpls.2023.1269995.

Han, Y., Wang, H., Wang, X. et al. (2019). Mechanism of floral scent production in *Osmanthus fragrans* and the production and regulation of its key floral constituents, β-ionone and linalool. *Horticulture Research* 6(1), 106. https://doi.org/10.1038/s41438-019-0189-4.

Harrison, P.J. and Bugg, T.D.H. (2014). Enzymology of the carotenoid cleavage dioxygenases: reaction mechanisms, inhibition and biochemical roles. *Archives of Biochemistry and Biophysics* 544, 105–111. https://doi.org/10.1016/j.abb.2013.10.005.

Huang, F., Molnár, P. and Schwab, W. (2009). Cloning and functional characterization of carotenoid cleavage dioxygenase 4 genes. *Journal of Experimental Botany* 60(11), 3011–3022. https://doi.org/10.1093/jxb/erp137.

Huang, Y., Guo, Y., Liu, Y. et al. (2018). 9-*cis*-Epoxycarotenoid dioxygenase 3 regulates plant growth and enhances multi-abiotic stress tolerance in rice. *Frontiers in Plant Science* 9, 162. https://doi.org/10.3389/fpls.2018.00162.

Huang, Y., Jiao, Y., Xie, N. et al. (2019). *OsNCED5*, a 9-*cis*-epoxycarotenoid dioxygenase gene, regulates salt and water stress tolerance and leaf senescence in rice. *Plant Science* 287, 110188. https://doi.org/10.1016/j.plantsci.2019.110188.

Hwang, S.G., Chen, H.C., Huang, W.Y. et al. (2010). Ectopic expression of rice *OsNCED3* in Arabidopsis increases ABA level and alters leaf morphology. *Plant Science* 178(1), 12–22. https://doi.org/10.1016/j.plantsci.2009.09.014.

Ilg, A., Bruno, M., Beyer, P. et al. (2014). Tomato carotenoid cleavage dioxygenases 1A and 1B: relaxed double bond specificity leads to a plenitude of dialdehydes, mono-apocarotenoids and isoprenoid volatiles. *FEBS Open Bio* 4(1), 584–593. https://doi.org/10.1016/j.fob.2014.06.005.

Imtiaz, H., Arif, Y., Alam, P. et al. (2023). Apocarotenoids biosynthesis, signaling regulation, crosstalk with phytohormone, and its role in stress tolerance. *Environmental and Experimental Botany* 210, 105337. https://doi.org/10.1016/j.envexpbot.2023.105337.

Iuchi, S., Kobayashi, M., Yamaguchi-Shinozaki, K. et al. (2000). A stress-inducible gene for 9-cis-epoxycarotenoid dioxygenase involved in abscisic acid biosynthesis under water stress in drought-tolerant cowpea. *Plant Physiology* 123(2), 553–562. https://doi.org/10.1104/pp.123.2.553.

Jamil, M., Lin, P.Y., Berqdar, L. et al. (2023). New series of zaxinone mimics (MiZax) for fundamental and applied research. *Biomolecules* 13(8), 1206. https://doi.org/10.3390/biom13081206.

Kloer, D.P. and Schulz, G.E. (2006). Structural and biological aspects of carotenoid cleavage. *Cellular and Molecular Life Sciences* 63 (19–20), 2291–2303. https://doi.org/10.1007/s00018-006-6176-6.

Ledger, S., Janssen, B.J., Karunairetnam, S. et al. (2010). Modified *CAROTENOID CLEAVAGE DIOXYGENASE8* expression correlates with altered branching in kiwifruit (*Actinidia chinensis*). *New Phytologist* 188(3), 803–813. https://doi.org/10.1111/j.1469-8137.2010.03394.x.

Lefebvre, V., North, H., Frey, A. et al. (2006). Functional analysis of Arabidopsis *NCED6* and *NCED9* genes indicates that ABA synthesized in the endosperm is involved in the induction of seed dormancy. *The Plant Journal* 45(3), 309–319. https://doi.org/10.1111/j.1365-313x.2005.02622.x.

Leroux, J., Truong, T.T., Pogson, B.J. et al. (2022). Detection and analysis of novel and known plant volatile apocarotenoids. In: *Methods in Enzymology: Volume 670* (ed. E.T. Wurtzel), 311–368. Academic Press. https://doi.org/10.1016/bs.mie.2022.03.020.

Li, F., Gong, X., Liang, Y. et al. (2022). Characteristics of a new carotenoid cleavage dioxygenase NtCCD10 derived from *Nicotiana tabacum*. *Planta* 256(5), 100. https://doi.org/10.1007/s00425-022-04013-y.

Liang, M., Zhu, J. and Jiang, J. (2018). Carotenoids biosynthesis and cleavage related genes from bacteria to plants. *Critical Reviews in Food Science and Nutrition* 58(14), 2314–2333. https://doi.org/10.1080/10408398.2017.1322552.

Liang, M.H., He, Y.J., Liu, D.M. et al. (2021). Regulation of carotenoid degradation and production of apocarotenoids in natural and engineered organisms. *Critical Reviews in Biotechnology* 41(4), 513–534. https://doi.org/10.1080/07388551.2021.1873242.

Lu, C., Qu, J., Deng, C. et al. (2022). The transcription factor complex CmAP3-CmPI-CmUIF1 modulates carotenoid metabolism by directly regulating the carotenogenic gene *CmCCD4a-2* in chrysanthemum. *Horticulture Research* 9, uhac020. https://doi.org/10.1093/hr/uhac020.

Lu, S., Ye, J., Zhu, K. et al. (2021). A fruit ripening-associated transcription factor CsMADS5 positively regulates carotenoid biosynthesis in citrus. *Journal of Experimental Botany* 72(8), 3028–3043.

Luo, B., Zhang, G., Yu, T. et al. (2024). Genome-wide association studies dissect low-phosphorus stress response genes underling field and seedling traits in maize. *Theoretical and Applied Genetics* 137(7), 172. https://doi.org/10.1007/s00122-024-04681-2.

Meng, N., Wei, Y., Gao, Y. et al. (2020). Characterization of transcriptional expression and regulation of carotenoid cleavage dioxygenase 4b in grapes. *Frontiers in Plant Science* 11, 483. https://doi.org/10.3389/fpls.2020.00483.

Messing, S., Gabelli, S. B., Echeverria, I. et al. (2010). Structural insights into maize viviparous14, a key enzyme in the biosynthesis of the phytohormone abscisic acid. *The Plant Cell* 22(9), 2970–2980. https://doi.org/10.1105/tpc.110.074815.

Moreno, J.C., Mi, J., Alagoz, Y. et al. (2021). Plant apocarotenoids: from retrograde signaling to interspecific communication. *The Plant Journal* 105(2), 351–375. https://doi.org/10.1111/tpj.15102.

Ohmiya, A. (2009). Carotenoid cleavage dioxygenases and their apocarotenoid products in plants. *Plant Biotechnology* 26(4), 351–358. https://doi.org/10.5511/plantbiotechnology.26.351.

Ohmiya, A., Kato, M., Shimada, T. et al. (2019). Molecular basis of carotenoid accumulation in horticultural crops. *The Horticulture Journal* 88(2), 135–149. https://doi.org/10.2503/hortj.utd-r003.

Paparella, A., Shaltiel-Harpaza, L. and Ibdah, M. (2021). β-Ionone: its occurrence and biological function and metabolic engineering. *Plants* 10(4), 754. https://doi.org/10.3390/plants10040754.

Priya, R., Sneha, P., Dass, J.F.P. et al. (2019). Exploring the codon patterns between *CCD* and *NCED* genes among different plant species. *Computers in Biology and Medicine* 114, 103449. https://doi.org/10.1016/j.compbiomed.2019.103449.

Priya, R., Sneha, P., Rivera Madrid, R. et al. (2017). Molecular modeling and dynamic simulation of *Arabidopsis thaliana* carotenoid cleavage dioxygenase gene: a comparison with *Bixa orellana* and *Crocus sativus*. *Journal of Cellular Biochemistry* 118(9), 2712–2721. https://doi.org/10.1002/jcb.25919.

Qin, X. and Zeevaart, J.A.D. (1999). The 9-*cis*-epoxycarotenoid cleavage reaction is the key regulatory step of abscisic acid biosynthesis in water-stressed bean. *Proceedings of the National Academy of Sciences of the United States of America* 96(26), 15354–15361. https://doi.org/10.1073/pnas.96.26.15354.

Ren, H., Fan, Y., Gao, Z. et al. (2007). Roles of a sustained activation of *NCED3* and the synergistic regulation of ABA biosynthesis and catabolism in ABA signal production in *Arabidopsis. Chinese Science Bulletin* 52(4), 484–491. https://doi.org/10.1007/s11434-007-0072-9.

Rodrigo, A., Bertels, F., Heled, J. et al. (2008). The perils of plenty: what are we going to do with all these genes. *Philosophical Transactions of the Royal Society B: Biological Sciences* 363(1512), 3893–3902.

Rodríguez-Ávila, N.L., Narváez-Zapata, J.A., Aguilar-Espinosa, M. et al. (2011). Regulation of pigment-related genes during flower and fruit development of *Bixa orellana. Plant Molecular Biology Reporter* 29(1), 43–50. https://doi.org/10.1007/s11105-010-0207-z.

Rubio, Á., Rambla, J.L., Santaella, M et al. (2008). Cytosolic and plastoglobule-targeted carotenoid dioxygenases from *Crocus sativus* are both involved in β-ionone release. *Journal of Biological Chemistry* 283(36), 24816–24825. https://doi.org/10.1074/jbc.m804000200.

Rubio-Moraga, Á., Ahrazem, O. Pérez-Clemente, R.M. et al. (2014a). Apical dominance in saffron and the involvement of the branching enzymes CCD7 and CCD8 in the control of bud sprouting. *BMC Plant Biology* 14(1), 171. https://doi.org/10.1186/1471-2229-14-171.

Rubio-Moraga, Á., Ahrazem, O., Rambla, J. L. et al. (2013). Crocins with high levels of sugar conjugation contribute to the yellow colours of early-spring flowering crocus tepals. *PLoS ONE* 8(9), e71946. https://doi.org/10.1371/journal.pone.0071946.

Rubio-Moraga, Á., Rambla, J.L. Fernández-de-Carmen, A. et al. (2014b). New target carotenoids for CCD4 enzymes are revealed with the characterization of a novel stress-induced carotenoid cleavage dioxygenase gene from *Crocus sativus. Plant Molecular Biology* 86, 555–569. https://doi.org/10.1007/s11103-014-0250-5.

Rubio-Moraga, Á., Trapero, A., Ahrazem, O. et al. (2010). Crocins transport in *Crocus sativus*: the long road from a senescent stigma to a newborn corm. *Phytochemistry* 71(13), 1506–1513. https://doi.org/10.1016/j.phytochem.2010.05.026.

Schwartz, S.H., Tan, B.C., Gage, D.A. et al. (1997). Specific oxidative cleavage of carotenoids by VP14 of maize. *Science* 276(5320), 1872–1874. https://doi.org/10.1126/science.276.5320.1872.

Schwartz, S.H., Tan, B.C., McCarty, D.R. et al. (2003). Substrate specificity and kinetics for VP14, a carotenoid cleavage dioxygenase in the ABA biosynthetic pathway. *Biochimica et Biophysica Acta (BBA)—General Subjects* 1619(1), 9–14. https://doi.org/10.1016/s0304-4165(02)00422-1.

Shi, J., Cao, C., Xu, J. et al. (2020). Research advances on biosynthesis, regulation, and biological activities of apocarotenoid aroma in horticultural plants. *Journal of Chemistry* 2020, 2526956. https://doi.org/10.1155/2020/2526956.

Shu, H., Guo, S. and Ni, W. (2020). Comprehensive transcriptome analyses of different *Crocus* flower tissues uncover genes involved in crocin biosynthesis. *Biologia Plantarum* 64, 504–511. https://doi.org/10.32615/bp.2020.068.

Sierra, J., McQuinn, R.P. and León, P. (2022). The role of carotenoids as a source of retrograde signals: impact on plant development and stress responses. *Journal of Experimental Botany* 73(21), 7139–7154. https://doi.org/10.1093/jxb/erac292.

Simkin, A.J. (2021). Carotenoids and apocarotenoids in planta: their role in plant development, contribution to the flavour and aroma of fruits and flowers, and their nutraceutical benefits. *Plants* 10(11), 2321. https://doi.org/10.3390/plants10112321.

Simkin, A.J., Schwartz, S.H. Auldridge, M. et al. (2004a). The tomato *carotenoid cleavage dioxygenase 1* genes contribute to the formation of the flavor volatiles β-ionone, pseudoionone, and geranylacetone. *The Plant Journal* 40(6), 882–892. https://doi.org/10.1111/j.1365-313x.2004.02263.x.

Simkin, A.J., Underwood, B.A. Auldridge, M. et al. (2004b). Circadian regulation of the PhCCD1 carotenoid cleavage dioxygenase controls emission of β-ionone, a fragrance volatile of petunia flowers. *Plant Physiology* 136(3), 3504–3514. https://doi.org/10.1104/pp.104.049718.

Snowden, K.C., Simkin, A.J., Janssen, B.J. et al. (2005). The *Decreased apical dominance1/Petunia hybrida CAROTENOID CLEAVAGE DIOXYGENASE8* gene affects branch production and plays a role in leaf senescence, root growth, and flower development. *The Plant Cell* 17(3), 746–759. https://doi.org/10.1105/tpc.104.027714.

Su, W., Zhang, C., Feng, J. et al. (2021). Genome-wide identification, characterization and expression analysis of the carotenoid cleavage oxygenase (*CCO*) gene family in *Saccharum*. *Plant Physiology and Biochemistry* 162, 196–210. https://doi.org/10.1016/j.plaphy.2021.02.041.

Sui, X., Kiser, P.D., von Lintig, J. et al. (2013). Structural basis of carotenoid cleavage: from bacteria to mammals. *Archives of Biochemistry and Biophysics* 539(2), 203–213. https://doi.org/10.1016/j.abb.2013.06.012.

Sui, X., Zhang, J., Golczak, M. et al. (2016). Key residues for catalytic function and metal coordination in a carotenoid cleavage dioxygenase. *Journal of Biological Chemistry* 291(37), 19401–19412. https://doi.org/10.1074/jbc.m116.744912.

Sun, T., Rao, S., Zhou, X. et al. (2022). Plant carotenoids: recent advances and future perspectives. *Molecular Horticulture* 2(1), 3. https://doi.org/10.1186/s43897-022-00023-2.

Tan, B., Joseph, L.M., Deng, W.T. et al. (2003). Molecular characterization of the *Arabidopsis* 9-*cis* epoxycarotenoid dioxygenase gene family. *The Plant Journal* 35(1), 44–56. https://doi.org/10.1046/j.1365-313x.2003.01786.x.

Tan, H., Chen, X., Liang, N. et al. (2019). Transcriptome analysis reveals novel enzymes for apo-carotenoid biosynthesis in saffron and allows construction of a pathway for crocetin synthesis in yeast. *Journal of Experimental Botany* 70(18), 4819–4834. https://doi.org/10.1093/jxb/erz211.

Teng, K., Li, J., Liu, L. et al. (2014). Exogenous ABA induces drought tolerance in upland rice: the role of chloroplast and ABA biosynthesis-related gene expression on photosystem II during PEG stress. *Acta Physiologiae Plantarum* 36(8), 2219–2227. https://doi.org/10.1007/s11738-014-1599-4.

Timmins, J.J., Kroukamp, H., Paulsen, I.T. et al. (2020). The sensory significance of apocarotenoids in wine: importance of carotenoid cleavage dioxygenase 1 (CCD1) in the production of β-ionone. *Molecules* 25(12), 2779. https://doi.org/10.3390/molecules25122779.

Umehara, M., Cao, M., Akiyama, K. et al. (2015). Structural requirements of strigolactones for shoot branching inhibition in rice and Arabidopsis. *Plant and Cell Physiology* 56(6), 1059–1072. https://doi.org/10.1093/pcp/pcv028.

Vogel, J.T., Tan, B.C., McCarty, D.R. et al. (2008). The carotenoid cleavage dioxygenase 1 enzyme has broad substrate specificity, cleaving multiple carotenoids at two different bond positions. *Journal of Biological Chemistry* 283(17), 11364–11373. https://doi.org/10.1074/jbc.m710106200.

Votta, C., Fiorilli, V., Haider, I. et al. (2022). Zaxinone synthase controls arbuscular mycorrhizal colonization level in rice. *The Plant Journal* 111(6), 1688–1700. https://doi.org/10.1111/tpj.15917.

Wang, J.Y., Haider, I., Jamil, M. et al. (2019). The apocarotenoid metabolite zaxinone regulates growth and strigolactone biosynthesis in rice. *Nature Communications* 10(1), 810. https://doi.org/10.1038/s41467-019-08461-1.

Wang, Y., Xu, J. and Liu, A. (2022). Identification of the carotenoid cleavage dioxygenase genes and functional analysis reveal *DoCCD1* is potentially involved in beta-ionone formation in *Dendrobium officinale*. *Frontiers in Plant Science* 13, 967819. https://doi.org/10.3389/fpls.2022.967819.

Wei, Y., Wan, H., Wu, Z. et al. (2016). A comprehensive analysis of carotenoid cleavage dioxygenases genes in *Solanum lycopersicum*. *Plant Molecular Biology Reporter* 34, 512–523.

Xu, Z., Chen, S., Wang, Y. et al. (2024). *Crocus* genome reveals the evolutionary origin of crocin biosynthesis. *Acta Pharmaceutica Sinica B* 14(4), 1878–1891. https://doi.org/10.1016/j.apsb.2023.12.013.

Xue, G., Hu, L., Zhu, L. et al. (2023). Genome-wide identification and expression analysis of *CCO* gene family in *Liriodendron chinense*. *Plants* 12(10), 1975. https://doi.org/10.3390/plants12101975.

Yao, Y., Jia, L., Cheng, Y. et al. (2022). Evolutionary origin of the carotenoid cleavage oxygenase family in plants and expression of pepper genes in response to abiotic stresses. *Frontiers in Plant Science* 12, 792832. https://doi.org/10.3389/fpls.2021.792832.

Zhang, J., He, L., Dong, J. et al. (2023). Integrated metabolic and transcriptional analysis reveals the role of carotenoid cleavage dioxygenase 4 (IbCCD4) in carotenoid accumulation in sweetpotato tuberous roots. *Biotechnology for Biofuels and Bioproducts* 16(1), 45. https://doi.org/10.1186/s13068-023-02299-y.

Zhang, S., Guo, Y., Zhang, Y. et al. (2021). Genome-wide identification, characterization and expression profiles of the *CCD* gene family in *Gossypium* species. *3 Biotech* 11(5), 249. https://doi.org/10.1007/s13205-021-02805-9.

Zheng, X., Xie, Z., Zhu, K. et al. (2015). Isolation and characterization of *carotenoid cleavage dioxygenase 4* genes from different citrus species. *Molecular Genetics and Genomics* 290(4), 1589–1603. https://doi.org/10.1007/s00438-015-1016-8.

Zhong, Y., Pan, X., Wang, R. et al. (2020). *ZmCCD10a* encodes a distinct type of carotenoid cleavage dioxygenase and enhances plant tolerance to low phosphate. *Plant Physiology* 184(1), 374–392. https://doi.org/10.1104/pp.20.00378.

Zia-Ul-Haq, M. (2021). Past, present and future of carotenoids research. In: *Carotenoids: Structure and Function in the Human Body* (eds. M. Zia-Ul-Haq, S. Dewanjee and M. Riaz), 827–854. Springer. https://doi.org/10.1007/978-3-030-46459-2_28.

17

Impact of AM Symbiosis on Apocarotenoid Production in Plants

Ishrat Mehmood[1,], Abdullah[2], Kaiser Iqbal Wani[3,5], Muhammad Aamir Manzoor[3,5], Tariq Aftab[2], and Nayaab Shahir Pandit[4]*

[1] *Department of Biophysics, Panjab University, Chandigarh, India*
[2] *Department of Botany, Aligarh Muslim University, Aligarh, Uttar Pradesh, India*
[3] *Key Laboratory for Quality Regulation of Tropical Horticultural Crops of Hainan Province, School of Tropical Agriculture and Forestry, Hainan University, Haikou, Hainan, China*
[4] *School of Computational and Integrative Sciences (SC&IS), Jawaharlal Nehru University, New Delhi, India*
[5] *National Key Laboratory for Tropical Crop Breeding, School of Breeding and Multiplication (Sanya Institute of Breeding and Multiplication), Hainan University, Sanya, Hainan, China*

*Corresponding author: ishratmehmood943@gmail.com

17.1 Introduction

The intricate relationships between plants and soil microorganisms have long fascinated biologists, with arbuscular mycorrhizal (AM) fungi emerging as pivotal players in enhancing plant health, nutrient acquisition, and stress resilience. These ancient symbionts, belonging to the phylum *Glomeromycota*, form mutualistic associations with over 80% of terrestrial plant species, facilitating bidirectional nutrient exchange and modulating physiological processes critical for survival (Smith and Read 2008). While the benefits of AM symbiosis in improving phosphorus (P) uptake and drought tolerance are well documented, emerging research highlights its profound influence on plant secondary metabolism, particularly the production of apocarotenoids—a class of bioactive compounds derived from carotenoid precursors. This chapter will explore the dynamic interplay between AM symbiosis and apocarotenoid biosynthesis, unraveling how these fungi orchestrate molecular and biochemical pathways to enhance plant adaptation to environmental challenges.

AM fungi colonize plant root cortical cells, forming intricate structures such as arbuscules and vesicles that serve as hubs for nutrient exchange. By extending their hyphal networks into the soil, AM fungi amplify the root system's absorptive capacity, delivering essential minerals like P and nitrogen (N) to the host plant in exchange for photosynthetic carbon (Bonfante and Genre 2010). Beyond nutrient transfer, AM symbiosis primes plants

Plant Apocarotenoids: Biosynthesis, Signaling and Agricultural Applications, First Edition. Edited by Kaiser Iqbal Wani and Tariq Aftab.
Published 2026 by John Wiley & Sons Ltd.

for systemic resistance against biotic and abiotic stresses, including pathogen attacks, salinity, and drought (Pozo and Azcón-Aguilar 2007). This priming effect is mediated through complex signaling cascades involving phytohormones, reactive oxygen species (ROS), and secondary metabolites, positioning AM fungi as master regulators of plant metabolic plasticity. Carotenoids, the vibrant lipid-soluble pigments responsible for the red, orange, and yellow hues in plants, are synthesized in plastids via the methylerythritol phosphate (MEP) pathway. These compounds, including β-carotene and lutein, not only are critical for photosynthesis and photoprotection but also serve as precursors for apocarotenoids—a diverse group of metabolites generated through enzymatic cleavage (Walter and Strack 2011). Carotenoid cleavage dioxygenases (CCDs) catalyze the oxidative cleavage of carotenoids, yielding apocarotenoids such as strigolactones (SLs), abscisic acid (ABA), and volatile β-ionone (Hou et al. 2016a). These derivatives play multifaceted roles in plant development and stress responses: SLs regulate root architecture and attract AM fungi, ABA governs stomatal closure during drought, and β-cyclocitral signals oxidative stress (Al-Babili and Bouwmeester 2015; Dickinson et al. 2019).

The intersection of AM symbiosis and apocarotenoid production represents a fascinating mutualistic dialog. SLs, exuded by plant roots, function as rhizospheric signals that stimulate hyphal branching in AM fungi, initiating colonization (Akiyama et al. 2005). Intriguingly, AM symbiosis reciprocally modulates carotenoid biosynthesis pathways, enhancing substrate availability for apocarotenoid production. For instance, mycorrhizal colonization upregulates genes encoding phytoene synthase (PSY) and CCDs, amplifying the synthesis of SLs and ABA (Walter and Strack 2010a; López-Ráez et al. 2017). This feedback loop not only reinforces fungal recruitment but also primes plants for stress adaptation, as apocarotenoids like ABA synergize with AM-induced antioxidants to mitigate oxidative damage (Ruiz-Lozano et al. 2016). At the molecular level, AM fungi orchestrate apocarotenoid dynamics through transcriptional reprogramming and hormonal crosstalk. Colonization activates symbiotic signaling pathways, including the conserved symbiosis pathway, which triggers calcium-dependent gene expression (Gutjahr and Parniske 2013). Transcriptomic studies reveal that AM symbiosis upregulates carotenoid biosynthetic genes (e.g., *PSY*, *LCYB*) while modulating *CCD* expression to balance SL and ABA synthesis (Fiorilli et al. 2018). Hormonal interplay further fine-tunes this process: jasmonic acid (JA) enhances SL production to sustain fungal symbiosis, while ABA primes systemic drought responses without antagonizing AM colonization (Wasternack and Hause 2013; de Vries et al. 2020). Such crosstalk underscores the evolutionary optimization of apocarotenoid pathways to meet the dual demands of symbiosis and stress resilience.

Under environmental duress, the AM fungi–apocarotenoid alliance becomes indispensable. Drought-stressed mycorrhizal plants exhibit elevated ABA levels, which coordinate stomatal closure and osmotic adjustment, while SLs enhance soil moisture foraging via extended hyphal networks (Aroca et al. 2013). Similarly, under nutrient deficiency, AM fungi induce apocarotenoid-mediated root remodeling, optimizing resource acquisition. Pathogen challenge elicits β-cyclocitral emissions, which deter herbivores and recruit beneficial microbes (Jia et al. 2019). These adaptive responses highlight the dual role of AM-induced apocarotenoids as both endogenous regulators and ecological mediators.

The ecological ramifications of AM fungi–apocarotenoid interactions extend to ecosystem stability and crop productivity. In natural ecosystems, AM fungi enhance plant diversity by modulating apocarotenoid-driven allelopathy and resource partitioning (Bennett and Groten 2022). In agriculture, AM inoculation offers a sustainable strategy to bolster crop resilience, reducing reliance on synthetic fertilizers and pesticides. For example, AM-enhanced SL production improves P uptake in rice, while ABA-mediated drought tolerance benefits maize in arid regions (Rillig et al. 2019; Thirkell et al. 2020). However, challenges remain in optimizing AM fungi–plant combinations for diverse agroecosystems and understanding genotype-specific responses. This chapter aims to illuminate the untapped potential of AM fungi–apocarotenoid networks in shaping plant health and ecosystem resilience.

17.2 Unveiling the Symbiotic Relationship: AM Fungi and Plant Health

AM fungi, belonging to the phylum *Glomeromycota*, form one of the most ancient and widespread symbiotic relationships with terrestrial plants, colonizing the roots of over 80% of plant species (Smith and Read 2008). This mutualistic association is pivotal for plant health, enhancing nutrient acquisition, stress resilience, and disease resistance. AM fungi colonize plant root cortical cells, forming specialized structures called arbuscules that facilitate nutrient exchange. Fungi enhance the host's access to immobile soil nutrients, particularly P and N, while receiving photosynthetic carbohydrates in return (Smith and Read 2008). Beyond nutrient uptake, AM symbiosis bolsters plant health by improving water-use efficiency, mitigating oxidative stress, and priming systemic defenses against pathogens and herbivores (Evelin et al. 2009). These benefits are partly mediated by fungal-induced changes in plant secondary metabolism, including the production of apocarotenoids. In recent years, research has increasingly highlighted the role of AM symbiosis in modulating the production of apocarotenoids—a class of bioactive metabolites derived from carotenoid cleavage—that influence plant development, stress responses, and communication. This chapter explores the mechanisms by which AM fungi regulate apocarotenoid biosynthesis and the implications for plant health.

17.3 Apocarotenoids: Multifunctional Metabolites in Plant Physiology

Apocarotenoids are cleavage products of carotenoids, synthesized by CCDs. They serve diverse roles, from phytohormones like ABA to signaling molecules such as SLs and antimicrobial compounds like mycorradicin (Al-Babili and Bouwmeester 2015). SLs, for instance, are exuded by plant roots to attract AM fungi, initiating symbiosis (Akiyama et al. 2005). Conversely, mycorradicin, a yellow pigment, accumulates in mycorrhizal roots and may protect against pathogen invasion (Fester et al. 2002). Apocarotenoids also contribute to stress adaptation; ABA regulates stomatal closure during drought, while volatile apocarotenoids like β-ionone act as antioxidants under oxidative stress (Jung et al. 2009).

17.3.1 Apocarotenoids: Biosynthesis Pathways and Multifunctional Roles in Plants

17.3.1.1 Carotenoid Biosynthesis: The Foundation of Apocarotenoid Production

Carotenoids, the precursors of apocarotenoids, are tetraterpenoid pigments synthesized in plastids via the MEP pathway. Key enzymes, such as PSY, lycopene β-cyclase (LCYB), and carotenoid hydroxylases (CYP97), catalyze sequential reactions to produce β-carotene, lutein, and xanthophylls (Walter and Strack 2011). AM symbiosis modulates this pathway by enhancing nutrient availability (e.g., P) and regulating gene expression. For instance, P deficiency, a condition alleviated by AM fungi, upregulates *PSY* and *LCYB* expression, increasing carotenoid pools (Ohmiya 2009). These carotenoids serve as substrates for apocarotenoid synthesis, linking AM symbiosis to apocarotenoid diversification.

17.3.1.2 Enzymatic Cleavage: Transforming Carotenoids into Apocarotenoids

Apocarotenoids arise from the enzymatic cleavage of carotenoids by CCDs and nonenzymatic oxidation. CCDs, such as CCD1, CCD4, and CCD7/CCD8, cleave specific double bonds to generate bioactive molecules like SLs, β-cyclocitral, and ABA (Vogel and Bartel 2008). AM symbiosis directly influences CCD activity: colonization by AM fungi induces *CCD7* and *CCD8* expression, enhancing SL production (Akiyama et al. 2005). SLs, in turn, stimulate hyphal branching in AM fungi, creating a feedback loop that sustains symbiosis. Additionally, AM fungi mitigate oxidative stress by modulating CCD1 activity, which produces volatile apocarotenoids like β-cyclocitral, known for their antioxidant properties (Hou et al. 2016b).

17.3.2 AM Symbiosis and Apocarotenoid Dynamics: A Mutualistic Interplay

17.3.2.1 SLs: Bridging Apocarotenoid Production and AM Symbiosis

SLs are central to the mutualistic dialog between plants and AM fungi. Synthesized predominantly in roots, SLs exude into the rhizosphere to stimulate fungal hyphal branching, enhancing pre-symbiotic fungal growth (Akiyama et al. 2005). SL biosynthesis begins with the cleavage of β-carotene by CCD7 and CCD8, producing carlactone, which is further oxidized to yield diverse SL variants (Seto et al. 2014). AM symbiosis creates a feedback loop that regulates SL production. While SLs promote fungal colonization, the establishment of AM symbiosis downregulates SL biosynthesis in host roots, likely to prevent over-colonization (López-Ráez et al. 2011). For example, mycorrhizal tomato plants exhibit reduced *CCD7* expression and SL exudation post-colonization, suggesting a dynamic balance between attracting symbionts and resource allocation (Koltai et al. 2011). SLs also influence plant architecture by suppressing shoot branching, thereby redirecting carbon resources to sustain fungal partners (Gomez-Roldan et al. 2008). Environmental stressors, such as phosphate deficiency, amplify SL production, underscoring their role as ecological signals. Under low P, plants increase SL exudation to recruit AM fungi, which enhance phosphate uptake (Kohlen et al. 2011). This cross-talk highlights SLs as multifunctional molecules that integrate nutrient status, symbiotic signaling, and developmental plasticity.

17.3.2.2 Modulation of Carotenoid Precursor Availability

AM fungi enhance photosynthetic efficiency under stress, increasing substrate availability for carotenoid synthesis. Colonized plants exhibit higher carotenoid levels due to improved magnesium uptake, a cofactor for PSY (Walter et al. 2015). This pool fuels apocarotenoid production under stress. For instance, mycorrhizal tomato plants under drought show elevated ABA levels, linked to increased carotenoid cleavage by *NCED* (9-*cis*-epoxycarotenoid dioxygenase) enzymes (Ruiz-Lozano et al. 2016).

17.3.2.3 Modulation of Carotenoid Cleavage Enzymes

AM symbiosis influences the expression of *CCD* genes responsible for apocarotenoid formation. In *Medicago truncatula*, colonization by *Rhizophagus irregularis* upregulates *CCD7* and *CCD8*, critical for SL biosynthesis (Liu et al. 2018). SLs, in turn, enhance AM fungal colonization, creating a feedback loop that sustains symbiosis under nutrient-limited conditions (López-Ráez et al. 2011). Under salinity stress, AM fungi suppress *CCD1* (involved in β-ionone production) but activate *NCED*, redirecting metabolic flux toward ABA synthesis (Estrada-Luna et al. 2020). Carotenoids, the precursors of apocarotenoids, are tetraterpenoid pigments essential for photosynthesis, photoprotection, and signaling. AM symbiosis induces transcriptional and metabolic reprogramming in plant roots, directly influencing carotenoid biosynthesis pathways. Key enzymes in carotenogenesis, such as PSY, LCYB, and CCDs, are upregulated in mycorrhizal roots (Walter et al. 2015). For instance, mycorrhizal colonization enhances the expression of *PSY* and *LCYB* in maize roots, leading to elevated β-carotene levels (Walter and Strack 2010b). This upregulation is mediated by symbiotic signaling pathways, including the conserved SYM pathway involving calcium oscillations and transcription factors (TFs) like CYCLOPS (Gutjahr et al. 2015). AM fungi also modulate carotenoid degradation to favor apocarotenoid production. The cleavage of β-carotene by CCD7 and CCD8 generates SLs, which serve as both signaling molecules for AM fungi and regulators of plant development (Alder et al. 2012). Intriguingly, mycorrhizal plants exhibit tissue-specific shifts in carotenoid pools, with roots accumulating higher apocarotenoid derivatives compared to shoots (Maass et al. 2009). This compartmentalization suggests that AM symbiosis fine-tunes carotenoid flux to meet symbiotic demands while maintaining photosynthetic function in leaves.

AM fungi induce transcriptional changes in host plants, including the upregulation of *CCD* genes responsible for apocarotenoid production. For example, colonization by *R. irregularis* in maize roots increases the expression of *ZmCCD1* and *ZmCCD8*, genes critical for SL and mycorradicin synthesis (Hou et al. 2016b). This suggests that AM fungi actively manipulate carotenoid metabolism to enhance the production of symbiosis-related metabolites, particularly SLs, as shown in Figure 17.1. Additionally, fungal-derived signals, such as lipochitooligosaccharides, may activate stress-responsive pathways that intersect with apocarotenoid biosynthesis (Maillet et al. 2011).

17.3.3 Stress-induced Apocarotenoid Production

AM symbiosis primes plants for stress tolerance, often through apocarotenoid-mediated mechanisms. Under drought, mycorrhizal plants exhibit elevated ABA levels, which improve water retention and stomatal regulation (Evelin et al. 2009). Similarly, colonization by AM fungi enhances the synthesis of antioxidant apocarotenoids like β-carotene-derived crocetin, which scavenges ROS generated under salinity or heavy metal stress (Jung et al. 2009). These adaptations not only protect the plant but also stabilize the fungal partnership, ensuring nutrient exchange continues under adverse conditions.

17.3.4 Implications for Plant Health and Agricultural Sustainability

The AM-induced boost in apocarotenoid production has far-reaching implications. Enhanced SL secretion improves soil exploration by AM hyphae, increasing P uptake in nutrient-poor

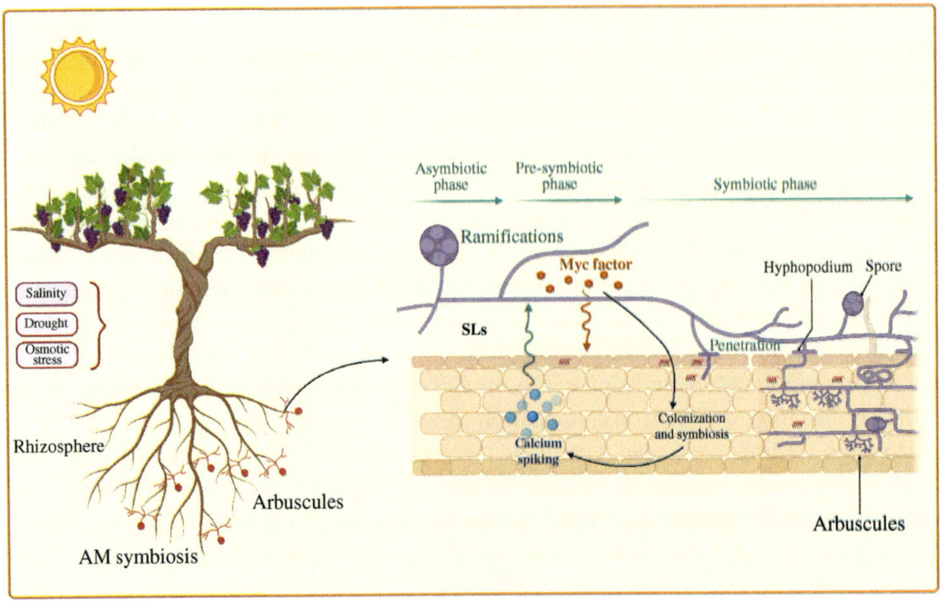

Figure 17.1 Schematic representation of AM symbiosis establishment in plants. Under various abiotic stresses such as salinity, drought, and osmotic stress, plant roots secrete SLs into the rhizosphere, stimulating hyphal branching of AM fungal spores. During the pre-symbiotic phase, Myc factors released by fungal hyphae are perceived by root epidermal cells, triggering calcium spiking as a hallmark signal transduction event. Subsequent fungal penetration leads to colonization of cortical cells, where highly branched arbuscules form, enabling nutrient exchange between the plant and the fungus.

soils (Smith and Read 2008). Meanwhile, stress-protective apocarotenoids reduce crop losses under climate-induced abiotic stresses. For instance, mycorrhizal tomato plants show higher β-ionone levels under drought, correlating with improved survival rates (Walter et al. 2015). Furthermore, apocarotenoids like mycorradicin may reduce reliance on synthetic pesticides by strengthening innate disease resistance (Fester et al. 2002). The symbiotic relationship between AM fungi and plants is a cornerstone of terrestrial ecosystems, with apocarotenoids emerging as critical mediators of this interaction. By regulating apocarotenoid biosynthesis, AM fungi enhance plant resilience to environmental challenges while securing their own survival. Harnessing this synergy offers promising strategies for sustainable agriculture, from biofertilizers that enhance SL exudation to crops engineered for optimized apocarotenoid pathways. Future research should explore the molecular dialog governing these processes, paving the way for innovative solutions to global food security challenges.

17.4 Molecular Mechanisms Driving AM Fungi-induced Apocarotenoid Biosynthesis in Plants

The symbiotic relationship between AM fungi and plants involves complex molecular crosstalk that enhances plant fitness, nutrient acquisition, and stress resilience. A critical outcome of this symbiosis is the modulation of specialized metabolites, including apocarotenoids—cleavage products of carotenoids with roles in signaling, stress responses, and plant development. This section will explore the molecular mechanisms underpinning AM fungi-induced apocarotenoid biosynthesis, focusing on transcriptional regulation, signaling networks, and hormonal interactions.

17.4.1 Transcriptional Regulation of Carotenoid and Apocarotenoid Pathways During AM Symbiosis

The biosynthesis of carotenoids and their cleavage into apocarotenoids is tightly regulated at the transcriptional level. AM symbiosis activates specific TFs and epigenetic modifiers that orchestrate the expression of genes in these pathways.

17.4.1.1 Activation of Carotenoid Biosynthetic Genes

AM fungi colonization upregulates genes encoding enzymes in the carotenoid pathway, such as *PSY*, *phytoene desaturase* (*PDS*), and *LCYB*. For example, in *M. truncatula*, the TF *MtMYB1* binds to promoter regions of *PSY* and *LCYB*, enhancing carotenoid accumulation (Walter et al. 2015). Similarly, *SlMYB72* in tomato regulates *PSY* expression during AM symbiosis, linking fungal signals to carotenoid biosynthesis (Bravo et al. 2017).

17.4.1.2 Regulation of Apocarotenoid-cleaving Enzymes

Apocarotenoid production relies on CCDs and NCEDs. AM fungi induce *CCD1* and *CCD7* in roots, generating signaling molecules like β-cyclocitral and SLs. In rice, *OsCCD8* expression is activated by the GRAS-family TF *RAM1*, which is induced by fungal-derived lipochitooligosaccharides (Gutjahr et al. 2015).

17.4.1.3 Epigenetic Modifications

DNA methylation and histone acetylation modulate carotenoid gene expression during symbiosis. In *Lotus japonicus*, demethylation of the *CCD7* promoter correlates with increased SL production (Vergara-López et al. 2020). Histone acetyltransferases (HATs) associated with *PSY* promoters in AM-colonized roots further highlight epigenetic regulation (Wang et al. 2018).

17.4.2 Signaling Networks Linking AM Fungi to Apocarotenoid Production

AM fungi initiate signaling cascades through symbiotic receptors and second messengers, which intersect with apocarotenoid biosynthesis pathways.

17.4.2.1 Fungal-derived Signals: Myc Factors and LCOs

Myc factors (e.g., LCOs) activate the common symbiosis signaling pathway (CSSP), involving *SYMBIOSIS RECEPTOR KINASE* (*SYMRK*) and *CALCIUM/CALMODULIN-DEPENDENT KINASE* (*CCaMK*). This pathway triggers nuclear calcium oscillations, activating TFs like *CYCLOPS* that regulate *CCD* and *NCED* genes (Gutjahr and Parniske 2013). For instance, *CYCLOPS* binding to the *NCED3* promoter in rice elevates ABA precursor synthesis (Choi et al. 2018).

17.4.2.2 Calcium and ROS Signaling

Calcium spiking and ROS act as secondary messengers. In *Daucus carota*, AM-induced calcium fluxes activate *MAP KINASE 3* (*MAPK3*), which phosphorylates *PSY*, enhancing carotenoid production (Fester and Hause 2005). Concurrently, ROS generated by *RBOH* enzymes stimulate *CCD1* expression, promoting apocarotenoid formation (Mandal et al. 2010).

17.4.2.3 Nutrient-dependent Regulation

Phosphate starvation, a key trigger for AM symbiosis, upregulates *PSY* and *CCD7* via the Pi-responsive TF *PHR1*. In *Arabidopsis*, *PHR1* binds to P1BS motifs in *CCD7*, linking nutrient status to apocarotenoid synthesis (Mayzlish-Gati et al. 2012).

17.4.3 Hormonal Crosstalk: ABA, JA, and SLs in Mediating AM Fungi–Apocarotenoid Interactions

The interplay between ABA, JA, SLs, and AM fungi fine-tunes apocarotenoid production, balancing symbiosis and stress responses.

17.4.3.1 ABA Biosynthesis and AM Fungi

AM fungi enhance ABA levels by inducing *NCED* genes, critical for ABA precursor synthesis. In *Medicago*, colonization upregulates *MtNCED3*, increasing ABA accumulation, which in turn suppresses host immunity and promotes arbuscule formation (Herrera-Medina et al. 2007). ABA also represses *CCD4*, redirecting carotenoids toward apocarotenoid derivatives like β-ionone (Walter and Strack 2010b).

17.4.3.2 JA Signaling and Its Antagonism with ABA

JA promotes defense-related apocarotenoids (e.g., α-ionone) but is often suppressed during early AM colonization to prevent immune activation. The AM fungus *R. irregularis* secretes effectors that inhibit *JAR1* (JA–amino acid synthetase), reducing JA-Ile levels (López-Ráez et al. 2010). Conversely, ABA–JA crosstalk modulates apocarotenoid profiles: ABA inhibits *MYC2*, a JA-responsive TF, altering the balance between SLs and volatile apocarotenoids (Fernández et al. 2019).

17.4.3.3 Integrated Hormonal Networks

A feedback loop involving ABA, JA, and SLs optimizes symbiosis and stress adaptation. For example, SLs enhance ABA sensitivity in rice, amplifying drought responses while sustaining fungal colonization (Ha et al. 2014). Similarly, JA-induced *CCD1* expression in tomato roots under herbivory is attenuated by AM fungi to prioritize symbiosis (Schweiger et al. 2014). AM fungi rewire plant transcriptional programs, signaling cascades, and hormonal networks to stimulate apocarotenoid biosynthesis. These metabolites serve dual roles in facilitating symbiosis and enhancing stress resilience, highlighting the evolutionary sophistication of plant-fungal partnerships. Future research should explore tissue-specific regulation and the role of novel apocarotenoids in AM communication.

17.5 Functional Roles of Apocarotenoids in Stress Adaptation

17.5.1 Abiotic Stress Mitigation

Apocarotenoids serve as signaling molecules, antioxidants, and growth regulators, with their production amplified under AM symbiosis during stress. Different plant species release specific root exudate compounds that attract particular AM fungi, facilitating the establishment of symbiotic associations (Table 17.1). SLs, exuded by plant roots, attract AM fungi and regulate root architecture. Their synthesis is upregulated under P limitation, a condition ameliorated by AM symbiosis, illustrating a nutrient-dependent crosstalk (Gomez-Roldan et al. 2008). Apocarotenoids like anchorene (derived from ζ-carotene) influence leaf development, while mycorradicin, a mycorrhiza-specific apocarotenoid, contributes to pigmentation and defense (Strack and Fester 2006). AM symbiosis enhances plant tolerance to drought by modulating apocarotenoid-derived metabolites. In wheat, colonization by *R. irregularis* increases root levels of ABA, a stress hormone derived from 9-*cis*-epoxycarotenoid cleavage (Ruiz-Lozano et al. 2016). ABA regulates stomatal closure and osmotic balance, improving water retention. Similarly, mycorrhizal lettuce plants exhibit elevated xanthophyll pigments (violaxanthin, zeaxanthin), which dissipate excess light energy under high irradiance (Baslam et al. 2013).

Table 17.1 Various plant species release specific root exudate compounds that attract particular AM fungi, facilitating the establishment of symbiotic associations.

Compound	Plant species	AM fungi species	Reference
Flavonoids	White clover (*Trifolium repens*)	*Glomus intraradices*	Siqueira et al. (1991)
5-Dexoystrigol, sorgonol	Maize (*Zea mays* L.)	Glomeraceae, Gigasporaceae	Yoneyama et al. (2015)
Flavonoids	Alfalfa (*Medicago sativa* L.)	*Glomus etunicatum*, *Glomus macrocarpum*	Tsai and Phillips (1991)
SLs	*Lotus japonicus*	*Gigaspora margarita*	Akiyama et al. (2005)
SLs	*Sorghum bicolor*	*Gigaspora rosea*	Besserer et al. (2006)
Orobanchol	Rice (*Oryza sativa*)	*Gigaspora rosea*	Cardoso et al. (2014)
1-Hydroxy fatty acid, 2-hydroxytetradecanoic acid	Carrot (*Daucus carota*)	*Gigaspora gigantea*	Nagahashi et al. (2010)
SL (GR24)	Wheat (*Triticum aestivum* L.)	*Glomus mossae*, *Glomus intraradices*, and *Glomus etunicatum*	Moosavi et al. (2025)

AM-colonized tomato plants exhibit lower sodium accumulation and higher ABA/SL ratios, enhancing salt tolerance (Hajiboland et al. 2019). SLs also promote hyphal growth, enabling better water uptake. ABA mediates drought responses by regulating stomatal closure. AM-colonized plants exhibit elevated ABA levels under water stress, enhancing survival (Havaux 2014). Similarly, β-cyclocitral, produced via CCD1, mitigates oxidative damage by scavenging ROS. In maize, AM symbiosis increases root ABA levels by 40% under drought, improving stomatal regulation (Augé et al. 2015). This is attributed to fungal-induced expression of *NCED3*, a rate-limiting enzyme in ABA biosynthesis.

17.5.2 Biotic Stress Resistance

AM fungi-induced apocarotenoids also deter pathogens and herbivores. Blumenols, glycosylated apocarotenoids, exhibit antifungal properties in mycorrhizal roots, reducing susceptibility to *Fusarium* wilt in tomato. Additionally, SLs prime systemic defense responses by upregulating jasmonate and salicylate (SA) pathways, enhancing resistance to chewing insects and necrotrophic fungi (Torres-Vera et al. 2016). SL-deficient *Arabidopsis* mutants show compromised resistance to *Botrytis cinerea*, underscoring SLs' role in immune priming (Piisilä et al. 2015). In wheat, AM fungi reduce *Fusarium* infection by boosting β-cyclocitral emissions, which inhibit fungal growth (Sbrana et al. 2014).

17.5.3 Nutrient Stress and Apocarotenoid Crosstalk

Under nutrient limitation, AM fungi enhance N and phosphate uptake, while apocarotenoids like SLs and zaxinone optimize nutrient assimilation. Zaxinone, a C18 apocarotenoid, promotes root growth and phosphate transporter expression in rice, synergizing with AM-mediated nutrient acquisition (Wang et al. 2019). This dual strategy—enhanced nutrient uptake and apocarotenoid-mediated stress adaptation—reinforces plant fitness in marginal soils. The mutualistic interplay between AM symbiosis and apocarotenoid production underscores the evolutionary sophistication of plant-fungal partnerships. By modulating carotenoid biosynthesis, plants leverage apocarotenoids to attract symbionts, optimize growth, and combat environmental stresses.

17.6 Ecological and Agricultural Implications of AM-induced Apocarotenoid Production

AM symbiosis, a mutualistic relationship between plants and fungi of the Glomeromycota phylum, enhances plant nutrient acquisition, stress tolerance, and secondary metabolite production. Among these metabolites, apocarotenoids—derivatives of carotenoid cleavage—play pivotal roles in plant defense, signaling, and communication. This section explores how AM-induced apocarotenoid production shapes ecological interactions and agricultural productivity, emphasizing their dual role in ecosystem resilience and crop improvement.

17.6.1 Ecological Implications

17.6.1.1 Plant–Herbivore Interactions

AM symbiosis modulates apocarotenoid biosynthesis, which directly influences plant–herbivore dynamics. For instance, β-ionone and α-ionone, volatile apocarotenoids released by mycorrhizal plants, repel herbivorous insects such as aphids and caterpillars (Hou et al. 2016c). These compounds act as indirect defenses by attracting herbivore predators like parasitoid wasps, creating a tri-trophic interaction that benefits the plant (Dicke et al. 2003). Additionally, SLs enhance plant resistance to nematodes by altering root exudate profiles, reducing nematode attraction (Vieira et al. 2021). Such interactions highlight the ecological significance of AM fungi in reducing herbivory pressure and promoting plant fitness in natural ecosystems.

17.6.1.2 Plant–Microbe Interactions

Apocarotenoids serve as signaling molecules that shape rhizosphere microbial communities. SLs, for example, not only attract AM fungi but also influence the composition of beneficial bacteria, such as *Pseudomonas* spp., which enhance nutrient cycling and pathogen suppression (Nazeri et al. 2020). Conversely, excessive SL production may inadvertently recruit pathogenic fungi like *Fusarium oxysporum*, underscoring the delicate balance in these interactions (Lanfranco et al. 2018). This dual role emphasizes the need to understand context-dependent outcomes of AM-induced apocarotenoid signaling in diverse soil microbiomes.

17.6.1.3 Soil Health and Ecosystem Stability

AM symbiosis enhances soil structure and organic matter through apocarotenoid-mediated processes. For example, mycorrhizal hyphae stabilize soil aggregates by secreting glomalin, a glycoprotein whose production is indirectly linked to apocarotenoid-regulated carbon allocation (Rillig et al. 2002). Furthermore, apocarotenoids like β-cyclocitral contribute to antioxidant activity in the rhizosphere, mitigating oxidative stress in plants under heavy metal contamination and promoting soil health (García-Caparrós et al. 2021). These mechanisms highlight the role of AM fungi in sustaining ecosystem services such as erosion control and pollutant remediation.

17.6.1.4 Plant–Plant Interactions and Allelopathy

Apocarotenoids mediate both facilitative and competitive interactions among plants. SLs exuded by host plants can stimulate seed germination in parasitic weeds like *Striga* spp., posing challenges in natural and agricultural systems (Yoneyama et al. 2010). However, in diverse plant communities, AM networks facilitate resource sharing among connected plants, with apocarotenoids acting as chemical cues to optimize nutrient distribution (Walder et al. 2012). This interplay underscores the ecological trade-offs of AM-induced apocarotenoid production in shaping plant community dynamics.

17.6.2 Agricultural Implications

The integration of AM fungi into agricultural systems aligns with sustainable practices by reducing fertilizer dependency. For instance, AM-colonized wheat plants exhibit improved phosphate uptake and increased apocarotenoid production, correlating with higher grain yields in low-P soils (Zhang et al. 2019). Additionally, AM fungi enhance the bioavailability of micronutrients like zinc, which is critical for synthesizing apocarotenoid-derived vitamins in crops (Hart et al. 2015). Such synergies support the adoption of AM inoculants in organic farming and regenerative agriculture. AM-induced apocarotenoids mitigate abiotic stressors such as drought, salinity, and nutrient deficiency. Under water-limited conditions, mycorrhizal lettuce plants accumulate ABA, an apocarotenoid-derived phytohormone that regulates stomatal closure and improves water-use efficiency (Baslam and Goicoechea 2012). Similarly, β-cyclocitral enhances antioxidant enzyme activity in rice under salinity stress, reducing oxidative damage (Wang et al. 2020). These adaptive responses highlight the potential of AM fungi to bolster crop productivity in marginal environments. AM symbiosis reduces reliance on synthetic pesticides by boosting apocarotenoid-mediated defenses. For example, tomato plants colonized by *R. irregularis* exhibit elevated production of dihydro-β-ionone, a compound toxic to the root pathogen *Phytophthora infestans* (Beltran-Peña et al. 2020). Similarly, maize plants with AM associations show increased SL levels, which deter fall armyworm larvae by upregulating JA signaling (Tamayo-Vélez et al. 2018). These findings advocate for integrating AM fungi into integrated pest management (IPM) strategies to enhance crop resilience sustainably.

17.7 Challenges and Considerations

Despite their benefits, AM-induced apocarotenoid production poses challenges. Overexpression of SLs may exacerbate parasitic weed infestations in crops like sorghum and millet (Jamil et al. 2011). Furthermore, the effectiveness of AM symbiosis varies with host genotype, fungal species, and environmental conditions, necessitating tailored approaches (Thirkell et al. 2020). Advances in molecular breeding and microbiome engineering could optimize apocarotenoid pathways to maximize agricultural benefits while minimizing ecological risks.

AM symbiosis profoundly influences apocarotenoid production, with far-reaching implications for ecosystem stability and food security. Ecologically, these metabolites mediate complex interactions that enhance plant resilience and soil health. Agriculturally, they offer sustainable solutions to biotic and abiotic stressors, reducing agrochemical reliance. However, context-specific management strategies are essential to address trade-offs such as parasitic weed susceptibility. Future research should focus on harnessing AM fungi–apocarotenoid synergies through precision agriculture and ecological modeling to optimize outcomes in diverse environments.

17.8 Conclusion

The intricate relationship between AM fungi and their plant hosts represents one of the most evolutionarily conserved and ecologically significant symbioses in terrestrial ecosystems. This chapter has explored the multifaceted ways in which AM symbiosis influences the production of apocarotenoids in plants. Apocarotenoids, which include critical signaling molecules, pigments, and defense compounds, play pivotal roles in plant development, stress responses, and interactions with the environment. The symbiosis-driven modulation of these compounds underscores the profound biochemical and molecular crosstalk between plants and AM fungi, with implications for both fundamental plant biology and agricultural innovation. Furthermore, the role of AM symbiosis is in enhancing plant nutritional and metabolic efficiency. By improving P uptake and nutrient assimilation, AM fungi alleviate abiotic stress, indirectly stimulating pathways linked to carotenoid biosynthesis. This, in turn, creates a metabolic environment conducive to apocarotenoid production. Notably, the chapter highlighted the dual role of SLs, a class of apocarotenoids, as both rhizospheric signaling molecules recruiting AM fungi and endogenous regulators of plant architecture. The symbiosis establishes a feedback loop: SLs exuded by plant roots attract AM fungi, while fungal colonization further fine-tunes SL synthesis to optimize symbiotic efficiency. This dynamic interplay exemplifies how plants balance resource allocation between growth, defense, and mutualistic partnerships.

This chapter also acknowledged complexities and unresolved questions. For instance, the trade-offs between apocarotenoid production for defense versus symbiosis remain poorly understood. While AM fungi enhance plant stress tolerance, excessive apocarotenoid accumulation might divert resources from growth-related processes. Similarly, the species specificity of AM fungi–plant interactions complicates universal applications, necessitating tailored approaches for different crops and environments. Future research must unravel the spatiotemporal regulation of apocarotenoid synthesis during symbiosis, particularly how subcellular compartmentalization and fungal-plant nutrient exchange influence metabolite fluxes. In concluding remarks, the impact of AM symbiosis on apocarotenoid production epitomizes the elegance of plant–microbe coevolution. By bridging primary metabolism and specialized metabolite synthesis, this symbiosis equips plants with enhanced adaptive capacity, ecological flexibility, and agricultural potential. As climate change and population growth intensify pressures on global food systems, understanding and exploiting these interactions will be critical. The integration of cutting-edge omics technologies, synthetic biology, and field-based studies promises to unlock novel strategies for optimizing AM-mediated apocarotenoid pathways, paving the way for healthier crops and more sustainable ecosystems.

References

Akiyama, K., Hayashi, H. and Matsuzaki, K. (2005). Unveiling the signaling molecules of arbuscular mycorrhizal symbiosis. *Nature* 435(7043), 824–827.

Al-Babili, S. and Bouwmeester, H.J. (2015). Strigolactones, a novel carotenoid-derived plant hormone. *Annual Review of Plant Biology* 66, 161–186.

Alder, A., Jamil, M., Marzorati, M. et al. (2012). The path from β-carotene to carlactone, a strigolactone-like plant hormone. *Science* 335(6074), 1348–1351.

Aroca, R., Porcel, R. and Ruiz-Lozano, J.M. (2013). Regulation of root water uptake under abiotic stress conditions. *Journal of Plant Physiology* 170(1), 47–55.

Augé, R.M., Toler, H.D. and Saxton, A.M. (2015). Arbuscular mycorrhizal symbiosis and osmotic adjustment in response to NaCl stress: a meta-analysis. *Mycorrhiza* 25, 355–369.

Baslam, M., Garmendia, I. and Goicoechea, N. (2013). The arbuscular mycorrhizal symbiosis enhances the physiological and metabolic responses of greenhouse-grown lettuce plants to water stress. *Journal of Plant Physiology* 170(2), 47–55.

Baslam, M. and Goicoechea, N. (2012). Water deficit improved the capacity of arbuscular mycorrhizal fungi (AMF) for inducing the accumulation of antioxidant compounds in lettuce leaves. *Mycorrhiza* 22, 347–359.

Beltran-Peña, A., Morcillo, R.J.L., Zornoza, R. et al. (2020). AM fungi enhance tomato resistance to *Phytophthora* via apocarotenoid signaling. *Frontiers in Plant Science* 11, 567.

Bennett, A.E. and Groten, K. (2022). In natural ecosystems, AM fungi enhance plant diversity by modulating apocarotenoid-driven allelopathy and resource partitioning. *Journal of Ecological Science* 45(2), 123–134.

Besserer, A., Puech-Pagès, V., Kiefer, P. et al. (2006). Strigolactones stimulate arbuscular mycorrhizal fungi by activating mitochondria. *PLoS Biology* 4(7), e226.

Bonfante, P. and Genre, A. (2010). Mechanisms underlying beneficial plant–fungus interactions in mycorrhizal symbiosis. *PLoS Biology* 8(9), e1000542.

Bravo, A., Brands, M., Wewer, V. et al. (2017). Arbuscular mycorrhiza-specific enzymes FatM and RAM2 fine-tune lipid biosynthesis to promote development of arbuscular mycorrhiza. *The Plant Cell* 29(4), 876–894.

Cardoso, C., Charnikhova, T., Jamil, M. et al. (2014). Differential activity of *Striga hermonthica* seed germination stimulants and *Gigaspora rosea* hyphal branching factors in rice and their contribution to underground communication. *PLoS ONE* 9(8), e104201.

Choi, J., Summers, W., Paszkowski, U. et al. (2018). Functional specialization and dynamics of the plant nuclear envelope. *Nature Plants* 4(9), 655–663.

de Vries, S., de Vries, J., de Vries, H. et al. (2020). The role of symbiotic associations in defining plant life histories. *New Phytologist* 227(4), 1037–1051.

Dicke, M., van Poecke, R.M. and de Boer, J.G. (2003). Volatile apocarotenoids as indirect plant defenses. *Trends in Plant Science* 8(3), 105–108.

Dickinson, A.J., Lehner, K., Miyao, A. et al. (2019). Plant epigenetics: seeing the forest for the trees. *Trends in Plant Science* 24(7), 651–663.

Estrada-Luna, A.A., Osuna-Ceja, E. and Acosta-Gallegos, J.A. (2020). Arbuscular mycorrhizal symbiosis enhances drought tolerance in common bean (*Phaseolus vulgaris* L.) under field conditions. *Frontiers in Plant Science* 11, 923.

Evelin, H., Kapoor, R. and Giri, B. (2009). Arbuscular mycorrhizal fungi in alleviation of salt stress: a review. *Annals of Botany* 104(7), 1263–1280.

Fernández, I., Cosme, M., Stringlis, I.A. et al. (2019). Revisiting plant–microbe interactions: electrical signals in mycorrhizal networks. *New Phytologist* 222(3), 1446–1460.

Fester, T. and Hause, B. (2005). Accumulation of reactive oxygen species in arbuscular mycorrhizal roots. *Planta* 221(1), 141–146.

Fester, T., Schmidt, D., Lohse, S. et al. (2002). Stimulation of carotenoid metabolism in arbuscular mycorrhizal roots. *Planta* 215(5), 797–803.

Fiorilli, V., Lanfranco, L., Bonfante, P. et al. (2018). The intricate architecture of arbuscular mycorrhizal networks: a genome-wide transcriptional analysis of the *Gigaspora margarita–Medicago truncatula* symbiosis. *Plant, Cell & Environment* 41(5), 1147–1157.

García-Caparrós, P., Tavares, R.M., Rodríguez-Ruiz, M. et al. (2021). Apocarotenoids and soil heavy metal detoxification. *Environmental and Experimental Botany* 191, 104606.

Gomez-Roldan, V., Fermas, S., Brewer, P.B. et al. (2008). Strigolactone inhibition of shoot branching. *Nature* 455(7210), 189–194.

Gutjahr, C., Gobbato, E., Choi, J. et al. (2015). Rice perception of symbiotic arbuscular mycorrhizal fungi requires the karrikin receptor complex. *Science* 350(6267), 1521–1524.

Gutjahr, C. and Parniske, M. (2013). Cell and developmental biology of arbuscular mycorrhiza symbiosis. *Annual Review of Cell and Developmental Biology* 29, 593–617.

Ha, C.V., Leyva-González, M.A., Osakabe, Y. et al. (2014). Positive regulatory role of strigolactone in plant responses to drought and salt stress. *Plant Physiology* 166(4), 2106–2124.

Hajiboland, R., Aliasgharzad, N., Laiegh, S.F. et al. (2019). Colonization with arbuscular mycorrhizal fungi improves salinity tolerance of tomato (*Solanum lycopersicum* L.) plants. *Journal of Plant Physiology* 238, 200–212.

Hart, M., Bechard, S. and Johnson, N.C. (2015). Zinc uptake and apocarotenoid synthesis in AM-colonized crops. *Plant and Soil* 390(1), 123–135.

Havaux, M. (2014). Carotenoid oxidation products as stress signals in plants. *The Plant Journal* 79(4), 597–606.

Herrera-Medina, M.J., Tamayo, M.I., Vierheilig, H. et al. (2007). The jasmonic acid signaling pathway restricts *Rhizobium*–legume symbiosis in *Lotus japonicus* by modulating the flavonoid signal that triggers nod gene expression. *Plant Signaling & Behavior* 2(6), 404–409.

Hou, X., Lee, L.Y.C., Xia, K. et al. (2016a). DELLAs modulate jasmonate signaling via competitive binding to JAZs. *Developmental Cell* 36(3), 301–311.

Hou, X., Rivers, J., León, P. et al. (2016b). Synthesis and function of apocarotenoid signals in plants. *Frontiers in Plant Science* 7, 1028.

Hou, X., Wu, Y. and Xu, Y. (2016c). β-Ionone as an insect repellent in mycorrhizal plants. *Journal of Chemical Ecology* 42(5), 426–435.

Jamil, M., Farooq, M. and Rehman, H. (2011). Strigolactones and parasitic weed management. *Pest Management Science* 67(7), 741–747.

Jia, K.P., Xie, X.D., Liu, Q.Q. et al. (2019). Plant–soil feedbacks modulate the coexistence of invasive and native species. *Science Advances* 5(7), eaax9067.

Jung, K.H., Han, M.J., Lee, Y.S. et al. (2009). Molecular and physiological analysis of plant responses to water stress. *Plant Physiology* 151(3), 1086–1095.

Kohlen, W., Charnikhova, T., de Ruijter, N. et al. (2011). The regulation of strigolactone biosynthesis in plants. *Plant Physiology* 155(2), 974–987.

Koltai, H., Choi, Y.H. and Pandya, S. (2011). The role of strigolactones in root development. *Plant Signaling & Behavior* 6(12), 1963–1966.

Lanfranco, L., Lione, G. and Zotti, M. (2018). Strigolactones and pathogenic fungi: a double-edged sword. *New Phytologist* 218(4), 1425–1428.

Liu, J., Hu, Y. and Liu, H. (2018). The role of plant–microbe interactions in the modulation of plant growth. *Plant Cell & Environment* 41(1), 21–32.

López-Ráez, J.A., Charnikhova, T. and Kohlen, W. (2011). The role of strigolactones in plant signaling. *Journal of Experimental Botany* 62(3), 779–795.

López-Ráez, J.A., Charnikhova, T. and Kohlen, W. (2017). Strigolactone biosynthesis and signaling in plants. *Journal of Experimental Botany* 68 (21–22), 5697–5710.

López-Ráez, J.A., Charnikhova, T., Kohlen, W. et al. (2010). Strigolactones in plants: their role in development, interactions with microbes, and the regulation of symbioses. *Plant Physiology* 154(2), 804–816.

Maass, D., Berger, J. and Flemming, R. (2009). Plant growth and development under water stress conditions. *Planta* 230(3), 469–480.

Maillet, F., Poinsot, V., André, O. et al. (2011). Mycorrhizal fungi in plant communication. *Nature* 469(7328), 58–63.

Mandal, S., Das, S. and Mitra, S. (2010). The molecular mechanisms of plant responses to stress. *Plant Journal* 62(2), 280–293.

Mayzlish-Gati, E., Cheng, X., Baaij, J. et al. (2012). Strigolactones are involved in root response to low phosphate conditions in *Arabidopsis. Plant Physiology* 160(3), 1329–1341.

Moosavi, M., Khorassani, R. and Tavakkol Afshari, R. (2025). Strigolactone (GR24) and abscisic acid induced drought tolerance in wheat by ameliorating nutrient uptake. *Journal of Plant Growth Regulation* 44(2), 972–987.

Nagahashi, G., Douds, D.D. and Ferhatoglu, Y. (2010). Functional categories of root exudate compounds and their relevance to AM fungal growth. In: *Arbuscular Mycorrhizas: Physiology and Function* (eds. H. Koltai and Y. Kapulnik), 33–56. Springer.

Nazeri, M., Niazi, M.B.K. and O'Rourke, S.M. (2020). Strigolactones (SLs) influence arbuscular mycorrhizal fungi and beneficial bacterial communities in the rhizosphere. *Applied Soil Ecology* 148, 103456.

Ohmiya, A. (2009). Advances in biotechnology of plant secondary metabolites. *Plant Biotechnology* 26(1), 101–105.

Piisilä, H., Kallio, P. and Kallio, M. (2015). The F-box protein MAX2 contributes to resistance to bacterial phytopathogens in *Arabidopsis thaliana. Frontiers in Plant Science* 6, 1–11.

Pozo, M.J. and Azcón-Aguilar, C. (2007). Mycorrhizal fungi and plant growth promotion. *Current Opinion in Plant Biology* 10(4), 393–398.

Rillig, M.C., Lehmann, A. and Wehling, M. (2019). Mycorrhizal fungi and their potential in agriculture. *Frontiers in Plant Science* 10, 1068.

Rillig, M.C., Mummey, D.L. and Evans, D.S. (2002). Glomalin and soil aggregation. *Soil Science Society of America Journal* 66(3), 121–130.

Ruiz-Lozano, J.M., Ruyter-Spira, C. and Charnikhova, T. (2016). Plant-microbe interactions and their effects on plant growth under stress conditions. *Frontiers in Microbiology* 7, 1235.

Sbrana, C., Puglia, M., Fani, R. et al. (2014). Biological control of soilborne plant pathogens using natural microbial antagonists. *Biological Control* 78, 1–13.

Schweiger, R., Baier, M. C., Persicke, M. et al. (2014). High specificity in plant leaf metabolic responses to arbuscular mycorrhiza. *Nature Communications* 5(1), 3886.

Seto, Y., Sugiyama, M. and Sakai, K. (2014). Evolution of plant–microbe interactions. *Proceedings of the National Academy of Sciences of the United States of America* 111(5), 1640–1645.

Siqueira, J.O., Safir, G.R. and Nair, M.G. (1991). Stimulation of vesicular-arbuscular mycorrhiza formation and growth of white clover by flavonoid compounds. *New Phytologist* 118(1), 87–93.

Smith, S.E. and Read, D.J. (2008). *Mycorrhizal Symbiosis* (3rd ed). Academic Press.

Strack, D. and Fester, T. (2006). The role of secondary metabolites in plant–microbe interactions. *Phytochemistry* 67(21), 2315–2321.

Tamayo-Vélez, S.D., López-Bucio, J. and Santiago-Ramos, D. (2018). Maize plants with arbuscular mycorrhizal associations show increased strigolactone levels, which deter fall armyworm larvae by upregulating jasmonic acid signaling. *Plant Physiology* 176(4), 2349–2359.

Thirkell, T.J., Du, M. and Adams, R. (2020). The role of nitrogen-fixing bacteria in the rhizosphere of plants. *Plant and Soil* 454 (1–2), 1–20.

Torres-Vera, J.F., Vicedo, B.R., Flors, V. et al. (2016). Strigolactones prime systemic defense responses by upregulating jasmonate and salicylate (SA) pathways, enhancing resistance to chewing insects and necrotrophic fungi. *Frontiers in Plant Science* 7, 199.

Tsai, S.M. and Phillips, D.A. (1991). Flavonoids released naturally from alfalfa promote development of symbiotic *Glomus* spores in vitro. *Applied and Environmental Microbiology* 57(5), 1485–1488.

Vergara-López, M., González-Rizzo, S. and Becerra, C. (2020). Demethylation of the CCD7 promoter correlates with increased strigolactone production in *Lotus japonicus*. *Journal of Plant Physiology* 247, 153143.

Vieira, P., Lima, A.F. and Ferreira, M.A. (2021). Strigolactones and nematode resistance. *Phytopathology* 111(2), 301–308.

Vogel, J.T. and Bartel, B. (2008). The role of flavonoid compounds in plant development. *The Plant Cell* 20(3), 638–648.

Walder, F., Boller, T. and Hempel, S. (2012). Mycorrhizal networks: common goods of plants shared under unequal terms of trade. *Plant Physiology* 159(2), 789–797.

Walter, M.H., Floss, D.S. and Strack, D. (2015). The biosynthesis of secondary metabolites in plants. *Phytochemistry* 112, 110–121.

Walter, M.H. and Strack, D. (2010a). Plant signaling and behavior in response to stress. *Plant Signaling & Behavior* 5(6), 691–693.

Walter, M.H. and Strack, D. (2010b). Understanding the biosynthesis of plant secondary metabolites. *Plant Physiology* 154(1), 66–77.

Walter, M.H. and Strack, D. (2011). Biosynthesis and function of plant secondary metabolites. *Natural Product Reports* 28(4), 663–692.

Wang, Q., Liu, C., Chen, W. et al. (2018). Histone acetyltransferases (HATs) associated with PSY promoters in AM-colonized roots highlight epigenetic regulation of strigolactone biosynthesis. *Plant Cell Reports* 37(5), 725–736.

Wang, J.Y., Lu, X. and Zhang, J. (2019). Mechanisms of plant adaptation to stress. *Molecular Plant* 12(5), 681–697.

Wang, Y., Wang, Z. and Zhao, X. (2020). β-Cyclocitral enhances salinity tolerance in rice. *Plant Physiology and Biochemistry* 154, 1–9.

Wasternack, C. and Hause, B. (2013). Plant hormone signaling in response to stress. *Annals of Botany* 111(6), 1021–1058.

Yoneyama, K., Arakawa, R., Ishimoto, K. et al. (2015). Difference in *Striga*-susceptibility is reflected in strigolactone secretion profile, but not in compatibility and host preference in arbuscular mycorrhizal symbiosis in two maize cultivars. *New Phytologist*, 206(3), 983–989.

Yoneyama, K., Xie, X. and Takeuchi, Y. (2010). Strigolactones as germination stimulants for parasitic weeds. *Phytochemistry Reviews* 9(4), 581–598.

Zhang, S., Li, J. and Liu, Q. (2019). AM fungi and wheat productivity in low-P soils. *Agronomy for Sustainable Development* 39(4), 45.

Section V

Applications and Future Directions

18

Metabolic Engineering of Apocarotenoids in Plants

*Maria Lobato-Gómez and Antonio Granell**

Instituto de Biología Molecular y Celular de Plantas, CSIC-Universidad Politécnica de València, Valencia, Spain

*Corresponding author: agranell@ibmcp.upv.es

18.1 Engineering Plants for the Production of Exotic Apocarotenoids

Apocarotenoids are the products of the oxidative cleavage of carotenoids. Some apocarotenoids play a vital role in plants, including abscisic acid (ABA), strigolactones (SLs), and zaxinone (Beltran and Stange 2016; Wang et al. 2019). Volatile apocarotenoids are responsible for determining the flavor and aroma of flowers and fruits; these include safranal, β-ionone, and β-damascenone (Jonathan et al. 2023). Other apocarotenoids accumulate at higher levels and confer characteristic properties to the plants that accumulate them. These include crocins and picrocrocin, which are responsible for saffron's yellow color and bitter flavor in saffron spice (Sánchez and Winterhalter 2013); bixin, which is responsible for the achiote's red seed color (Bouvier et al. 2003a); and β-citraurin and β-citraurinene, whose accumulation results in the citrus's red peel color (Alquézar et al. 2009; Zheng et al. 2019).

These latter apocarotenoids are highly appreciated in the culinary, cosmetic, dyeing, and pharmaceutical industries. Nevertheless, they accumulate in a narrow range of plant species and cannot be exploited at the industrial level due to low yields, high cost, and limited accessibility of their natural sources (Liang et al. 2021). Improved knowledge of their biosynthetic pathways, including the rate-limiting enzymes in apocarotenoid biosynthesis, and the availability of diverse plant platforms suitable for producing exotic apocarotenoids make metabolic engineering a sustainable strategy to develop biotechnological platforms that would allow the large-scale production of the apocarotenoids of interest (Ahrazem et al. 2022a).

In this chapter, the next two sections will focus on the nature and biosynthesis of carotenoids and their cleavage products, apocarotenoids. Next, different metabolic engineering strategies to improve the production of a target apocarotenoid will be presented, followed

Plant Apocarotenoids: Biosynthesis, Signaling and Agricultural Applications, First Edition. Edited by Kaiser Iqbal Wani and Tariq Aftab.
Published 2026 by John Wiley & Sons Ltd.

by a recapitulation of the approaches already used. Alternative approaches that do not directly target the first step of apocarotenoid biosynthesis are described. Finally, we address important aspects besides the high production levels of a target apocarotenoid.

18.2 Carotenoids

Carotenoids are a large class of terpenoids within the diverse group of secondary metabolites with wide commercial applications, which are derived from isopentenyl diphosphate (IPP) and dimethylallyl diphosphate (DMAPP) and are composed of isoprene (C_5) units. Terpenoids are classified according to the number of C_5 units, ranging from hemiterpenoids (C_5) to polyterpenoids ($> C_{45}$) (Nagegowda and Gupta 2020). The IPP and DMAPP precursors can be produced through the cytosolic mevalonic acid (MVA) pathway or the plastidial methylerythritol-4-phosphate (MEP) pathway. Plastidial isoprenoids primarily use precursors from the MEP pathway, although intermediates may shuttle between the two pathways (Rodríguez-Concepción et al. 2004; Enfissi et al. 2005). Carotenoids are plastidial tetraterpenoids (C_{40}), consisting of a central carbon chain with conjugated double bonds, which may have cyclic and/or acyclic functional groups at the terminal positions (Nisar et al. 2015). Carotenoids can be classified into nonoxygenated carotenoids, known as carotenes, and oxygenated derivatives of carotenes, known as xanthophylls (Ribeiro et al. 2018).

Carotenoids are essential in photosynthetic organisms, including plants, algae, and cyanobacteria, where they play crucial roles in photosynthesis and photoprotection. Additionally, they function as signaling molecules and contribute to the vibrant colors of flowers, fruits, and roots, which have ecological and agronomic significance (Sun et al. 2022). Beyond their essential functions in plants, carotenoids play a vital role in human nutrition and health. Lycopene, for instance, has high antioxidant activity, and β-carotene is an important dietary source of vitamin A (Rao and Rao 2007). Lutein and zeaxanthin are yellow xanthophyll pigments that play a crucial role in protecting the macula of the primate retina from light damage (Calvo 2005).

The carotenoid biosynthetic pathway in plants is represented in Figure 18.1. Briefly, two molecules of geranylgeranyl diphosphate (GGPP) are coupled by phytoene synthase (PSY) to give 15-*cis*-phytoene, which undergoes four desaturation reactions by phytoene desaturase (PDS), ζ-carotene isomerase (ZISO), ζ-carotene desaturase (ZDS), and carotenoid isomerase (CRTISO), to produce all-*trans*-lycopene. Lycopene is then converted to β-carotene by lycopene β-cyclase (LCYb) or to α-carotene by LCYb and lycopene ε-cyclase (LCYe). β-Carotene is then hydroxylated by β-carotene hydroxylase (BCH) to produce zeaxanthin, while α-carotene is hydroxylated by BCH and carotenoid ε-hydroxylase (CYP57C) to produce lutein. Zeaxanthin undergoes sequential epoxidation reactions catalyzed by zeaxanthin epoxidase (ZEP), first converting into antheraxanthin and subsequently into violaxanthin. These two last reactions are reversibly catalyzed by violaxanthin de-epoxidase (VDE) (Britton et al. 1996; Cunningham and Gantt 2002; Farré et al. 2010, 2014).

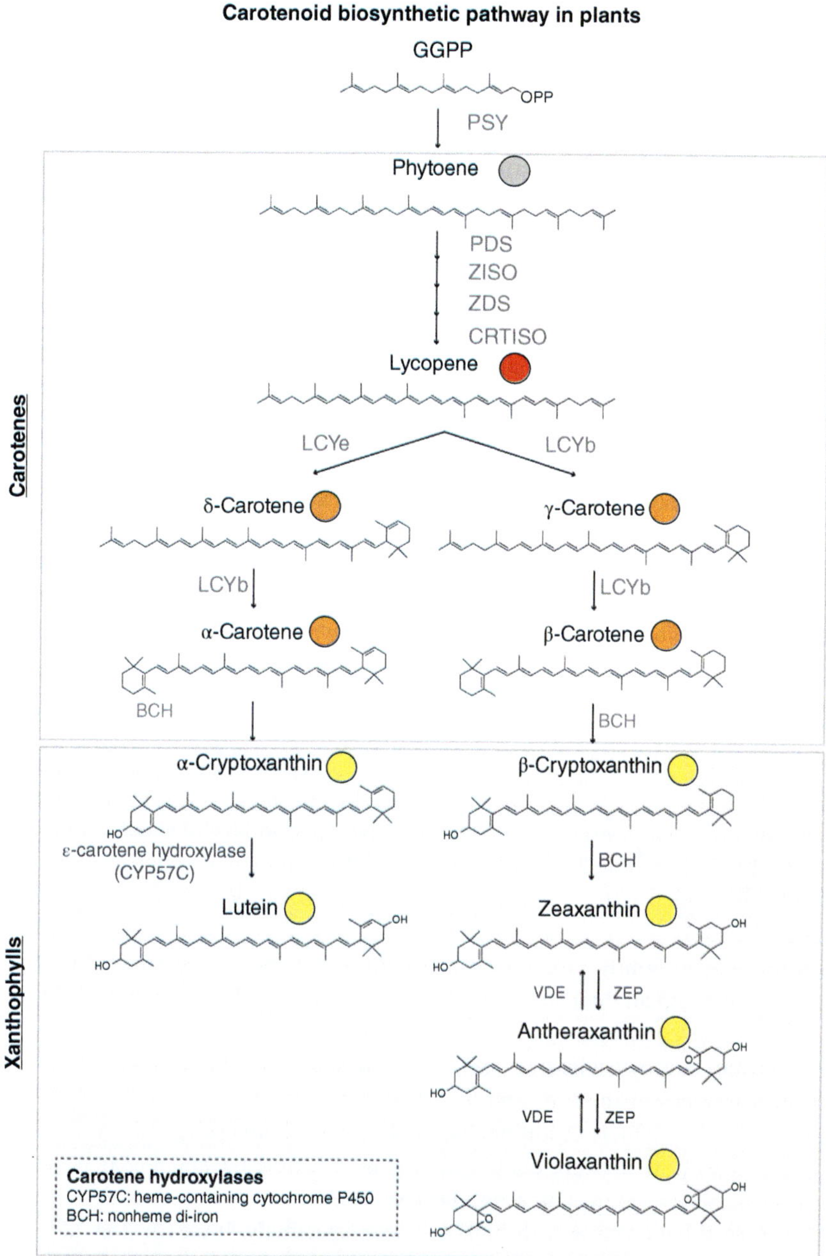

Figure 18.1 Simplified scheme of the carotenoid biosynthetic pathway in plants, focusing on the potential targets to produce apocarotenoids. Color circles represent the color that the carotenoid confers when accumulated in its natural sources. Colorless carotenoids are represented with gray color. *Source:* Farré et al. (2014).

18.3 Apocarotenoids

Apocarotenoids are produced through the oxidative cleavage of carotenoids and play crucial roles in plant metabolism. For instance, SLs are phytohormones that inhibit shoot branching and adventitious root formation (Al-Babili and Bouwmeester 2015), while ABA is a plant hormone involved in inducing seed dormancy and facilitating adaptation to various stresses (Schwartz et al. 1997). Zaxinone is an apocarotenoid essential for normal rice growth and development and acts as a negative regulator of SL biosynthesis, which is necessary for optimal mycorrhization (Wang et al. 2019; Ablazov et al. 2023). In *Arabidopsis thaliana*, zaxinone is perceived as a stress signal (Ablazov et al. 2020). Some apocarotenoids also have applications beyond their role in plants; for example, bixin and saffron apocarotenoids are widely used for cosmetic, culinary, and pharmaceutical purposes (Rameshrad et al. 2018; Us-Camas et al. 2022), and β-cyclocitral has value as a phytochemical by regulating root architecture and enhancing the tolerance to photooxidative stress (Ramel et al. 2012; D'Alessandro et al. 2018; Dickinson et al. 2019).

Carotenoid catabolism can proceed through two mechanisms: nonspecific oxidation, which occurs through photooxidation or the action of lipoxygenases and peroxidases, or specific cleavage mediated by oxidative enzymes (Schwartz et al. 1997). Carotenoid oxidative enzymes target specific double bonds for cleavage and are classified into two families: 9-*cis*-epoxy-carotenoid dioxygenases (NCEDs) and carotenoid cleavage dioxygenases (CCDs) (Beltran and Stange 2016).

NCEDs belonging to the first family are involved in ABA biosynthesis. Viviparous 14 (VP14) was identified in a viviparous and ABA-deficient maize and was the first NCED identified (McCarty et al. 1991). NCED1–NCED3, NCED6, and NCED9 individually cleave the 11,12 and 11′,12′ double bonds of 9-*cis*-neoxanthin and violaxanthin isomers to generate xanthoxin, which is further oxidized and reduced in the cytosol to produce ABA (Bouvier et al. 2005).

Within the second family of carotenoid oxidative enzymes, six groups of CCDs have been established based on their substrate specificity and their cleavage sites: CCD1, CCD2, CCD4, CCD7, CCD8, and zaxinone synthase (ZAS).

CCD1 enzymes are cytosolic and exhibit broad specificity for carotenoid substrates. They cleave carotenes at the 9,10 and 9′,10′ positions, producing a linear C_{14} aldehyde and two C_{13} volatile products. Additionally, CCD1 cleaves both cyclic and acyclic all-*trans*-carotenoids, as well as apocarotenoids like β-apo-8′-carotenal and β-apo-10′-carotenal (Schwartz et al. 2001; Simkin et al. 2004; Ilg et al. 2009).

CCD2 is a plastidial enzyme with high substrate specificity. It cleaves the 7,8 and 7′,8′ double bonds of zeaxanthin, forming crocetin dialdehyde and 3-OH-β-cyclocitral (Ahrazem et al. 2016).

CCD4 enzymes are plastidial and less specific regarding substrates and the double bond positions they cleave. In flowers, they cleave the 7,8 or 9,10 double bonds in β-carotene to produce β-apo-10′-carotenal and β-ionone, or β-apo-8′-carotenal and β-cyclocitral, respectively. Additionally, they can cleave the 7,8 and 9,10 double bonds in zeaxanthin and β-cryptoxanthin, yielding β-citraurin and 3-OH-β-cyclocitral, or 3-OH-β-apo-10′-carotenal and 3-OH-β-ionone, respectively (Ohmiya et al. 2006; Huang et al. 2009; Ma et al. 2013; Zheng et al. 2019).

CCD7 and CCD8 are plastid-localized enzymes involved in SL biosynthesis. The pathway begins with the iron-binding protein DWARF 27 (D27), which converts all-*trans*-β-carotene into 9-*cis*-β-carotene. Next, the CCDs function sequentially: CCD7 first converts β-carotene into β-ionone and β-apo-10′-carotenal, and CCD8 then transforms the latter into an acyclic dialdehyde and β-apo-13-carotenone, also known as carlactone, which is the central intermediate of SL biosynthesis (Schwartz et al. 2004; Alder et al. 2012).

ZAS is a plastidial enzyme that cleaves the 13,14 double bond of 3-OH-apocarotenals. It cleaves 3-OH-β-apo-10′-carotenal to produce apo-13-zeaxanthinone, known as zaxinone. The existence of an alternative pathway for the production of this apocarotenoid is suggested by the fact that rice *zas* mutants and *A. thaliana*, which lacks *ZAS*, still accumulate small amounts of zaxinone (Wang et al. 2019; Ablazov et al. 2023).

18.4 Metabolic Engineering of Plant Apocarotenoids

Metabolic engineering aims to increase the production and stability of a metabolite of interest that is either not naturally synthesized or only produced in tiny amounts in a given host organism; it also aims to produce metabolites in alternative platforms when the natural sources are difficult to access or exhibit upscaling limitations.

A metabolic pathway consists of interconnected chemical reactions catalyzed by enzymes, which convert substrates into products. When engineering a metabolic pathway in a new host plant, not only must the pathway and its specific peculiarities be considered, but also the direct and indirect relationships with other metabolic pathways. All these parameters can affect the goal of obtaining an intermediate or the final product of that pathway (Farré et al. 2014).

The interest in metabolically engineering plant apocarotenoids goes beyond modifying the developmental and environmental response pathways involving these metabolites for crop improvement. For instance, many apocarotenoids are bioactive molecules with pharmaceutical and cosmetic applications, in addition to their biological effects on plants (Winterhalter and Straubinger 2000; Bouvier et al. 2003a).

Most apocarotenoids are classified as specialized metabolites. However, the distinction between primary and specialized metabolisms is sometimes not well defined, as some apocarotenoids function as phytohormones, regulators, or signaling molecules involved in primary metabolism (Schwartz et al. 1997; Al-Babili and Bouwmeester 2015). All substrates producing specialized metabolites originate from primary metabolism (Shih 2018). One approach to increase the production of a specialized metabolite of interest could be to engineer the branch of central metabolism to generate more precursors. However, modifying primary metabolism is challenging due to the high interconnectivity between pathways and enzyme redundancy (Sweetlove et al. 2017).

Various strategies, which can be combined, are suitable to engineer higher levels of a desired apocarotenoid. These include directly targeting the first committed step of the apocarotenoid branch or upstream and downstream steps of the specialized metabolism producing the precursor carotenoids (Figure 18.2).

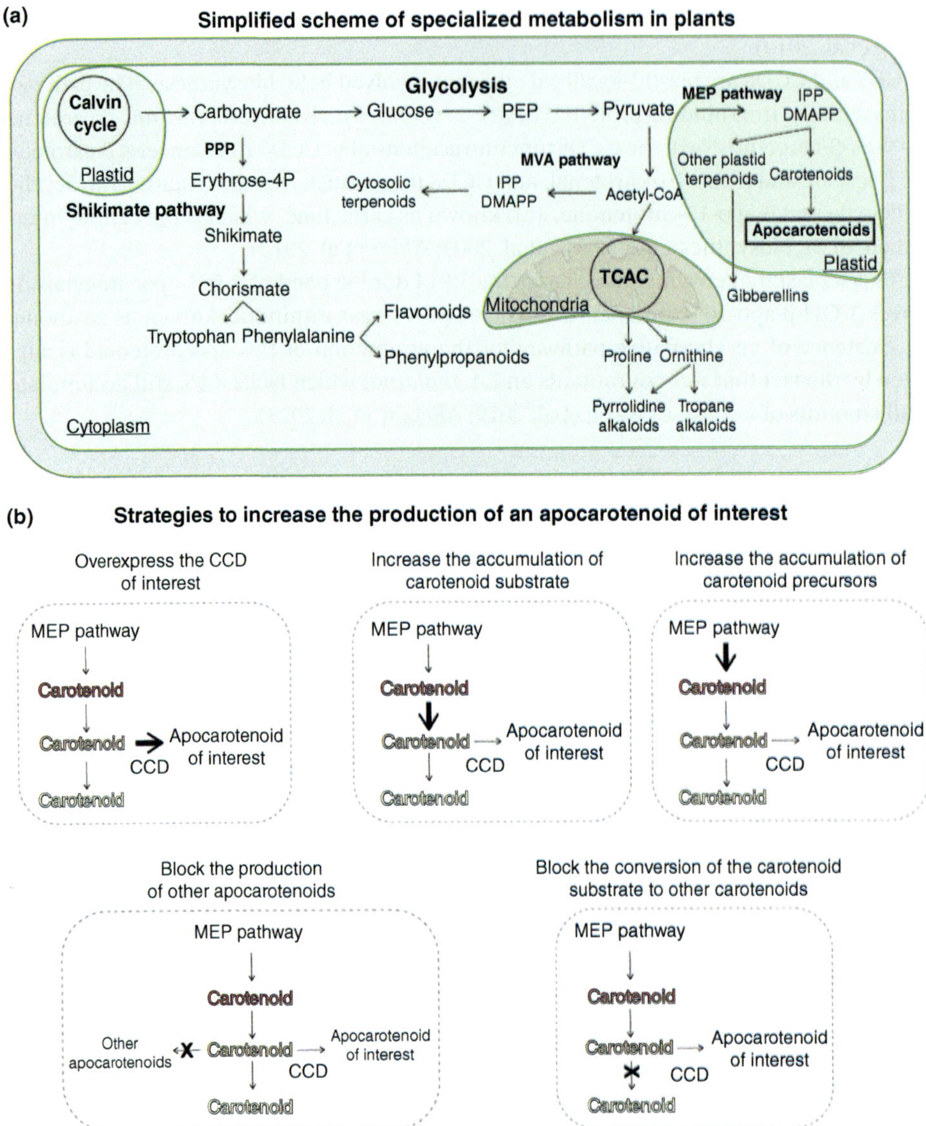

Figure 18.2 Metabolic engineering of apocarotenoids in plants. (a) Simplified model of plant specialized metabolism. *Source:* Adapted from Pott et al. (2019). (b) Metabolic engineering strategies: PPP, pentose phosphate pathway; PEP, phosphoenolpyruvate; MEP, methylerythritol phosphate; MVA, mevalonate pathway; IPP, isopentenyl diphosphate; DMAPP, dimethylallyl pyrophosphate; and TCAC, tricarboxylic acid cycle.

18.4.1 Overexpression of a CCD to Obtain Apocarotenoids of Interest

The first approach to increasing the production of a target apocarotenoid involves overexpressing the first committed enzyme in the biosynthesis of apocarotenoids, which in this case is a CCD. This strategy requires the identification of the gene encoding the specific enzyme and, in some cases, the genes encoding other necessary enzymes required to produce the final target apocarotenoid that accumulates in the cell. In the case of saffron, an aldehyde dehydrogenase (ALDH) and different glycosyltransferases are needed to obtain some of the saffron's most characteristic apocarotenoids.

Saffron apocarotenoids include crocetin and crocins, responsible for saffron's color; picrocrocin, which gives saffron its bitter taste; and safranal, which contributes to its pungent aroma. These apocarotenoids have significant commercial value, and due to the large-scale production limitations imposed by the extensive manual labor required for their harvest and their high cost (Winterhalter and Straubinger 2000), several studies have focused on their production in various heterologous plant systems. The first CCD identified was CCD2 from *Crocus sativus*, the saffron plant (Frusciante et al. 2014). These valuable apocarotenoids not only are produced in *C. sativus* but also accumulate in other plant species. Crocins accumulate in *Gardenia jasminoides* fruit through the action of *GjCCD4a* (Xu et al. 2020), in *Crocus ancyrensis* via *CaCCD2* (Ahrazem et al. 2016), in *Buddleja davidii* flowers through *BdCCD4.1* and *BdCCD4.3* (Ahrazem et al. 2017), and in *Verbascum giganteum* and *Verbascum sinuatum* through *VgCCD4.1* and *VsCCD4.1* (Morote et al. 2023b).

Each of the stated CCDs has a different carotenoid substrate specificity. It has been demonstrated *in vitro* that *CsCCD2* can use zeaxanthin, lutein, and 3-OH-β-apocarotenals as substrates to produce crocetin dialdehyde and 4-hydroxy-2,6,6-trimethyl-1-cyclohexene-1-carboxaldehyde (HTCC) or 3-OH-β-cyclocitral, but it cannot use β-carotene or lycopene as substrates (Frusciante et al. 2014). *In planta* and *Escherichia coli*, *GjCCD4a* can use β-carotene as a substrate, in addition to zeaxanthin and lutein. In *E. coli*, *GjCCD4a* can also cleave lycopene to produce crocetin dialdehyde (Xu et al. 2020; Zheng et al. 2022). Both *VgCCD4.1* and *VsCCD4.1* also cleaved β-carotene *in planta* (Morote et al. 2023b), while *BdCCD4.1*, *BdCCD4.3*, and *CaCCD2* exhibited substrate specificity similar to that of *CsCCD2* (Ahrazem et al. 2016, 2017).

The biosynthetic pathway for saffron apocarotenoids is shown in Figure 18.3. Briefly, the first and most important step of the pathway is the cleavage of the 7,8 and 7′,8′ double bonds of zeaxanthin by the corresponding CCD, resulting in two molecules of HTCC and one molecule of crocetin dialdehyde (Frusciante et al. 2014). When β-carotene is used as the carotenoid substrate for the CCD, two molecules of β-cyclocitral are obtained instead of HTCC.

Crocetin dialdehyde is converted to crocetin by an ALDH. In *C. sativus* and *G. jasminoides*, this is done by *CsALDH3* and *GjALDH2C3*, respectively (Demurtas et al. 2018; Xu et al. 2020). The glycosylation of crocetin by different UDP-glycosyltransferases (UGTs) results in the production of a series of crocins, which can be classified based on the type and number of sugar moieties at each of the two ends of the molecule. In *C. sativus* and *G. jasminoides*, *CsUGT2*, *CsUGT91P3*, *GjUGT74F8*, and *GjUGT94E13* were identified as the UGTs responsible for this step (Moraga et al. 2004; Xu et al. 2020; López et al. 2021).

The two molecules of HTCC are glycosylated by a UGT to produce picrocrocin; in *C. sativus*, this is done by *CsUGT709G1* (Diretto et al. 2019). Picrocrocin is converted to safranal during spice processing or cooking (Frusciante et al. 2014). Interestingly, the apocarotenoids derived from zeaxanthin cleavage, including picrocrocin, and the volatile compounds obtained from its degradation, are not detected in *G. jasminoides* fruits (Carmona et al. 2006).

Identifying the different CCDs able to synthesize saffron apocarotenoids has allowed their production on various heterologous platforms with different levels of achievement.

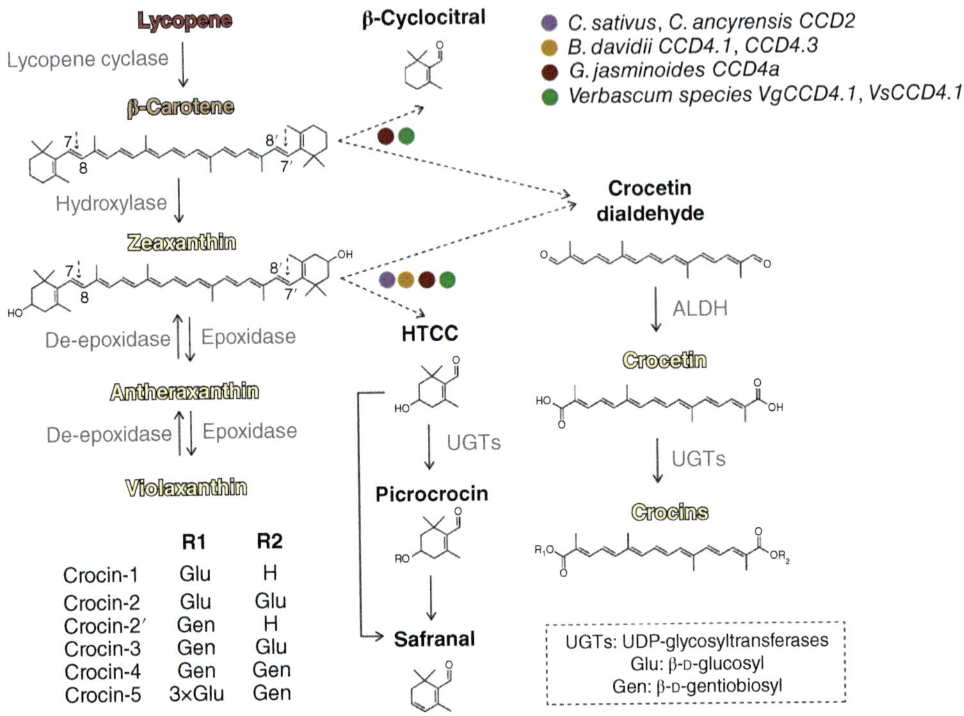

Figure 18.3 Saffron apocarotenoid biosynthetic pathway: CCD, carotenoid cleavage dioxygenase. *Source:* Adapted from Zheng et al. (2022).

Thus, only the crocin precursor crocetin was obtained in *E. coli* expressing *CsCCD2*, even when the cleavage enzyme was expressed with two *G. jasminoides UGTs* and one fungal *ALDH*. The addition of three additional bacterial *UGTs* to the system was needed to produce the glycosylated crocins (Wang et al. 2019). Crocetin has also been obtained in *Saccharomyces cerevisiae* by the expression of *CsCCD2* (Liu et al. 2020; Song et al. 2020). These results indicate that the two microbial systems lack the UGTs needed to convert crocetin to crocins; therefore, plants could be more adequate expression systems.

These valuable apocarotenoids have been produced in various heterologous plant platforms (Table 18.1). Contrary to what happens in microorganisms, in plants, adding UGTs or ALDHs is not strictly required to produce crocins in many species. For instance, *Nicotiana benthamiana*, *Nicotiana tabacum*, and *Solanum lycopersicum* have produced crocins by the heterologous expression of the CCD alone (Martí et al. 2020; Ahrazem et al. 2022b; Frusciante et al. 2022; Martí et al.,et al. 2020; Morote et al. 2023b). In plant cell suspensions of *N. benthamiana* and *N. tabacum* cv. BY-2, the expression of CCDs from either *C. sativus* or *G. jasminoides*, together with *PaCrtB* to boost carotenoid accumulation and *CsUGT93P1* to enhance crocin stability, resulted in high levels of crocins (Lobato-Gómez et al. 2025a). This was not the case for the transient transformation of maize kernel and the stable transformation of rice calluses, where the expression of *CsCCD2* and *CaCCD2*, respectively, resulted in the accumulation of low levels of crocetin, but not crocins (Frusciante et al. 2014; Ahrazem et al. 2016).

Table 18.1 Production of saffron apocarotenoids in heterologous plant platforms: CCD, carotenoid cleavage dioxygenase; PSY, plant phytoene synthase; CrtB, bacterial phytoene synthase; BCH, plant ß-carotene hydroxylase; CrtZ, bacterial ß-carotene hydroxylase; OR, ORANGE; ALDH, aldehyde dehydrogenase; UGT, UDP-glycosyltransferase; DXS, 1-deoxy-d-xylulose-5-phosphate synthase; 35S, CaMV 35S promoter; T7 RNA poly, T7 RNA polymerase/promoter system; UBQ10, *A. thaliana* ubiquitin 10 promoter; and stable (chl), chloroplast stable transformation.

Species	Expression	Trans genes	Saffron apocarotenoid production (DW)	Promoter	Reference
Nicotiana benthamiana leaf	Transient, virus-driven	*CsCCD2*	1.9-mg/g Crocins, 6.9-mg/g picrocrocin	35S	Marti et al. (2020)
		CsCCD2, PaCrtB	3.5-mg/g Crocins, 6.9-mg/g picrocrocin		
		CsCCD2, CsBCH2	2.2-mg/g Crocins, 4.2-mg/g picrocrocin		
		VgCCD4.1	9.8-mg/g Crocins		Morote et al. (2023b)
		VsCCD4.1	14.38-mg/g Crocins		
		CsCCD2	483.1-µg/g Crocins, 80.8-µg/g picrocrocin		Demurtas et al. (2023)
		CsCCD2	883.7-µg/g Crocins, 62.5-µg/g picrocrocin		
	Transient	*CsCCD2*	20.3-µg/g Crocins, 9.2-µg/g picrocrocin		
		CsCCD2	115.3-µg/g Crocins, 17.8-µg/g picrocrocin		
		CsCCD2	0.7-mg/g Crocins		Zheng et al. (2022)
		GjCCD4a	1.6-mg/g Crocins		

(*Continued*)

Table 18.1 (Continued)

Species	Expression	Trans genes	Saffron apocarotenoid production (DW)	Promoter	Reference
		BoCCD4-3	39.4-µg/g Crocins ND picrocrocin		Frusciante et al. (2022)
		CsCCD2	33.4-µg/g Crocins 0.07-µg/g picrocrocin		
	Stable	*GjCCD4a*, *GjALDH2*, *GjUGT4*, and *GjUGT94*	1-mg/g Crocins	35S, ubiquitin	Xie et al. (2023)
	Stable	*CsCCD2*	36-µg/g Crocins	35S	Ahrazem et al. (2022b)
Nicotiana tabacum leaf 		*CsCCD2*, *BrCrtZ*, and *AtORMut*	136-µg/g Crocins		
	Stable (chl)	*BoCCD4-3*	58.9-µg/g Crocins	T7 RNA poly	Frusciante et al. (2022)
Nicotiana glauca leaf 	Stable	*CsCCD2*	400-µg/g Crocins	35S	Ahrazem et al. (2022b)
		BdCCD4.1	302-µg/g Crocins		Huang et al. (2022)
Citrus paradise callus 	Stable	*GjCCD4a*, *PaCrtB*	2.7-µg/g Crocetin, 7-µg/g Crocins, 16.3-µg/g picrocrocin	35S	Zheng et al. (2022)
		GjCCD4a, *PaCrtB*, and *OsBCH*	9.9-µg/g Crocetin, 12-µg/g Crocins 42.5-µg/g picrocrocin		
		CsCCD2, *PaCrtB*	0.6-µg/g Crocetin 3.6-µg/g picrocrocin		

		CsCCD2, PaCrtB, and *OsBCH*	3-μg/g Crocetin 3-μg/g Crocins 25.8-μg/g picrocrocin		
Solanum lycopersicum fruit	Stable	*BoCCD4-3*	110.3-μg/g Crocins	35S	Frusciante et al. (2022)
		CsCCD2, CsUGT2, CsUGT709, and *TF*	14.48-mg/g Crocins 2.92-mg/g picrocrocin	E8	Ahrazem et al. (2022a)
		CsCCD2, CsUGT2, and *CsUGT709*	6× More Crocins than TF, 15× More picrocrocin than TF	E8	Lobato-Gómez et al. (2024)
Solanum tuberosum tuber	Stable	*CsCCD2, CsUGT2,* and *CsUGT709*	360-μg/g Crocins 800-μg/g picrocrocin	Batatin	Gómez-Gómez et al. (2022)
			3.7-mg/g Crocins 2.4-mg/g picrocrocin		Gómez-Gómez et al. (2024)
Zea mays endosperm	Transient	*CsCCD2*	0.01-μg/g Crocetin	35S	Frusciante et al. (2014)
Oryza sativa callus	Stable	*CaCCD2, ZmPSY1, PaCrtl,* and *AtDXS*	1-μg/g Crocetin	Ubiquitin, hordein, prolamin	Ahrazem et al. (2016)
Nicotiana benthamiana cell suspension	Stable	*CsCCD2, PaCrtB,* and *CsUGT93P1*	375-μg/g Crocins, 3-μg/g picrocrocin	35S	Lobato-Gómez et al. (2025a)
Nicotiana tabacum cell suspension	Stable	*GjCCD4a, PaCrtB,* and *CsUGT93P1*	770-μg/g Crocetin, 55-μg/g picrocrocin	35S	Lobato-Gómez et al. (2025a)

The most commonly used enzyme to produce saffron apocarotenoids in transgenic plants is *CsCCD2*. It has been introduced in *N. benthamiana* (Martí et al. 2020; Zheng et al. 2022; Demurtas et al. 2023; Lobato-Gómez et al. 2025a), *N. tabacum* (Ahrazem et al. 2022b; Lobato-Gómez et al. 2025a), *Nicotiana glauca* (Ahrazem et al. 2022b), *Citrus paradise* (Zheng et al. 2022), *S. lycopersicum* (Ahrazem et al. 2022a; Morote et al. 2023a; Lobato-Gómez et al. 2024), *Solanum tuberosum* (Gómez-Gómez et al. 2022, 2024), and *Zea mays* (Frusciante et al. 2014).

The second most used enzyme is *GjCCD4a*, probably to take advantage of its broader substrate specificity. It has been introduced in *N. benthamiana* (Zheng et al. 2022; Xie et al. 2023; Lobato-Gómez et al. 2025a), *N. tabacum* (Lobato-Gómez et al. 2025a), and *C. paradise* (Zheng et al. 2022). The CCDs from *Verbascum* species are also promising enzymes, as they can also use β-carotene as a substrate, and their overexpression in *N. benthamiana* resulted in high levels of crocin accumulation (Morote et al. 2023b).

The 35S promoter has been used to drive the expression of the CCD of interest throughout most of the plant platforms, except in the case of *N. benthamiana* stable transformation with *GjCCD4a*, where the *A. thaliana* ubiquitin 10 promoter has been used (Xie et al. 2023). In the case of tomato and potato, fruit- and tuber-specific promoters, respectively, were used to produce saffron apocarotenoids solely in the organs of interest (Ahrazem et al. 2022a; Gómez-Gómez et al. 2022, 2024; Morote et al. 2023a; Lobato-Gómez et al. 2024).

In transient experiments, the use of a viral vector derived from *the Tobacco etch virus* (TEV) resulted in significantly higher levels of crocins and picrocrocin than in conventional agroinfiltration experiments using T-DNA plasmids (Demurtas et al. 2023). TEV viral vectors produced the highest levels of crocins in *N. benthamiana* by transient expression of *CsCCD2* (Martí et al. 2020). The highest levels of transient expression overall have been obtained by expressing CCD4s from *Verbascum* species with the TEV viral system (Morote et al. 2023b).

The second most researched apocarotenoid with commercial interest after saffron apocarotenoids is bixin. Bixin is produced in the fruit of achiote (*Bixa orellana*). Bixin is a dicarboxylic monomethyl ester apocarotenoid responsible for the red color of *B. orellana* seeds, and it has been widely used in the cosmetic and culinary industries (Bouvier et al. 2003a). The suggested bixin metabolic pathway is represented in Figure 18.4. Briefly, bixin biosynthesis initiates with lycopene cleavage at 5,6 and 5′,6′ double bonds by *BoCCD4*, generating bixin aldehyde. Then, *BoALDH* oxidizes the aldehyde groups to carboxylate groups to produce norbixin. Subsequently, one of the carboxyl groups of norbixin is methylated by norbixin methyltransferase (*BoMTH*) to produce bixin (Us-Camas et al. 2022).

Unlike saffron apocarotenoids, there are no reports of bixin production in plant heterologous platforms. The expression of one CCD from *B. orellana* resulted in a limited amount of crocins in *N. benthamiana* transient expression, *N. tabacum* chloroplast transformation, and *S. lycopersicum* stable transformation. However, it failed to produce the expected bixin (Frusciante et al. 2022). Instead, bixin was produced in *E. coli* engineered to produce lycopene by the expression of *BoCCD4*, *BoALDH*, and *BoMTH* (Bouvier et al. 2003a). More recently, it was demonstrated that *BoCCD4-3* and *BoCCD1-1* can cleave lycopene both *in vitro* and in *E. coli* (Carballo-Uicab et al. 2019; Us-Camas et al. 2022).

An alternative to using heterologous systems for producing apocarotenoids of interest is to screen the diversity within the species and target genotypes that accumulate higher levels of the apocarotenoids of interest. Recently, it has been discovered that the red coloration

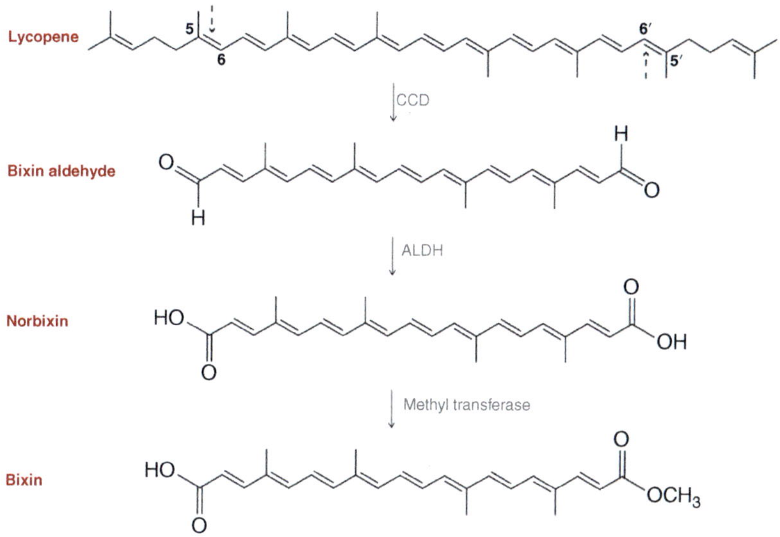

Figure 18.4 Bixin biosynthesis in *B. orellana*. *Source:* Adapted from Us-Camas et al. (2022).

in the citrus peel of some citrus varieties is the result of a cis-regulatory enhancer in the promoter of *Citrus CCD4b*, which produces β-citraurin and β-citraurinene by the cleavage of 7′,8′ double bonds of zeaxanthin and β-cryptoxanthin (Rodrigo et al. 2013; Zheng et al. 2019). Similar findings were also observed in tomato. The fruits of a *Solanum habrochaites* introgression line, IL8A, exhibited a dark orange color instead of the typical red color of the recurrent background and showed an increase in geranylacetone and β-ionone. These changes were associated with an 8,000-fold increase in *ShCCD4b* expression. It was demonstrated that *ShCCD4b* cleaves ζ-carotene and β-carotene 9,10 and 9′,10′ double bonds in an *E. coli* strain engineered to accumulate these carotenoids, which, therefore, are likely the substrates of *ShCCD4b* in IL8A fruits (Yoo et al. 2023).

18.4.2 Metabolic Engineering of Carotenoids to Provide Diverse and Higher Levels of Substrates for Apocarotenoid Production

A second approach to increase the levels of an apocarotenoid of interest is to increase the levels of the CCD substrates, carotenoids. This strategy requires identifying the specific carotenoid substrate for the CCD of interest. In some cases, CCDs can act on multiple substrates, while in others, they exhibit high specificity.

Table 18.1 shows some examples of the CCDs of interest expressed with carotenoid biosynthetic genes. In transient virus-driven experiments with *CsCCD2*, it was demonstrated that the co-expression of *CsCCD2* together with *Pantoea ananatis* PSY (*CrtB*) resulted in a significant increase in carotenoid content compared to the expression of *CsCCD2* alone (Martí et al. 2020). Similar results were observed in transgenic *N. tabacum*, where crocin accumulation was significantly higher when *CsCCD2* was co-expressed with *Bacillus* BCH (*CrtZ*) and the *A. thaliana ORANGE* (*OR*) mutant (Ahrazem et al. 2022b). In citrus callus, the co-expression of *GjCCD4a* or *CsCCD2* with *PaCrtB* and *OsBCH* resulted in significantly higher levels of crocins than the expression of the same genes, except *OsBCH*, which catalyzes the conversion from β-carotene to zeaxanthin, demonstrating that the hydroxylation of β-carotene to zeaxanthin is a rate-limiting step in this system (Zheng et al. 2022).

Another strategy to boost the production of our target apocarotenoid is to use cultivars or mutant plants that accumulate higher levels of the substrate carotenoid. This has already been exploited in combination with the expression of *CsCCD2*. In *N. benthamiana*, the transient expression of *CsCCD2* in a genotype that accumulates higher levels of zeaxanthin than the wild type resulted in an almost sixfold increase and a twofold increase with the conventional *Agrobacterium* infiltration system and the virus-driven system, respectively (Demurtas et al. 2023). In *S. tuberosum*, crocin accumulation was increased 10-fold by the use of a potato variety that accumulates high levels of zeaxanthin in the tuber (Gómez-Gómez et al. 2022, 2024). In tomato, which is the platform with the highest levels of crocins produced in plant heterologous systems, the use of tomato mutant plants that, instead of accumulating high levels of lycopene in the fruit, accumulate β-carotene or zeaxanthin, increased crocin levels by 2 to 6 and 15 times, respectively. These lines have been named Tomaffron (TF) (Morote et al. 2023a; Lobato-Gómez et al. 2024). However, the tomato plant with high levels of zeaxanthin exhibits many negative effects on plant growth, and the higher levels of crocins do not compensate for the fruit yield loss (Karniel et al. 2020; Lobato-Gómez et al. 2024).

18.4.3 Use of Platforms with Enhanced Carotenoid Levels to Produce the Apocarotenoids of Interest

Biotechnological platforms for producing apocarotenoids could benefit from strategies to increase the accumulation of a carotenoid of interest. Almost all known apocarotenoids with pharmacological, cosmetic, or other applications are derived from lycopene, β-carotene, and zeaxanthin cleavage. Table 18.2 lists different studies in which these three carotenoids have been increased using transgenic, genome editing, genome silencing, or breeding approaches. Only studies in which platforms used to produce heterologous apocarotenoids have been selected.

18.4.3.1 Lycopene

Tomato fruit is a source of high levels of lycopene. This high-antioxidant carotenoid is the substrate for bixin production (Us-Camas et al. 2022), and tomato fruit could be a promising heterologous platform to produce this red-colored apocarotenoid. Lycopene and β-carotene levels in ripe tomato fruit vary between genotypes, and even within the same genotype, accumulation may differ due to environmental and growing conditions. Several studies have aimed at enhancing lycopene levels in tomato fruit. Overexpression of 1-deoxy-D-xylulose-5-phosphate synthase (*DXS*), a key enzyme in the MEP pathway, led to lycopene accumulation of 6.7 mg/g DW (Enfissi et al. 2005). Overexpression of the *A. thaliana OR* mutant in tomato led to nearly a twofold increase in lycopene content (Yazdani et al. 2019). The silencing of stay-green 1 (*SGR1*) resulted in a 3.9-fold increase in lycopene production compared to the wild type (Luo et al. 2013). Similar results were observed by targeting *SlSGR1* with CRISPR/Cas9, which resulted in a 3.1-fold increase in lycopene content in ripe fruit (Gianoglio et al. 2022). A *cycB* mutant identified in a TILLING population accumulated 2.4 times more lycopene than the wild type (Silletti et al. 2013), as observed in the *old-gold* recessive mutants (Ronen et al. 2000).

Table 18.2 Selected examples of strategies to increase the carotenoid content in different plant systems: Crtl, bacterial phytoene desaturase; PSY, plant phytoene synthase; CrtB, bacterial phytoene synthase; LCYb, plant lycopene β-cyclase; CYCb, chromoplastic LCYb; CrtY, bacterial lycopene β-cyclase; CHY and BCH, β-carotene hydroxylases; *hp3*, *high pigment 3* mutation; $hp2^{dg}$, *high pigment 2* mutation; B^{Sh}, *BETA* mutation from *Solanum habrochaites; gs*, *green stripe* mutation; SGR1, stay-green 1; CCD, carotenoid cleavage dioxygenase; OR, ORANGE; ZEP, zeaxanthin epoxidase; DXS, 1-deoxy-d-xylulose-5-phosphate synthase; and Chl, chloroplast.

Species	Strategy	Target gene(s)	Carotenoid increase (DW)	Reference
Solanum lycopersicum fruit 	Overexpression	*EuCrtl*	3.5-Fold increase β-carotene (700 μg/g)	Romer et al. (2000)
		EcDXS	1.6-Fold increase lycopene (6.7 mg/g), 1.4-Fold increase β-carotene (700 μg/g)	Enfissi et al. (2005)
		$AtOR^{His}$	1.8-Fold increase lycopene (989 μg/g), 3.1-Fold increase β-carotene (359 μg/g FW)	Yazdani et al. (2019)
		SlPSY	1.4-Fold increase β-carotene (819.3 μg/g)	Fraser et al. (2007)
	Overexpression (chl)	*LCYb*	5-Fold increase β-carotene (980 μg/g)	Apel and Bock (2009)
	Silencing	*SlSGR1*	3.9-Fold increase lycopene (328.5 μg/g), 34.5-Fold increase β-carotene (69 μg/g)	Luo et al. (2013)
	TILLING	*SlCYCB*	2.3-Fold increase lycopene (1.4 mg/g)	Silletti et al. (2013)
$hp3/B^{Sh}$ Xantomato	Natural mutations	*hp3*, B^{Sh}	26-Fold increase β-carotene (287.2 μg/g)	Karniel et al. (2020)
		hp3, hp^{2d}, B^{Sh}, and *gs*	37-Fold increase β-carotene (474 μg/g), zeaxanthin detected (577 μg/g)	
Solanum tuberosum tuber 	Overexpression	*EuCrtB*, *EuCrtl*, and *EuCrtY*	3,600-Fold increase β-carotene (47 μg/g)	Diretto et al. (2007a)
		BoOR	β-Carotene detected (5 μg/g)	Lopez et al. (2008)

(Continued)

Table 18.2 (Continued)

Species	Strategy	Target gene(s)	Carotenoid increase (DW)	Reference
	Silencing	*StLCYe*	14-Fold increase β-carotene (43.6 ng/g)	Diretto et al. (2006)
		StCHY1, StCHY2	38-Fold increase β-carotene (85.3 ng/g)	Diretto et al. (2007b)
		StZEP	130-Fold increase zeaxanthin (40 μg/g)	Romer et al. (2002)
Zea mays endosperm 	Overexpression	*ZmPSY1, PaCrtl, Osdhar,* and *EcfolE*	Lycopene detected (23 μg/g), 169-Fold increase β-carotene (59.3 μg/g), 112-Fold increase zeaxanthin (35.8 μg/g)	Naqvi et al. (2009)
Golden Rice 1, Golden Rice 2	Overexpression	*NpPSY, EuCrtl*	β-Carotene detected (2 μg/g)	Ye et al. (2000)
		ZmPSY, and *EuCrtl*	23-Fold increase β-carotene (37 μg/g)	Paine et al. (2005)
		CaPSY, PaCrtl, and *CaBCH*	Zeaxanthin detected (0.8 μg/g)	Ha et al. (2019)
Oryza sativa endosperm 	CRISPR/Cas9	*OsOR*	β-Carotene detected (2.9 μg/g)	Endo et al. (2019)
	Silencing	*OsCCD1*	1.3-Fold increase β-carotene (1.5 μg/g), 1.4-Fold-increase zeaxanthin (0.1 μg/g)	Ko et al. (2018)
		OsCCD4a	Twofold increase β-carotene (2 μg/g), 1.7-Fold increase zeaxanthin (0.1 μg/g)	

18.4.3.2 β-Carotene

β-Carotene has been the target of many metabolic engineering approaches due to its importance in the human diet, among other applications. In this context, β-carotene is the substrate of *G. jasminoides* CCD4a and *Verbascum* species CCD4s to produce crocins and β-cyclocitral. In contrast to lycopene, different strategies have been used to produce higher levels of this carotenoid in all the selected plant platforms. The highest levels of β-carotene (almost 1 mg/g DW) have been obtained in tomato fruit by chloroplast transformation with *Erwinia herbicola LCYb* gene (Apel and Bock 2009). High levels of β-carotene, from 700 to 820 μg/g, have been achieved in tomato fruit by the overexpression of *Erwinia uredovora* carotene desaturase (*CrtI*) (Römer et al. 2000), *EcDXS* (Enfissi et al. 2005), or *SlPSY* (Enfissi et al. 2005), showing that β-carotene content can be engineered in different steps of the carotenoid biosynthetic pathway in tomato fruit.

In tomato fruit, the chromoplastic *LCYb* (*CycB*) expression is very low. However, plants with mutations in the promoter region of *CycB* exhibited a significantly higher expression of this gene in the fruit, and the β-carotene percentage in the fruit increased from 7% to 46% (Ronen et al. 2000). This mutation, named B^{Sh}, was combined with the *high pigment 3* mutation (*hp3*), which impaired ZEP and resulted in a 26-fold increase (287 μg/g DW) in β-carotene content compared to M82. The increase was even greater (37-fold increase, 474 μg/g DW) when these two mutations were combined with *high pigment 2* ($hp2^{dg}$) and *green stripe* (*gs*) mutations; this mutant plant was named Xantomato (Karniel et al. 2020).

The levels of β-carotene in the other platforms are significantly lower compared to tomato fruit. For instance, the highest levels of β-carotene obtained in potato tuber were achieved by the expression of *E. uredovora CrtB*, *CrtI*, and LCYb (*CrtY*). The transgenic potato tubers accumulated 3,600 times more β-carotene than the wild-type tubers, and the final concentration was 47 μg/g DW (Diretto et al. 2007a). The biofortification of *Z. mays* by expressing *ZmPSY1*, *PaCrtI*, *Oryza sativa dhar*, and *E. coli folE* resulted in plants accumulating 169 times more β-carotene in the kernels, with a final concentration of 59.3 μg/g DW (Naqvi et al. 2009). In rice, the highest β-carotene levels (37 μg/g DW) were obtained in Golden Rice 2 by the expression of *ZmPSY* and *EuCrtI* (Paine et al. 2005).

18.4.3.3 Zeaxanthin

Zeaxanthin is the only carotenoid substrate that can be utilized by all the identified CCDs producing saffron apocarotenoids. The accumulation of zeaxanthin in maize kernels was increased 112 times in biofortified *Z. mays*, reaching 38 μg/g DW (Naqvi et al. 2009). In rice, a slight increase in zeaxanthin was achieved by silencing *OsCCD1* and *OsCCD4a* independently in the lines overexpressing *CaPSY* and *PaCrtI*. The levels reached were 0.1 μg/g DW (Ko et al. 2018). Interestingly, zeaxanthin does not accumulate in conventional red-ripe tomato fruit. However, up to 577 μg/g DW of zeaxanthin was accumulated in Xantomato fruit by stacking the four *hp3*, $hp2^{dg}$, B^{Sh}, and *gs* mutations (Karniel et al. 2020). Xantomato represents an interesting platform for expressing *CsCCD2*.

18.4.3.4 Targeting Endogenous CCDs Competing for the Same Substrate

Another strategy to increase the apocarotenoids of interest involves downregulating endogenous CCDs that utilize the same carotenoid substrate as the CCD of interest. In this case, the CCDs targeted for downregulation cleave the carotenoid at a different double bond than the one involved in producing the desired apocarotenoid. The endogenous CCDs should first be identified and characterized in order to use this approach.

18.4.4 Selection of an Appropriate Matrix to Produce the Apocarotenoids of Interest

From a biotechnological point of view, achieving high levels of the target apocarotenoid is as important as defining the organs where the metabolite will be produced, the organelle where it will be stored, whether it will negatively impact plant growth and development, how stable the apocarotenoid will be in the matrix, how scalable the platform is, and the level of containment needed (Figure 18.5).

The apocarotenoids that accumulate in high amounts in their natural sources are mainly water-soluble, suggesting that they are stored in vacuoles. It has been demonstrated that crocins are synthesized in the chromoplast and subsequently sequestered into vacuoles (Moraga et al. 2004). Bixin, β-citraurin, and β-citraurinene are most likely sequestered in the vacuole due to their polar nature (Bouvier et al. 2003a; Zheng et al. 2021). Other volatile apocarotenoids accumulate at lower levels in the cells because they are not stored in storage organelles and are, therefore, more challenging to produce in heterologous platforms at high levels since they are readily released.

Nicotiana benthamiana is a highly suitable platform for performing transient expression experiments and testing the effect of the enzymes of interest, as it has been done by the expression of *GjCCD4a*, *CsCCD2*, *Verbascum* species *CCD4s*, and *BoCCD4-3*. The transient expression process is fast, allowing the rapid evaluation of different CCDs' ability to produce saffron apocarotenoids using the substrates available in the leaves (Table 18.1). *Nicotiana tabacum* can also be used to perform transient experiments, although it has not yet been tested for any CCD.

Stable transformation of *N. benthamiana* and *N. tabacum* is slower than in transient experiments but faster than in many other crop plants, and it would facilitate the upscaling of production. However, using constitutive promoters could lead to the depletion of carotenoids, and the accumulation of high levels of apocarotenoids could negatively affect plant growth. Additionally, the use of strong constitutive promoters can result in post-transcriptional silencing of the transgenes (Matsunaga et al. 2019).

In contrast to whole plants, plant cell suspensions could be the ideal platform for containment, sterility, and environmental control (Lobato-Gómez et al. 2025a). Plant cell packs (PCPs) could be used to transiently express and test different combinations of genes of interest by infusing *Agrobacterium tumefaciens* carrying the appropriate construct. Compared to transient expression experiments in *Nicotiana*, this method allows for the simultaneous multiplex analysis of more constructs (Rademacher et al. 2019). PCPs are a valuable tool for identifying the best transgene combination and serve as a first step before transforming cell suspensions, which would allow scaling up the production of the apocarotenoid of interest. Plant cell suspensions can be grown in bioreactors, and the optimal conditions for producing the metabolite of interest can be easily optimized (Huang and McDonald 2012). One disadvantage of plant cell suspensions compared to transgenic whole-plant systems is that they require continuous growth and frequent cell refreshment; one solution to this problem is the cryopreservation of cells (Mustafa et al. 2011). However, the main disadvantage of producing apocarotenoids in bioreactors is the high initial investment required

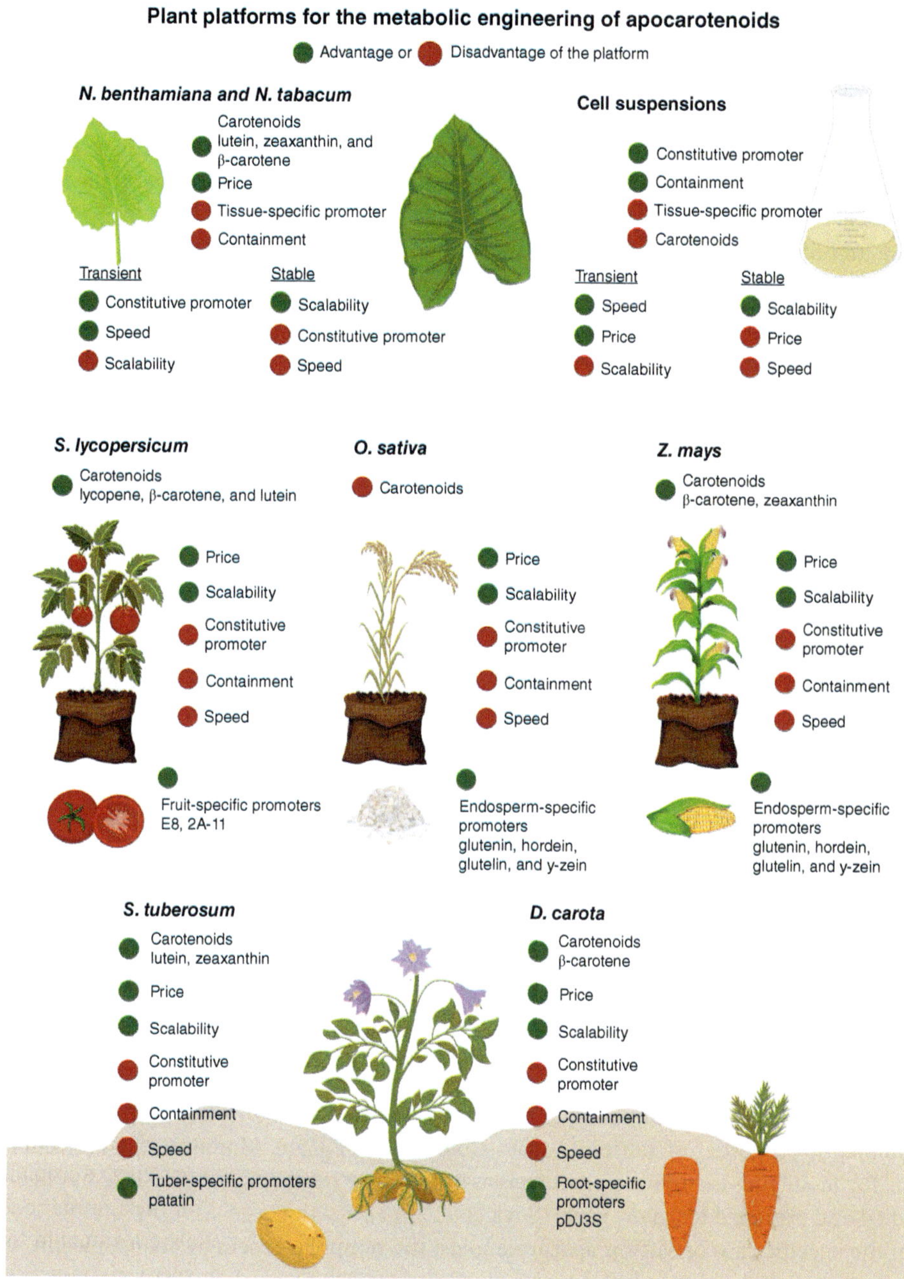

Figure 18.5 Proposed matrices to produce heterologous apocarotenoids.

to set them up. Nevertheless, various bioreactors are now available, including disposable bioreactors, which are more affordable (Lehmann et al. 2014).

The most suitable plant platforms for producing apocarotenoids direct their accumulation to storage organs using promoters expressed only in the targeted organ when this organ has already achieved the mature stage. This results in the production of apocarotenoids of interest without growth and developmental penalties. Tomato, rice, maize, potato, and carrot (*Daucus carota*) are food crops optimized for the cultivation and productivity of those target organs, and the harvesting of fruits, kernels, or tubers can be done mechanically. These crops have genetic backgrounds selected for the accumulation of substrate carotenoids, and the apocarotenoids of interest will be accumulated in a matrix selected for human consumption.

Rice and maize are two of the most cultivated staple crops (FAOSTAT 2024) that can be processed into different products for long-term storage. They can be engineered with CCDs using endosperm-specific promoters. The main advantage of maize over rice as a possible apocarotenoid biotechnological platform is the availability of varieties that accumulate high levels of carotenoids in the endosperm. In addition, even though carotenoid accumulation has been achieved in rice through metabolic engineering, the levels remain low compared to those in other available platforms (Table 18.2).

Potato is the most cultivated vegetable worldwide, followed by tomato (FAOSTAT 2024). It represents an interesting biotechnological platform for producing apocarotenoids, as tuber-specific promoters have been identified, and lutein and zeaxanthin accumulate in the tuber of some varieties. In addition, metabolic-engineered potato plants with higher levels of lutein, zeaxanthin, and even β-carotene in the tubers are available (Table 18.2).

Carrot has not yet been used to metabolically engineer the accumulation of exotic apocarotenoids. However, it is a promising platform due to its high levels of β-carotene (1 mg/g DW), the availability of root-specific promoters, and the characterization of some of its CCDs (Arango et al. 2010; Li et al. 2021).

Solanum lycopersicum has been proven to be the best platform for producing saffron apocarotenoids due to the availability of fruit-specific promoters acting during ripening, such as E8, and the activation of the carotenoid pathway in the fruit (Ahrazem et al. 2022a). In addition, the natural variation in fruit carotenoid composition allows for the selection of genotypes that accumulate higher amounts of the different carotenoid substrates for the CCDs (Karniel et al. 2020). An additional advantage of producing saffron apocarotenoids in tomato fruit is that the tomato processing and canning industry has developed cost-effective methods for the long-term preservation of tomato-based products, including dry powder and concentrates (Kelebek et al. 2017), with the latter being a suitable matrix for the storage of saffron apocarotenoids within a tomato matrix, which demonstrates the stability of these valuable compounds after processing and storage, while maintaining the organoleptic properties of saffron (Lobato-Gómez et al. 2025b). Moreover, the molecular basis for modifying the tomato plant chassis to facilitate mechanical harvesting has been studied and exploited for many years (Soyk et al. 2017).

In the specific case of saffron apocarotenoids, the nonpolar crocin aglycon, crocetin, is efficiently glycosylated to produce crocins in *C. sativus* (Dufresne et al. 1999). Crocins and picrocrocin are polar metabolites, and the interaction between chromoplasts and vacuoles observed in *C. sativus* at the stage of highest apocarotenoid accumulation suggests their sequestration in the vacuole (Bouvier et al. 2003b), which was confirmed by the identification of an ABCC transporter mediating the accumulation of crocins in the vacuole (Demurtas et al. 2019). In *N. benthamiana* leaves transiently expressing *CsCCD2*, crocins were localized in the vacuole, suggesting the presence of transporters that sequester crocins into the vacuole in heterologous plant systems (Martí et al. 2020).

The stability of saffron apocarotenoids post-harvest and during storage has been evaluated in *C. sativus* and *G. jasminoides*. In the saffron plant, the temperature and the drying time employed in the spice processing are detrimental to its final apocarotenoid profile (Yao et al. 2018). In *G. jasminoides*, crocins exhibited a decrease after thermal treatments ranging from 50 to 100 °C, while in *C. sativus*, the drying of saffron stigmata at temperatures up to 90 °C for short times resulted in higher crocin levels compared to non-processed samples (Rocchi et al. 2018). During storage, the HTCC and picrocrocin contents decrease while safranal content increases (Sereshti et al. 2018). The difference observed in crocin stability between saffron and gardenia reflects the importance of the matrix where these apocarotenoids are accumulated.

Taken together, to optimize the metabolic engineering of apocarotenoids, knowledge of the stepwise enzymes in the pathway and the availability of the carotenoid substrate are not the only important factors to consider. Other factors should be considered when aiming to develop a biotechnological platform, including the matrix in which the apocarotenoids will be produced, the effect that these apocarotenoids will have on the biotechnological plant platform, the stability of the compounds after processing the matrix, their stability during storage, the scalability and possibility of mechanization of the process, and the containment of the platform.

References

Ablazov, A., Mi, J., Jamil, M. et al. (2020). The apocarotenoid zaxinone is a positive regulator of strigolactone and abscisic acid biosynthesis in Arabidopsis roots. *Frontiers in Plant Science* 11, 578. https://doi.org/10.3389/FPLS.2020.00578.

Ablazov, A., Votta, C., Fiorilli, V. et al. (2023). *ZAXINONE SYNTHASE 2* regulates growth and arbuscular mycorrhizal symbiosis in rice. *Plant Physiology* 191(1), 382–399. https://doi.org/10.1093/PLPHYS/KIAC472.

Ahrazem, O., Diretto, G., Argandoña, J. et al. (2017). Evolutionarily distinct carotenoid cleavage dioxygenases are responsible for crocetin production in *Buddleja davidii*. *Journal of Experimental Botany* 68(16), 4663–4677. https://doi.org/10.1093/jxb/erx277.

Ahrazem, O., Diretto, G., Rambla, J.L. et al. (2022a). Engineering high levels of saffron apocarotenoids in tomato. *Horticulture Research* 9, uhac074. https://doi.org/10.1093/hr/uhac074.

Ahrazem, O., Rubio-Moraga, A., Berman, J. et al. (2016). The carotenoid cleavage dioxygenase CCD2 catalysing the synthesis of crocetin in spring crocuses and saffron is a plastidial enzyme. *New Phytologist* 209(2), 650–663. https://doi.org/10.1111/nph.13609.

Ahrazem, O., Zhu, C., Huang, X. et al. (2022b). Metabolic engineering of crocin biosynthesis in *Nicotiana* species. *Frontiers in Plant Science* 13, 861140. https://doi.org/10.3389/fpls.2022.861140.

Al-Babili, S. and Bouwmeester, H.J. (2015). Strigolactones, a novel carotenoid-derived plant hormone. *Annual Review of Plant Biology* 66, 161–186. https://doi.org/10.1146/ANNUREV-ARPLANT-043014-114759.

Alder, A., Jamil, M., Marzorati, M. et al. (2012). The path from β-carotene to carlactone, a strigolactone-like plant hormone. *Science* 335(6074), 1348–1351. https://doi.org/10.1126/SCIENCE.1218094.

Alquézar, B., Rodrigo, M. and Zacarías, L. (2009). Carotenoid biosynthesis and their regulation in citrus fruits. *Tree and Forestry Science and Biotechnology* 2(1), 23–37.

Apel, W. and Bock, R. (2009). Enhancement of carotenoid biosynthesis in transplastomic tomatoes by induced lycopene-to-provitamin A conversion. *Plant Physiology* 151(1), 59–66. https://doi.org/10.1104/PP.109.140533.

Arango, J., Salazar, B., Welsch, R. et al. (2010). Putative storage root specific promoters from cassava and yam: cloning and evaluation in transgenic carrots as a model system. *Plant Cell Reports* 29(6), 651–659. https://doi.org/10.1007/S00299-010-0851-7.

Beltran, J.C.M. and Stange, C. (2016). Apocarotenoids: a new carotenoid-derived pathway. *Sub-Cellular Biochemistry* 79, 239–272. https://doi.org/10.1007/978-3-319-39126-7_9.

Bouvier, F., Dogbo, O. and Camara, B. (2003a). Biosynthesis of the food and cosmetic plant pigment bixin (annatto). *Science* 300(5628), 2089–2091. https://doi.org/10.1126/SCIENCE.1085162.

Bouvier, F., Isner, J.C., Dogbo, O. et al. (2005). Oxidative tailoring of carotenoids: a prospect towards novel functions in plants. *Trends in Plant Science* 10(4), 187–194. https://doi.org/10.1016/j.tplants.2005.02.007.

Bouvier, F., Suire, C., Mutterer, J. et al. (2003b). Oxidative remodeling of chromoplast carotenoids: identification of the carotenoid dioxygenase *CsCCD* and *CsZCD* genes involved in Crocus secondary metabolite biogenesis. *Plant Cell* 15(1), 47–62. https://doi.org/10.1105/tpc.006536.

Britton, G., Liaaen-Jensen, S. and Pfander, H. (eds.) (1996). *Carotenoids:* Volume 2: *Synthesis* (1st ed). Birkhäuser Basel. https://doi.org/10.1007/978-3-0348-9323-7.

Calvo, M.M. (2005). Lutein: a valuable ingredient of fruit and vegetables. *Critical Reviews in Food Science and Nutrition* 45 (7–8), 671–696. https://doi.org/10.1080/10408690590957034.

Carballo-Uicab, V.M., Cárdenas-Conejo, Y., Vallejo-Cardona, A.A. et al. (2019). Isolation and functional characterization of two dioxygenases putatively involved in bixin biosynthesis in annatto (*Bixa orellana* L.). *PeerJ* 7(6), e7064. https://doi.org/10.7717/PEERJ.7064.

Carmona, M., Zalacain, A., Sánchez, A.M. et al. (2006). Crocetin esters, picrocrocin and its related compounds present in *Crocus sativus* stigmas and *Gardenia jasminoides* fruits. Tentative identification of seven new compounds by LC-ESI-MS. *Journal of Agricultural and Food Chemistry* 54(3), 973–979. https://doi.org/10.1021/JF052297W.

Cunningham, F.X. and Gantt, E. (2002). Genes and enzymes of carotenoid biosynthesis in plants. *Annual Review of Plant Physiology and Plant Molecular Biology* 49(1), 557–583. https://doi.org/10.1146/annurev.arplant.49.1.557.

D'Alessandro, S., Ksas, B. and Havaux, M. (2018). Decoding β-cyclocitral-mediated retrograde signaling reveals the role of a detoxification response in plant tolerance to photooxidative stress. *The Plant Cell* 30(10), 2495–2511. https://doi.org/10.1105/TPC.18.00578.

Demurtas, O.C., de Brito Francisco, R., Diretto, G. et al. (2019). ABCC transporters mediate the vacuolar accumulation of crocins in saffron stigmas. *The Plant Cell* 31(11), 2789–2804. https://doi.org/10.1105/tpc.19.00193.

Demurtas, O.C., Frusciante, S., Ferrante, P. et al. (2018). Candidate enzymes for saffron crocin biosynthesis are localized in multiple cellular compartments. *Plant Physiology* 177(3), 990–1006. https://doi.org/10.1104/pp.17.01815.

Demurtas, O.C., Sulli, M., Ferrante, P. et al. (2023). Production of saffron apocarotenoids in *Nicotiana benthamiana* plants genome-edited to accumulate zeaxanthin precursor. *Metabolites* 13(6), 729. https://doi.org/10.3390/metabo13060729.

Dickinson, A.J., Lehner, K., Mi, J. et al. (2019). β-Cyclocitral is a conserved root growth regulator. *Proceedings of the National Academy of Sciences of the United States of America* 116(21), 10563–10567. https://doi.org/10.1073/pnas.1821445116.

Diretto, G., Ahrazem, O., Rubio-Moraga, Á. et al. (2019). UGT709G1: a novel uridine diphosphate glycosyltransferase involved in the biosynthesis of picrocrocin, the precursor of safranal in saffron (*Crocus sativus*). *New Phytologist* 224(2), 725–740. https://doi.org/10.1111/nph.16079.

Diretto, G., Al-Babili, S., Tavazza, R. et al. (2007a). Metabolic engineering of potato carotenoid content through tuber-specific overexpression of a bacterial mini-pathway. *PLoS ONE* 2(4), e350. https://doi.org/10.1371/JOURNAL.PONE.0000350.

Diretto, G., Tavazza, R., Welsch, R. et al. (2006). Metabolic engineering of potato tuber carotenoids through tuber-specific silencing of lycopene epsilon cyclase. *BMC Plant Biology* 6(1), 13.

Diretto, G., Welsch, R., Tavazza, R., Mourgues, F., Pizzichini, D., Beyer, P., & Giuliano, G. (2007b). Silencing of beta-carotene hydroxylase increases total carotenoid and beta-carotene levels in potato tubers. BMC Plant Biology, 7(1), 11.

Dufresne, C., Cormier, F., Dorion, S. et al. (1999). Glycosylation of encapsulated crocetin by a *Crocus sativus* L. cell culture. *Enzyme and Microbial Technology* 24 (8–9), 453–462. https://doi.org/10.1016/S0141-0229(98)00143-4.

Endo, A., Saika, H., Takemura, M. et al. (2019). A novel approach to carotenoid accumulation in rice callus by mimicking the cauliflower Orange mutation via genome editing. *Rice* 12(1), 81.

Enfissi, E.M.A., Fraser, P.D., Lois, L.M. et al. (2005). Metabolic engineering of the mevalonate and non-mevalonate isopentenyl diphosphate-forming pathways for the production of health-promoting isoprenoids in tomato. *Plant Biotechnology Journal* 3(1), 17–27. https://doi.org/10.1111/J.1467-7652.2004.00091.X.

FAOSTAT (2024). Food and agriculture data. https://www.fao.org/faostat/en/.20 April, 2025.

Farré, G., Blancquaert, D., Capell, T. et al. (2014). Engineering complex metabolic pathways in plants. *Annual Review of Plant Biology* 65, 187–223. https://doi.org/10.1146/annurev-arplant-050213-035825.

Farré, G., Sanahuja, G., Naqvi, S. et al. (2010). Travel advice on the road to carotenoids in plants. *Plant Science* 179 (1–2), 28–48. https://doi.org/10.1016/j.plantsci.2010.03.009.

Fraser, P.D., Enfissi, E.M., Halket, J.M. et al. (2007). Manipulation of phytoene levels in tomato fruit: effects on isoprenoids, plastids, and intermediary metabolism. *The Plant Cell* 19(10), 3194–3211.

Frusciante, S., Diretto, G., Bruno, M. et al. (2014). Novel carotenoid cleavage dioxygenase catalyzes the first dedicated step in saffron crocin biosynthesis. *Proceedings of the National Academy of Sciences of the United States of America* 111(33), 12246–12251. https://doi.org/10.1073/pnas.1404629111.

Frusciante, S., Demurtas, O.C., Sulli, M. et al. (2022). Heterologous expression of *Bixa orellana* cleavage dioxygenase 4–3 drives crocin but not bixin biosynthesis. *Plant Physiology* 188(3), 1469–1482.

Gianoglio, S., Comino, C., Moglia, A. et al. (2022). In-depth characterization of greenflesh tomato mutants obtained by CRISPR/Cas9 editing: a case study with implications for breeding and regulation. *Frontiers in Plant Science* 13, 936089. https://doi.org/10.3389/fpls.2022.936089.

Gómez-Gómez, L., López-Jimenez, A.J., Martinez Fajardo, C. et al. (2024). Metabolic engineering of crocins and picrocrocin apocarotenoids in potato group phureja. *Engineering Reports* 6(12), e12997. https://doi.org/10.1002/ENG2.12997.

Gómez-Gómez, L., Morote, L., Frusciante, S. et al. (2022). Fortification and bioaccessibility of saffron apocarotenoids in potato tubers. *Frontiers in Nutrition* 9, 1045979. https://doi.org/10.3389/fnut.2022.1045979.

Ha, S.H., Kim, J.K., Jeong, Y.S. et al. (2019). Stepwise pathway engineering to the biosynthesis of zeaxanthin, astaxanthin and capsanthin in rice endosperm. *Metabolic Engineering* 52, 178–189.

Huang, T.K. and McDonald, K.A. (2012). Bioreactor systems for *in vitro* production of foreign proteins using plant cell cultures. *Biotechnology Advances* 30(2), 398–409. https://doi.org/10.1016/J.BIOTECHADV.2011.07.016.

Huang, F.C., Molnár, P. and Schwab, W. (2009). Cloning and functional characterization of carotenoid cleavage dioxygenase 4 genes. *Journal of Experimental Botany* 60(11), 3011–3022. https://doi.org/10.1093/JXB/ERP137.

Huang, X., Morote, L., Zhu, C. et al. (2022). The biosynthesis of non-endogenous apocarotenoids in transgenic *Nicotiana glauca*. *Metabolites* 12(7), 575.

Ilg, A., Beyer, P. and Al-Babili, S. (2009). Characterization of the rice carotenoid cleavage dioxygenase 1 reveals a novel route for geranial biosynthesis. *The FEBS Journal* 276(3), 736–747. https://doi.org/10.1111/J.1742-4658.2008.06820.X.

Jonathan, D.C., Rosa, U.C., Yair, C.C. et al. (2023). Carotenoid cleavage dioxygenases 4 from woodiness plant and their relationships with herbaceous plants. *Industrial Crops and Products* 205, 117529. https://doi.org/10.1016/J.INDCROP.2023.117529.

Karniel, U., Koch, A., Zamir, D. et al. (2020). Development of zeaxanthin-rich tomato fruit through genetic manipulations of carotenoid biosynthesis. *Plant Biotechnology Journal* 18, 2292–2303. https://doi.org/10.1111/pbi.13387.

Kelebek, H., Selli, S., Kadiroğlu, P. et al. (2017). Bioactive compounds and antioxidant potential in tomato pastes as affected by hot and cold break process. *Food Chemistry* 220, 31–41. https://doi.org/10.1016/j.foodchem.2016.09.190.

Ko, M.R., Song, M.H., Kim, J.K. et al. (2018). RNAi-mediated suppression of three carotenoid-cleavage dioxygenase genes, *OsCCD1*, *4a*, and *4b*, increases carotenoid content in rice. *Journal of Experimental Botany* 69(21), 5105–5116. https://doi.org/10.1093/jxb/ery300.

Lehmann, N., Dittler, I., Lämsä, M. et al. (2014). Disposable bioreactors for cultivation of plant cell cultures. In: *Production of Biomass and Bioactive Compounds Using Bioreactor Technology* (eds. K.Y. Paek, H. Murthy and J.J. Zhong), 17–46. Springer. https://doi.org/10.1007/978-94-017-9223-3_2.

Li, T., Deng, Y.J., Liu, J.X. et al. (2021). DcCCD4 catalyzes the degradation of α-carotene and β-carotene to affect carotenoid accumulation and taproot color in carrot. *The Plant Journal* 108(4), 1116–1130. https://doi.org/10.1111/TPJ.15498.

Liang, M.H., He, Y.J., Liu, D.M. et al. (2021). Regulation of carotenoid degradation and production of apocarotenoids in natural and engineered organisms. *Critical Reviews in Biotechnology* 41(4), 513–534. https://doi.org/10.1080/07388551.2021.1873242.

Liu, T., Dong, C., Qi, M. et al. (2020). Construction of a stable and temperature-responsive yeast cell factory for crocetin biosynthesis using CRISPR-Cas9. *Frontiers in Bioengineering and Biotechnology* 8, 653. https://doi.org/10.3389/FBIOE.2020.00653.

Lobato-Gómez, M., Drapal, M., Fernández-Muñoz, R. et al. (2024). Maximizing saffron apocarotenoid production in varied tomato fruit carotenoid contexts. *The Plant Journal* 120, 966–983. https://doi.org/10.1111/tpj.17030.

Lobato-Gómez, M., Laurel, M., Vázquez-Vilar, M. et al. (2025a). Novel plant cell suspension platforms for saffron apocarotenoid production and its impact on carotenoid and volatile profiles. *Plant Biotechnology Journal* 23, 3903–3918. https://doi.org/10.1111/pbi.70153.

Lobato-Gómez, M., Rambla, J.L., Zacarías-García, J. et al. (2025b). Valorization of transgenic Tomaffron-based products enriched in saffron apocarotenoids. *Food Chemistry* 488, 144828. https://doi.org/10.1016/J.FOODCHEM.2025.144828.

López, A.J., Frusciante, S., Niza, E. et al. (2021). A new glycosyltransferase enzyme from family 91, UGT91P3, is responsible for the final glucosylation step of crocins in saffron (*Crocus sativus* L.). *International Journal of Molecular Sciences* 22(16), 8815. https://doi.org/10.3390/ijms22168815.

Lopez, A.B., Van Eck, J., Conlin, B.J. et al. (2008). Effect of the cauliflower or transgene on carotenoid accumulation and chromoplast formation in transgenic potato tubers. *Journal of Experimental Botany* 59(2), 213–223.

Luo, Z., Zhang, J., Li, J. et al. (2013). A STAY-GREEN protein SlSGR1 regulates lycopene and β-carotene accumulation by interacting directly with SlPSY1 during ripening processes in tomato. *New Phytologist* 198(2), 442–452. https://doi.org/10.1111/nph.12175.

Ma, G., Zhang, L., Matsuta, A. et al. (2013). Enzymatic formation of β-citraurin from β-cryptoxanthin and zeaxanthin by carotenoid cleavage dioxygenase4 in the flavedo of citrus fruit. *Plant Physiology* 163(2), 682–695. https://doi.org/10.1104/PP.113.223297.

Martí, M., Diretto, G., Aragonés, V. et al. (2020). Efficient production of saffron crocins and picrocrocin in *Nicotiana benthamiana* using a virus-driven system. *Metabolic Engineering* 61, 238–250. https://doi.org/10.1016/j.ymben.2020.06.009.

Matsunaga, W., Shimura, H., Shirakawa, S. et al. (2019). Transcriptional silencing of 35S driven-transgene is differentially determined depending on promoter methylation heterogeneity at specific cytosines in both plus- and minus-sense strands. *BMC Plant Biology* 19(1), 24. https://doi.org/10.1186/S12870-019-1628-Y.

McCarty, D.R., Hattori, T., Carson, C.B. et al. (1991). The viviparous-1 developmental gene of maize encodes a novel transcriptional activator. *Cell* 66(5), 895–905. https://doi.org/10.1016/0092-8674(91)90436-3.

Moraga, A.R., Nohales, P.F., Pérez, J.A.F. et al. (2004). Glucosylation of the saffron apocarotenoid crocetin by a glucosyltransferase isolated from *Crocus sativus* stigmas. *Planta* 219(6), 955–966. https://doi.org/10.1007/s00425-004-1299-1.

Morote, L., Lobato-Gómez, M., Ahrazem, O. et al. (2023a). Crocins-rich tomato extracts showed enhanced protective effects *in vitro*. *Journal of Functional Foods* 101, 105432. https://doi.org/10.1016/j.jff.2023.105432.

Morote, L., Rubio-Moraga, Á., López Jiménez, A.J. et al. (2023b). *Verbascum* species as a new source of saffron apocarotenoids and molecular tools for the biotechnological production of crocins and picrocrocin. *The Plant Journal* 14, 1–15. https://doi.org/10.1111/tpj.16589.

Mustafa, N.R., de Winter, W., van Iren, F. et al. (2011). Initiation, growth and cryopreservation of plant cell suspension cultures. *Nature Protocols* 6(6), 715–742. https://doi.org/10.1038/nprot.2010.144.

Nagegowda, D.A. and Gupta, P. (2020). Advances in biosynthesis, regulation, and metabolic engineering of plant specialized terpenoids. *Plant Science* 294, 110457. https://doi.org/10.1016/J.PLANTSCI.2020.110457.

Naqvi, S., Zhu, C., Farre, G. et al. (2009). Transgenic multivitamin corn through biofortification of endosperm with three vitamins representing three distinct metabolic pathways. *Proceedings of the National Academy of Sciences of the United States of America* 106(19), 7762–7767. https://doi.org/10.1073/PNAS.0901412106.

Nisar, N., Li, L., Lu, S. et al. (2015). Carotenoid metabolism in plants. *Molecular Plant* 8(1), 68–82. https://doi.org/10.1016/J.MOLP.2014.12.007.

Ohmiya, A., Kishimoto, S., Aida, R. et al. (2006). Carotenoid cleavage dioxygenase (CmCCD4a) contributes to white color formation in chrysanthemum petals. *Plant Physiology* 142(3), 1193–1201. https://doi.org/10.1104/PP.106.087130.

Paine, J.A., Shipton, C.A., Chaggar, S. et al. (2005). Improving the nutritional value of golden rice through increased pro-vitamin A content. *Nature Biotechnology* 23(4), 482–487. https://doi.org/10.1038/nbt1082.

Pott, D.M., Osorio, S. and Vallarino, J.G. (2019). From central to specialized metabolism: an overview of some secondary compounds derived from the primary metabolism for their role in conferring nutritional and organoleptic characteristics to fruit. *Frontiers in Plant Science* 10, 835. https://doi.org/10.3389/FPLS.2019.00835.

Rademacher, T., Sack, M., Blessing, D. et al. (2019). Plant cell packs: a scalable platform for recombinant protein production and metabolic engineering. *Plant Biotechnology Journal* 17(8), 1560–1566. https://doi.org/10.1111/PBI.13081.

Ramel, F., Birtic, S., Ginies, C. et al. (2012). Carotenoid oxidation products are stress signals that mediate gene responses to singlet oxygen in plants. *Proceedings of the National Academy of Sciences of the United States of America* 109(14), 5535–5540. https://doi.org/10.1073/PNAS.1115982109.

Rameshrad, M., Razavi, B.M. and Hosseinzadeh, H. (2018). Saffron and its derivatives, crocin, crocetin and safranal: a patent review. *Expert Opinion on Therapeutic Patents* 28(2), 147–165. https://doi.org/10.1080/13543776.2017.1355909.

Rao, A.V. and Rao, L.G. (2007). Carotenoids and human health. *Pharmacological Research* 55(3), 207–216. https://doi.org/10.1016/J.PHRS.2007.01.012.

Ribeiro, D., Freitas, M., Silva, A.M.S. et al. (2018). Antioxidant and pro-oxidant activities of carotenoids and their oxidation products. *Food and Chemical Toxicology* 120, 681–699. https://doi.org/10.1016/j.fct.2018.07.060.

Rocchi, R., Mascini, M., Sergi, M. et al. (2018). Crocins pattern in saffron detected by UHPLC-MS/MS as marker of quality, process and traceability. *Food Chemistry* 264, 241–249. https://doi.org/10.1016/j.foodchem.2018.04.111.

Rodrigo, M.J., Alquézar, B., Alós, E. et al. (2013). A novel carotenoid cleavage activity involved in the biosynthesis of *Citrus* fruit-specific apocarotenoid pigments. *Journal of Experimental Botany* 64(14), 4461–4478. https://doi.org/10.1093/JXB/ERT260.

Rodríguez-Concepción, M., Forés, O., Martínez-García, J.F. et al. (2004). Distinct light-mediated pathways regulate the biosynthesis and exchange of isoprenoid precursors during Arabidopsis seedling development. *The Plant Cell* 16(1), 144–156. https://doi.org/10.1105/TPC.016204.

Römer, S., Fraser, P.D., Kiano, J.W. et al. (2000). Elevation of the provitamin A content of transgenic tomato plants. *Nature Biotechnology* 18(6), 666–669. https://doi.org/10.1038/76523.

Römer, S., Lübeck, J., Kauder, F. et al. (2002). Genetic engineering of a zeaxanthin-rich potato by antisense inactivation and co-suppression of carotenoid epoxidation. *Metabolic Engineering* 4(4), 263–272.

Ronen, G., Carmel-Goren, L., Zamir, D. et al. (2000). An alternative pathway to β-carotene formation in plant chromoplasts discovered by map-based cloning of *beta* and *old-gold* color mutations in tomato. *Proceedings of the National Academy of Sciences of the United States of America* 97(20), 11102–11107. https://doi.org/10.1073/pnas.190177497.

Sánchez, A.M. and Winterhalter, P. (2013). Carotenoid cleavage products in saffron (*Crocus sativus* L.). *ACS Symposium Series* 1134, 45–63. https://doi.org/10.1021/bk-2013-1134.ch005.

Schwartz, S.H., Qin, X. and Loewen, M.C. (2004). The biochemical characterization of two carotenoid cleavage enzymes from *Arabidopsis* indicates that a carotenoid-derived compound inhibits lateral branching. *Journal of Biological Chemistry* 279(45), 46940–46945. https://doi.org/10.1074/JBC.M409004200.

Schwartz, S.H., Qin, X. and Zeevaart, J.A.D. (2001). Characterization of a novel carotenoid cleavage dioxygenase from plants. *Journal of Biological Chemistry* 276(27), 25208–25211. https://doi.org/10.1074/JBC.M102146200.

Schwartz, S.H., Tan, B.C., Gage, D.A. et al. (1997). Specific oxidative cleavage of carotenoids by VP14 of maize. *Science* 276(5320), 1872–1874. https://doi.org/10.1126/SCIENCE.276.5320.1872.

Sereshti, H., Ataolahi, S., Aliakbarzadeh, G. et al. (2018). Evaluation of storage time effect on saffron chemical profile using gas chromatography and spectrophotometry techniques coupled with chemometrics. *Journal of Food Science and Technology* 55(4), 1350–1359. https://doi.org/10.1007/s13197-018-3046-9.

Shih, P.M. (2018). Towards a sustainable bio-based economy: redirecting primary metabolism to new products with plant synthetic biology. *Plant Science* 273, 84–91. https://doi.org/10.1016/J.PLANTSCI.2018.03.012.

Silletti, M.F., Petrozza, A., Stigliani, A.L. et al. (2013). An increase of lycopene content in tomato fruit is associated with a novel *Cyc-B* allele isolated through TILLING technology. *Molecular Breeding* 31(3), 665–674. https://doi.org/10.1007/S11032-012-9824-6.

Simkin, A.J., Schwartz, S.H., Auldridge, M. et al. (2004). The tomato *carotenoid cleavage dioxygenase 1* genes contribute to the formation of the flavor volatiles beta-ionone, pseudoionone, and geranylacetone. *The Plant Journal: For Cell and Molecular Biology* 40(6), 882–892. https://doi.org/10.1111/J.1365-313X.2004.02263.X.

Song, T., Wu, N., Wang, C. et al. (2020). Crocetin overproduction in engineered *Saccharomyces cerevisiae* via tuning key enzymes coupled with precursor engineering. *Frontiers in Bioengineering and Biotechnology* 8, 578005. https://doi.org/10.3389/FBIOE.2020.578005.

Soyk, S., Müller, N.A., Park, S.J. et al. (2017). Variation in the flowering gene *SELF PRUNING 5G* promotes day-neutrality and early yield in tomato. *Nature Genetics* 49(1), 162–168. https://doi.org/10.1038/ng.3733.

Sun, T., Rao, S., Zhou, X. et al. (2022). Plant carotenoids: recent advances and future perspectives. *Molecular Horticulture* 2(1), 3. https://doi.org/10.1186/s43897-022-00023-2.

Sweetlove, L.J., Nielsen, J. and Fernie, A.R. (2017). Engineering central metabolism—a grand challenge for plant biologists. *The Plant Journal* 90(4), 749–763. https://doi.org/10.1111/TPJ.13464.

Us-Camas, R., Aguilar-Espinosa, M., Rodríguez-Campos, J. et al. (2022). Identifying *Bixa orellana* L. new carotenoid cleavage dioxygenases 1 and 4 potentially involved in bixin biosynthesis. *Frontiers in Plant Science* 13, 829089. https://doi.org/10.3389/FPLS.2022.829089.

Wang, J.Y., Haider, I., Jamil, M. et al. (2019). The apocarotenoid metabolite zaxinone regulates growth and strigolactone biosynthesis in rice. *Nature Communications* 10(1), 810. https://doi.org/10.1038/s41467-019-08461-1.

Winterhalter, P. and Straubinger, M. (2000). Saffron—renewed interest in an ancient spice. *Food Reviews International* 16(1), 39–59. https://doi.org/10.1081/FRI-100100281.

Xie, L., Luo, Z., Jia, X. et al. (2023). Synthesis of crocin I and crocin II by multigene stacking in *Nicotiana benthamiana. International Journal of Molecular Sciences* 24(18), 14139. https://doi.org/10.3390/IJMS241814139.

Xu, Z., Xu, Z., Pu, X. et al. (2020). Tandem gene duplications drive divergent evolution of caffeine and crocin biosynthetic pathways in plants. *BMC Biology* 18(1), 63. https://doi.org/10.1186/s12915-020-00795-3.

Yao, C., Qian, X.D., Zhou, G.F. et al. (2018). A comprehensive analysis and comparison between vacuum and electric oven drying methods on Chinese saffron (*Crocus sativus* L.). *Food Science and Biotechnology* 28(2), 355–364. https://doi.org/10.1007/s10068-018-0487-x.

Yazdani, M., Sun, Z., Yuan, H. et al. (2019). Ectopic expression of *ORANGE* promotes carotenoid accumulation and fruit development in tomato. *Plant Biotechnology Journal* 17(1), 33–49. https://doi.org/10.1111/pbi.12945.

Ye, X., Al-Babili, S., Kloti, A. et al. (2000). Engineering the provitamin A (β-carotene) biosynthetic pathway into (carotenoid-free) rice endosperm. *Science* 287(5451), 303–305.

Yoo, H.J., Chung, M.Y., Lee, H.A. et al. (2023). Natural overexpression of *CAROTENOID CLEAVAGE DIOXYGENASE 4* in tomato alters carotenoid flux. *Plant Physiology* 192(2), 1289–1306. https://doi.org/10.1093/plphys/kiad049.

Zheng, X., Mi, J., Balakrishna, A. et al. (2022). *Gardenia* carotenoid cleavage dioxygenase 4a is an efficient tool for biotechnological production of crocins in green and non-green plant tissues. *Plant Biotechnology Journal* 20(11), 2202–2216. https://doi.org/10.1111/pbi.13901.

Zheng, X., Mi, J., Deng, X. et al. (2021). LC-MS-based profiling provides new insights into apocarotenoid biosynthesis and modifications in citrus fruits. *Journal of Agricultural and Food Chemistry* 69(6), 1842–1851. https://doi.org/10.1021/acs.jafc.0c06893.

Zheng, X., Zhu, K., Sun, Q. et al. (2019). Natural variation in CCD4 promoter underpins species-specific evolution of red coloration in citrus peel. *Molecular Plant* 12(9), 1294–1307. https://doi.org/10.1016/j.molp.2019.04.014.

Index

Note: Page numbers in *italics* and **bold** refers to figures and tables respectively.

Plant Apocarotenoids: Biosynthesis, Signaling and Agricultural Applications, First Edition. Edited by Kaiser Iqbal Wani and Tariq Aftab.
Published 2026 by John Wiley & Sons Ltd.

b

d

n

Printed and bound by CPI Group (UK) Ltd, Croydon, CR0 4YY
17/09/2026
14964297-0002